METHODS IN MOLECULAR BIOLOGY

Series Editor
John M. Walker
School of Life and Medical Sciences
University of Hertfordshire
Hatfield, Hertfordshire, AL10 9AB, UK

For further volumes:
http://www.springer.com/series/7651

Progranulin

Methods and Protocols

Edited by

Andrew Bateman and Hugh P. J. Bennett

Department of Medicine, McGill University MUHC, Montreal, QC, Canada

Siu Tim Cheung

Department of Surgery, The Chinese University of Hong Kong, Hong Kong, China

Humana Press

Editors
Andrew Bateman
Department of Medicine
McGill University MUHC
Montreal, QC, Canada

Hugh P. J. Bennett
Department of Medicine
McGill University MUHC
Montreal, QC, Canada

Siu Tim Cheung
Department of Surgery
The Chinese University of Hong Kong
Hong Kong, China

ISSN 1064-3745 ISSN 1940-6029 (electronic)
Methods in Molecular Biology
ISBN 978-1-4939-8557-9 ISBN 978-1-4939-8559-3 (eBook)
https://doi.org/10.1007/978-1-4939-8559-3

Library of Congress Control Number: 2018946638

Printed on acid-free paper

This Humana Press imprint is published by the registered company Springer Science+Business Media, LLC part of Springer Nature.
The registered company address is: 233 Spring Street, New York, NY 10013, U.S.A.

Preface

The molecular biology of progranulin has evolved from a handful of peaks on a chromatogram to a molecule of increasing significance in numerous cellular, physiological, and pathological contexts that include, but are by no means limited to, cancer, inflammation and repair, and neurodegeneration. Progranulin offers exciting translational possibilities in the development of therapies and diagnostic reagents. This volume of *Methods in Molecular Biology* collects a series of experimental strategies from experts in the field that have proven useful in the study of progranulin as a regulatory protein in a wide cross-section of biological and disease processes. The volume is divided into four parts; (I) *Introduction*, (II) *Analytical and In Vitro Methods for the Study of Progranulin Biology*, (III) *Methods to Investigate the Role of Progranulin in Cancer*, and (IV) *Animal Models to Study the Roles of Progranulin in Health and Disease*. There is, of course, no hard and fast dividing line between the parts, and we expect a large degree of crossover in the applicability of the methods between research areas.

The history of progranulin research has proven one of recurrent surprises. The identification of progranulin emerged not from studies on progranulin itself but rather on much smaller polypeptides called granulins, which are also sometimes called epithelins, and that were identified over 25 years ago. Upon sequencing the cDNA for each of the granulin polypeptides it was discovered, unexpectedly, that each granulin polypeptide arises from the same parental protein, which came to be called progranulin, proepithelin, or the granulin–epithelin precursor (GEP). The functional significance of the granulin polypeptides versus full-length progranulin is still unresolved as in some bioassays the granulin peptides stimulate the same biological actions as progranulin while in other assays the granulins trigger different responses from progranulin, often, in fact, opposing the action of progranulin. Soon after the primary sequences of the granulins and progranulin were reported it emerged that other proteins, notably an acrosome protein called acrogranin and a secreted growth factor called PC cell-derived growth factor (PCDGF), are identical to progranulin. A period of research followed that focused on progranulin as a growth factor acting through the extracellular regulated kinase (ERK) and protein kinase B/AKT signaling systems. Important roles for progranulin were identified in cancer both in terms of its elevated expression in a wide range of cancers and in its ability to promote tumor progression.

The discovery that progranulin is active in wound repair and in the regulation of the inflammatory response prompted a second major thrust in progranulin research. The role of progranulin and the granulin peptides in inflammation and repair is complex but in most instances it enhances the repair process through its growth factor-like properties and simultaneously suppresses an overactive inflammatory reaction that might cause bystander cell damage. The net effect is to promote tissue repair and limit tissue damage. The clinical importance of progranulin was further reinforced when it was discovered that humans that are haploinsufficient for the progranulin gene almost invariably succumb to a progressive and fatal neurodegenerative disease called frontotemporal dementia. Even more surprisingly, the loss of both alleles of progranulin results in a different neurodegenerative disease which is characterized by lysosome storage defects and is called neuronal ceroid lipofuscinosis. Moreover, the targeted gene delivery of progranulin to key regions of the brain dramatically inhibits the progression of neurodegenerative phenotypes associated with Parkinson's disease and Alzheimer's disease in mice.

In Part II, *Analytical and In Vitro Methods for the Study of Progranulin Biology*, the preparation of highly purified granulin peptides and full-length progranulin is discussed, as are methods to generate granulin peptides in large quantities and to analyze their spatial conformation. Each chapter presents specific techniques and examples but the protocols are transferable to granulin peptides or progranulin from any source as well as to synthetic or genetically modified forms of granulin or progranulin. As database resources become ever more powerful it is likely that in silico approaches to molecular biology and biomedical research will become increasingly relevant and a chapter that addresses data mining strategies for progranulin research in neurodegenerative diseases should, it is hoped, provide a useful point of entry for such studies.

In Part III, *Methods to Investigate the Role of Progranulin in Cancer*, readers will find methods and protocols for analyzing the roles of progranulin/granulin–epithelin precursor in diverse cancer types. Several groups from different parts of the world working on progranulin have contributed in this part. The authors have provided details of their expertise and tools in this competitive research field. These methods are representative studies aimed at examining the significance of progranulin in the clinical samples including blood and tissue specimens. Their functional roles on growth proliferation, properties of cancer stem cells, signaling, interaction with the tumor microenvironment, and the use of antibodies as therapeutics have been examined in cell and animal models. Emerging data shows that progranulin has even more complex functions than the already well-known role as a crucial growth factor. Further studies on the progranulin cell surface partner (receptor) and the intracellular protein partners are warranted. We hope that the current update protocols will contribute to further improve our knowledge of the role of progranulin in human cancers.

The final part, *Animal models to Study the Role of Progranulin in Health and Disease*, presents strategies to investigate the molecular biology of progranulin *in vivo*, backed up, in many of the chapters, by corresponding experiments that employ cellular models. This part includes protocols that employ diverse animal models, including the invertebrate worm *Caenorhabditis elegans*, a nonmammalian vertebrate the zebrafish *Danio rerio*, and mice. The neurobiology of progranulin is represented by chapters that employ *Progranulin* gene manipulation in *C. elegans*, and zebrafish embryos, *Progranulin* gene ablation in mice, and viral delivery of the *Progranulin* gene to targeted regions of the brain. This part includes methods to probe cell surface binding partners for progranulin, using the interaction of progranulin with sortilin as an example. Sortilin carries progranulin into the lysosome, a process that has great importance in the emerging field of progranulin in lysosomal biology. Progranulin is expressed from the earliest stages of life and a chapter on its role in the preimplantation embryo with its contribution to early development. Chapters dedicated to progranulin in arthritis, atherosclerosis, and the use of a tissue-targeted *Progranulin* transgene in the developing vasculature explore the complex molecular physiology of progranulin in the regulation of inflammation and tissue repair.

Progranulin offers a wealth of research opportunities. Despite our growing knowledge of its many roles in health and disease, much more remains to be discovered. It offers attractive possibilities for the development of new therapeutic agents across a range of pathologies. It is hoped that this volume will be useful not only to those who already include progranulin in their research but will also encourage newcomers to incorporate progranulin into their research programs. We would like to acknowledge the great assistance of Dr. Wing Yee Cheung in editing the manuscripts.

Montreal, QC, Canada — *Andrew Bateman*
Montreal, QC, Canada — *Hugh P. J. Bennett*
Hong Kong, China — *Siu Tim Cheung*

Contents

Contributors

FABIAN ARECHAVALETA-VELASCO • *Research Unit in Reproductive Medicine, UMAE Hospital de Gineco Obstetricia No. 4 "Luis Castelazo Ayala", Instituto Mexicano del Seguro Social, Mexico City, Mexico*

ANDREW BATEMAN • *Division of Experimental Medicine, Department of Medicine, McGill University, Montreal, QC, Canada; Program in Metabolic Diseases and Their Complications, Centre for Translational Biology, Research Institute of McGill University Health Centre, Montreal, QC, Canada*

ANTONINO BELFIORE • *Department of Health and Endocrinology, University Magna Graecia of Catanzaro, Catanzaro, Italy*

HUGH P. J. BENNETT • *Department of Medicine, McGill University, Montreal, QC, Canada; Program in Metabolic Diseases and Their Complications, Centre for Translational Biology, The Metabolic Disorders and Complications Program, Research Institute of McGill University Health Centre, Montreal, QC, Canada*

JAN BODDAERT • *Molecular Pathology Group, Laboratory of Cell Biology and Histology, Faculty of Medicine and Health Sciences, University of Antwerp, Antwerpen, Belgium*

SAMUEL BROOK • *Human Oncology and Pathogenesis Program (HOPP), Memorial Sloan Kettering Cancer Center, New York, NY, USA*

CHRISTINE VAN BROECKHOVEN • *Neurodegenerative Brain Diseases Group, Center for Molecular Neurology, VIB, University of Antwerp – CDE, Antwerp, Belgium; Institute Born-Bunge, University of Antwerp, Antwerp, Belgium*

SIMONE BURASCHI • *Cancer Cell Biology and Signaling Program, Department of Pathology, Anatomy and Cell Biology, Kimmel Cancer Center, Thomas Jefferson University, Philadelphia, PA, USA*

VICTORIA J. BUTLER • *Department of Neurology, University of California at San Francisco, San Francisco, CA, USA*

PHYLLIS F. Y. CHEUNG • *Department of Surgery, The Chinese University of Hong Kong, Hong Kong, China; Department of Anatomical and Cellular Pathology, The Chinese University of Hong Kong, Hong Kong, China*

SIU TIM CHEUNG • *Department of Surgery, The Chinese University of Hong Kong, Hong Kong, China; Li Ka Shing Institute of Health Sciences, The Chinese University of Hong Kong, Hong Kong, China*

BABYKUMARI P. CHITRAMUTHU • *Department of Medicine, McGill University, Montreal, QC, Canada; The Metabolic Disorders and Complications Program, Research Institute of the McGill University Health Centre, Montreal, QC, Canada; Centre for Translational Biology, Research Institute of McGill University Health Centre, Montreal, QC, Canada*

MARC CRUTS • *Neurodegenerative Brain Diseases Group, Center for Molecular Neurology, VIB, University of Antwerp – CDE, Antwerp, Belgium; Institute Born-Bunge, University of Antwerp, Antwerp, Belgium*

EUGENE DANIELS • *Department of Anatomy, McGill University, Montreal, QC, Canada*

LAURA DIAZ-CUETO • *Research Unit in Reproductive Medicine, UMAE Hospital de Gineco Obstetricia No. 4 "Luis Castelazo Ayala", Instituto Mexicano del Seguro Social, Mexico City, Mexico*

JAMES JULIAN DOYLE • *Division of Experimental Medicine, Faculty of Medicine, McGill University, Montreal, Canada; Metabolic Disorders and Complications, Research Institute of the McGill University Health Center, Montreal, Canada; Axe Neurosciences, Centre de Recherche du Centre Hospitalier de l'Université de Montréal, Montreal, Canada*

MOSHE ELKABETS • *The Shraga Segal Department of Microbiology, Immunology and Genetics, Faculty of Health Sciences, Ben-Gurion University of the Negev, Beer-Sheva, Israel*

SZE WAI FUNG • *Department of Surgery, The Chinese University of Hong Kong, Hong Kong, China; Department of Surgery, The University of Hong Kong, Hong Kong, China; School of Biomedical Sciences, The University of Hong Kong, Hong Kong, China*

GEORGE L. GERTON • *Center for Research on Reproduction and Women's Health, Perelman School of Medicine, University of Pennsylvania, Philadelphia, PA, USA*

DAVID HICKS • *A&G Pharmaceutical Inc., Columbia, MD, USA*

TSUNG-YUAN HSU • *Department of Neurology, University of California at San Francisco, San Francisco, CA, USA*

FENGHUA HU • *Department of Molecular Biology and Genetics, Weill Institute for Cell and Molecular Biology, Cornell University, Ithaca, NY, USA*

RENATO V. IOZZO • *Cancer Cell Biology and Signaling Program, Department of Pathology, Anatomy and Cell Biology, Kimmel Cancer Center, Thomas Jefferson University, Philadelphia, PA, USA*

AIMEE W. KAO • *Department of Neurology, University of California at San Francisco, San Francisco, CA, USA*

RYOTA KAWASE • *Department of Cardiology, Sousei Hospital, Osaka, Japan*

SAMIR KUMAR-SINGH • *Molecular Pathology Group, Laboratory of Cell Biology and Histology, Faculty of Medicine and Health Sciences, University of Antwerp, Antwerpen, Belgium; Department of Translational Neurosciences, Faculty of Medicine and Health Sciences, University of Antwerp, Antwerpen, Belgium*

CHUAN-JU LIU • *Department of Orthopaedic Surgery, New York University Medical Center, New York, NY, USA; Department of Cell Biology, New York University School of Medicine, New York, NY, USA*

ANDREA MORRIONE • *Biology of Prostate Cancer Program, Department of Urology, Thomas Jefferson University, Philadelphia, PA, USA*

TOHRU OHAMA • *Department of Cardiovascular Medicine, Osaka University Graduate School of Medicine, Osaka, Japan; Department of Dental Anesthesiology, Osaka University Graduate School of Dentistry, Osaka, Japan*

J. ALEX PARKER • *Axe Neurosciences, Centre de Recherche du Centre Hospitalier de l'Université de Montréal, Montreal, Canada; Département de Neurosciences, Université de Montréal, Montreal, Canada*

DANIEL H. PAUSHTER • *Department of Molecular Biology and Genetics, Weill Institute for Cell and Molecular Biology, Cornell University, Ithaca, NY, USA*

YASUSHI SAKATA • *Department of Cardiovascular Medicine, Osaka University Graduate School of Medicine, Osaka, Japan*

GINETTE SERRERO • *A&G Pharmaceutical Inc., Columbia, MD, USA*

MANUELA STEFANELLO • *Biology of Prostate Cancer Program, Department of Urology, Thomas Jefferson University, Philadelphia, PA, USA*

PETER M. SULLIVAN • *Department of Molecular Biology and Genetics, Weill Institute for Cell and Molecular Biology, Cornell University, Ithaca, NY, USA*

RYUTA TANIMOTO • *Biology of Prostate Cancer Program, Department of Urology, Thomas Jefferson University, Philadelphia, PA, USA*

HUISHI TOH • *Department of Medicine, McGill University, Montreal, QC, Canada; Centre for Translational Biology, Research Institute of McGill University, Montreal, QC, Canada*

DMITRI TOLKATCHEV • *The Gene and Linda Voiland School of Chemical Engineering and Bioengineering, Washington State University, Pullman, WA, USA*

JACKALINA M. VAN KAMPEN • *Neurodyn Life Sciences, Charlottetown, PE, Canada; Department of Biomedical Science, University of Prince Edward Island, Charlottetown, PE, Canada*

JIAN-LU WEI • *Department of Orthopaedic Surgery, New York University Medical Center, New York, NY, USA*

HANS WILS • *Molecular Pathology Group, Laboratory of Cell Biology and Histology, Faculty of Medicine and Health Sciences, University of Antwerp, Antwerpen, Belgium*

NICHOLAS C. L. WONG • *Department of Surgery, The University of Hong Kong, Hong Kong, China*

SHI-QIONG XU • *Biology of Prostate Cancer Program, Department of Urology, Thomas Jefferson University, Philadelphia, PA, USA*

SHIZUYA YAMASHITA • *Department of Cardiovascular Medicine, Osaka University Graduate School of Medicine, Osaka, Japan; Department of Community Medicine, Osaka University Graduate School of Medicine, Osaka, Japan; Rinku General Medical Center, Osaka, Japan*

CHI WAI YIP • *Department of Surgery, The Chinese University of Hong Kong, Hong Kong, China; Department of Surgery, The University of Hong Kong, Hong Kong, China*

CHUN YU ZHAO • *Division of Experimental Medicine, Department of Medicine, McGill University, Montreal, QC, Canada; Centre for Translational Biology, Research Institute of McGill University Health Centre, Montreal, QC, Canada*

XIAOLAI ZHOU • *Department of Molecular Biology and Genetics, Weill Institute for Cell and Molecular Biology, Cornell University, Ithaca, NY, USA*

Part I

Introduction

Chapter 1

A Brief Overview of Progranulin in Health and Disease

Andrew Bateman, Siu Tim Cheung, and Hugh P. J. Bennett

Abstract

The purpose of this brief overview of the progranulin protein is to provide a sense of the range and extent of the roles of progranulin in normal physiology and pathology. Progranulin has received attention due to its role in neurodegeneration, where mutation of a single copy of *GRN*, the gene encoding progranulin, results in frontotemporal dementia, whereas viral delivery of progranulin to the brains of mice exhibiting Parkinson's or Alzheimer's disease phenotypes inhibits the progression of the neurodegenerative phenotypes. Of equal importance, progranulin protects tissues against the harmful effects of poorly controlled inflammation and promotes tissue regeneration after injury at a multitude of sites throughout the body. Progranulin is overexpressed by many types of cancer and contributes to their progression. Given suitable analytical methods and model systems, progranulin offers a wealth of research possibilities.

Key words Progranulin, Granulin, Epithelin, Proepithelin, Granulin-epithelin precursor, PC-cell-derived growth factor, Acrogranin, Epithelial transforming growth factor (TGFe), Development, Wounds, Repair, Inflammation, Cancer, Neurodegeneration

1 Introduction

Progranulin is implicated in many different biological events from the development of the preimplantation embryo through to the maintenance and repair of adult tissues and offers many opportunities for further research. In the limited space available, it is impossible to discuss all the activities of progranulin. Therefore here we will highlight its protective and regenerative properties, its roles in cancer, and aspects of progranulin neurobiology. The protein and genomic structure of progranulin and the relationship between progranulin and granulin peptides will be summarized. Several alternate names for progranulin have been used, which can be confusing when first encountering the progranulin literature, and we will therefore outline these nomenclatures and their origin.

Andrew Bateman et al. (eds.), *Progranulin: Methods and Protocols*, Methods in Molecular Biology, vol. 1806,
https://doi.org/10.1007/978-1-4939-8559-3_1,

2 Progranulin: Structure, D-istribution, and Activities

2.1 The Structure of Progranulin

Progranulin is composed of a string of repeats of the granulin module aligned like beads on a chain (Fig. 1) [1]. The granulin module typically contains 12 cysteine residues positioned in a characteristic pattern that is found in no other protein [2, 3] and that forms six disulfide bridges [4] (Fig. 1). The spatial conformation of the granulin modules was solved by 2D solution nuclear magnetic resonance spectroscopy and found to consist of a stack of up to four beta-hairpins held together by parallel pairs of disulfide bonds that run down the axis of the fold from one hairpin to the next [4, 5]. Mammalian progranulin possesses seven and a half such granulin modules [1, 6], designated modules A through to G for the full granulin modules and p for the single half module called paragranulin (Fig. 1). Progranulin is a glycoprotein, with five possible *N*-glycosylation sites [7]. Two of these sites occur within granulin domains, namely, modules C and E, and both sites have been shown to be glycosylated. At the genomic level, each module is encoded by two exons [8, 9]. The N-type exons encode the amino

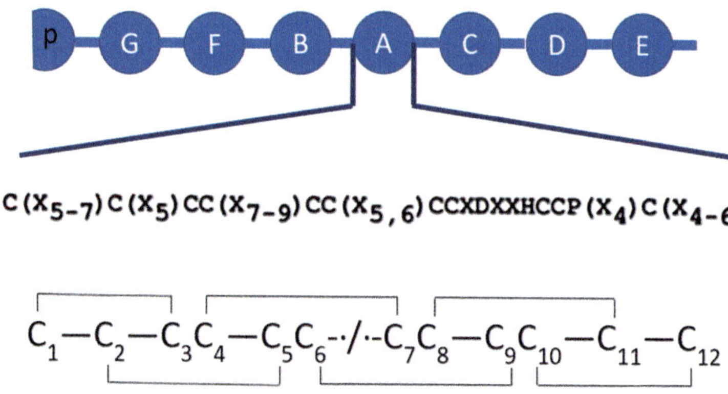

Fig. 1 The structure of progranulin. Progranulin consists of seven granulin modules (A to G) and one half granulin module (p) that corresponds to the region encompassing the amino-terminal first six cysteines of the full granulin modules. The modules A to D were initially identified and named as excised polypeptides before it was known that they are all contained within progranulin [2]. It is for this reason that the lettering of the modules is not in alphabetical order. The consensus sequence of a typical granulin module is given below, where X can be any amino acid, C is cysteine, D is aspartic acid, H is histidine, and P is proline. The connectivity of the 12 cysteines of a typical granulin module as six disulfide bridges is displayed in the lower panel. There is an approximate inverted symmetry between the disulfide alignments of the amino- and carboxyl-termini that centers around the –/– mark in the drawing. This position also corresponds to the location of the intronic splice site where the two exons (N type, C type, or CN type, as discussed in the text) are brought together to encode each granulin module

or N-terminal half of a granulin module, while the C-type exons encode the carboxyl or C-terminal half of a granulin module. Splicing together an N-type exon and a C-type exon generates mRNA that encodes a complete granulin module [8, 9]. A third type of exon exists, the CN type that encodes for the C-terminal half of a granulin module and then the N-terminal half of the next granulin module. Splicing together two CN exons (-C**N**-**C**N-) or a CN exon with either an N exon (-**N**-**C**N-) or a C exon (-C**N**-**C**-) will generate mRNAs encoding a full granulin module as indicated by the bold underlined text [8, 9]. The gene for progranulin is designated *Grn* (or *GRN* in humans, chromosome localization 17q21.31). Mammalian genomes and most other tetrapods possess only one *Grn* gene. In contrast, fish and invertebrate genomes often carry several *Grn* genes [10]. The zebrafish, for example, possesses four *Grn* genes, two of which encode for progranulins with only one and a half granulin modules each and two genes that encode for much larger progranulins with multiple modules [11]. The granulin module is evolutionarily ancient [10]. Thus, a modified form of the granulin module, with 14 rather than 12 cysteines, occurs in a family of plant cysteine proteases [10]. Slime molds such as *Dictyostelium* possess a progranulin with only one granulin domain [10]. The choanoflagellate, *Monosiga brevicollis*, a single-celled organism that is thought to be a living representative of the types of cells that were evolutionary precursors for the multicellular animals, possesses a progranulin with multiple modules that is similar in many respects to the multi-modular progranulins found in higher animals [10].

2.2 Nomenclature and History

The granulin modules can be excised from progranulin and occur in tissues and biological fluids as individual polypeptides of approximately 5 to 5.5 kDa [2]. These polypeptides were called granulins because they were isolated from the granular fraction of human granulocytes [2]. The murine homologs were first isolated from rodent kidneys and were called epithelins [3]. It was only following the sequencing of the cDNA for the granulins [1] and epithelins [6] that it became obvious that all the granulin-epithelin peptides that had been previously isolated were encoded within a single gene that came to be known as progranulin or proepithelin or the granulin-epithelin precursor (GEP). Further research soon revealed that progranulin had been isolated in other contexts although not identified as such until later. The guinea pig acrosome, for example, was shown to contain a protein, called acrogranin [12], that, once its cDNA structure was known, was found to be guinea pig progranulin [13]. In other studies, a protein called PC-cell-derived growth factor, or PCDGF, was also identified, upon sequencing of its cDNA, to be murine progranulin [14]. This protein was isolated from the conditioned medium of PC cells, a highly tumorigenic adipocytic teratoma that secretes PCDGF (progranulin) as an autocrine growth factor [14]. PCDGF is also sometimes referred to as GP88 for glycoprotein 88 kDa. A third group demonstrated that a partial amino-terminal protein sequence from a 25 kDa protein called epithelial transforming growth factor (TGFe) was identical in sequence to the amino terminus of the granulin A peptide [15]. TGFe was isolated from bovine kidneys and identified using procedures that were similar to those used to identify transforming growth factors alpha and beta but was distinguishable from other growth factors based on the spectrum of cells that respond to TGFe and its relative stability after treatment with acidic media [16, 17]. In summary, the protein progranulin, sometimes called proepithelin, the granulin-epithelin precursor (GEP), acrogranin, or PCDGF (GP88), contains all seven and a half granulin modules. TGFe is most probably a biologically active naturally occurring fragment of progranulin with the granulin A module at its amino terminus. Peptides that are derived from progranulin through proteolysis correspond to the 12-cysteine granulin modules which are called granulins or epithelins.

2.3 Progranulin Versus Granulins

As noted above, the proteolysis of progranulin by, for example, elastase releases the granulin modules as peptides, the granulins. The granulin peptides are biologically active, although generally less potent than the intact progranulin protein [3, 5]. Importantly, however, granulin peptide activity is not always the same as that of progranulin itself. Progranulin stimulates the growth of cells of diverse origins, including epithelial cells and fibroblasts. Some granulin peptides are also proliferative, for example, granulin F stimulates proliferation of MDA-BB-468 breast cancer cells, as does progranulin [5]. Other peptides are antiproliferative. Granulin A, for example, inhibits the growth of the MDA-MB-468 cells [5]. Most granulins however were only weakly active in this assay, and their production may therefore be a means to inactivate progranulin activity. The neurotrophic actions of progranulin can be reproduced by granulin E [18, 19]. Progranulin tends to suppress inflammation [20–22], whereas granulin B stimulates the production of the pro-inflammatory chemokine interleukin-8 (also known CXCL8) [20]. Given the different actions associated with different granulin peptides, and their lower potency relative to intact progranulin, it remains unclear when, and to what extent, the granulin peptides assume a significant role physiologically or in disease. The picture is further complicated by the existence of biologically active progranulin fragments, such as TGFe, that are intermediate in size between the granulin peptides and progranulin. Cell culture models have done little to clarify this issue. Cells that express progranulin in culture tend to produce the full-length progranulin protein, sometimes accompanied by a slightly smaller protein that is thought to be incompletely glycosylated progranulin. Little spontaneous fragmentation is observed in cell culture in vitro.

Progranulin can be stabilized from proteolysis by a protein called secretory leukocyte protease inhibitor (SLPI) [20]. This is a serine protease inhibitor found in mucosal secretions. It inhibits elastase, cathepsin G, and trypsin, among other enzymes. Interestingly, SLPI protects progranulin not only through its inhibitory effects on serine protease action but also by direct binding to progranulin in the intervening polypeptide sequences between granulin modules [20]. The granulin modules are very resistant to proteolysis so the binding of SLPI at the intervening sequences will block access of the proteases to the more susceptible cleavage sites between the modules. Progranulin is, in addition, subject to proteolysis by tissue matrix metalloproteinases [23]. The net activity of the progranulin system is therefore determined by the relative levels of progranulin, the presence of proteases, and protease inhibitors (summarized in Fig. 2).

2.4 Distribution of Progranulin Expression

Progranulin is strongly expressed in myeloid cells such as neutrophils [2], macrophages [24], and microglia [25]. It is strongly expressed in actively cycling epithelial cells such as the basal keratinocytes in the skin and in enterocytes of the villus crypt in the intestine [26]. There are exceptions to this pattern, however, as a few epithelial cells that are not highly mitotic such as the epithelia

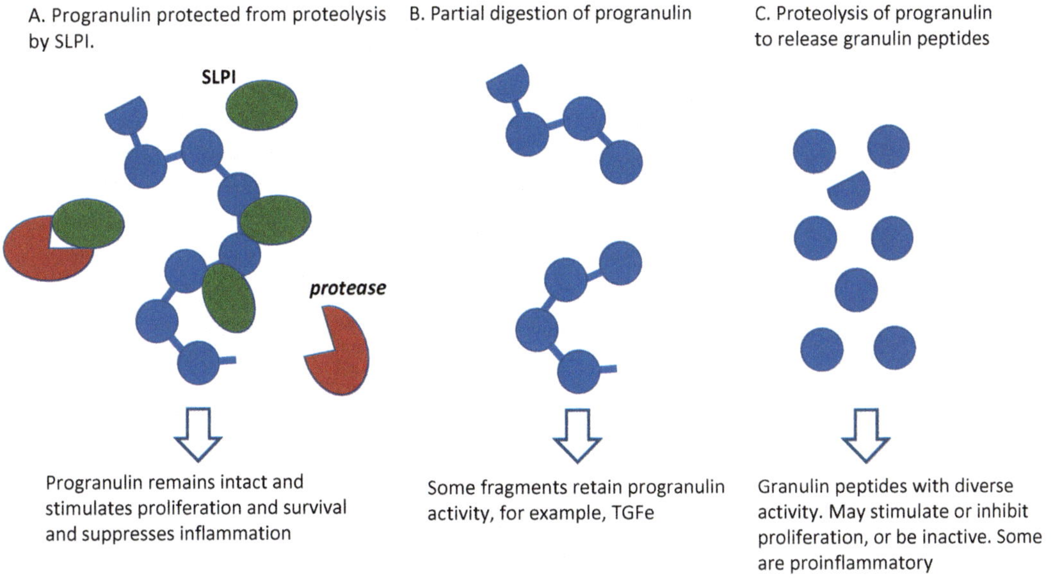

Fig. 2 Progranulin and granulin peptides. Proteolytic cleavage of progranulin releases the granulin peptides. In (**a**), progranulin is protected from proteolysis by the serine protease inhibitor (SLPI), which both inhibits the protease such as elastase and binds directly to progranulin to further block its proteolytic cleavage. The intact form of progranulin stimulates proliferation and survival and suppresses inflammation. (**b**) Intermediate-sized products of progranulin proteolysis have been isolated from tissues, and some, such as TGFe, retain progranulin-like biological activity. (**c**) The end products of progranulin digestion are the granulin peptides. Some of these retain progranulin-like proliferative activity, but others are antiproliferative, and some may stimulate inflammation

lining of the seminal vesicles express progranulin at high levels [26]. Progranulin is expressed in a subset of neurons that include some cerebral cortical neurons, cells in the hippocampus, Purkinje cells of the cerebellum, and motor neurons, among others [26]. It is sometimes said that progranulin expression is ubiquitous, but this is misleading. While progranulin can be detected in most tissues, some tissues, such as muscles, do not express progranulin [26], while in other cells progranulin expression is either absent or very low unless the cell is activated to induce progranulin production. Fibroblasts and endothelial cells, for example, are progranulin negative or, at most, only weakly positive [26], but upon activation, for example, following tissue injury, their progranulin expression is rapidly induced and may become prominent [27]. This was one of the first clues that suggested a role for progranulin in tissue repair and regeneration.

2.5 Physiological Roles for Progranulin

Progranulin has been implicated in physiological processes throughout life. It is involved in the growth of preimplantation embryos [28], in embryo implantation [29], and in the formation of the placenta [30]. In adulthood progranulin has growth factor-like roles in wound healing [27] and angiogenesis [27, 31]. It is involved in bone growth and repair [32]. It plays an important role in the regulation of inflammation, tending to suppress the inflammatory response [20–22, 33] by reducing pro-inflammatory cytokine production and increasing the production of anti-inflammatory cytokines. Moreover, progranulin suppresses the action of tumor necrosis factor-alpha possibly by a direct binding interaction of progranulin with tumor necrosis factor receptors [22]. Progranulin knockout mice are viable and fertile but develop a slowly progressing neurological phenotype [34]. The importance of progranulin becomes apparent when the mice are traumatized in some fashion. For example, progranulin knockout mice are considerably more susceptible to inflammatory arthritis than their wild-type counterparts [22]. Treating mice suffering from experimentally induced inflammatory arthritis with recombinant progranulin, or modified analogs of progranulin, reduces the severity of the arthritic phenotype. Similarly crossing progranulin knockout mice with ApoE mice that are prone to atherosclerosis markedly increases the severity of the atherosclerotic phenotype [35]. These are only two of many examples where the loss of progranulin renders progranulin knockout mice more susceptible to inflammatory situations. An important role for progranulin is, therefore, to protect tissues from injury by limiting the deleterious consequences of inflammation associated with tissue trauma [20]. It is likely, however, that once progranulin has exerted its protective role as a modulator of overactive inflammation, its growth factor-like properties will then contribute to tissue repair [27]. Clearly the dual protective and regenerative activities of progranulin make it a potential target for the development of novel therapies for conditions associated with tissue damage.

2.6 Progranulin and Cancer

Many tumors express progranulin at levels greatly above the corresponding non-diseased tissue, a feature that may have value in tumor diagnosis and prognosis and implies an important role for progranulin in cancer progression (reviewed in [36]). This includes cancers of the liver [37], breast [38], ovary [39, 40], endometrium [41], prostate [42], kidney [43], and bladder [44] as well as gliomas [45], multiple myelomas [46], and uterine leiomyosarcomas [47]. There is a correlation between progranulin-positive staining of cancer cells and progranulin levels circulating in the plasma and tumor prognosis. In breast cancer, for example, tissue levels of progranulin predict the risk of recurrence in estrogen receptor-positive invasive ductal carcinomas [48]. Serum PGRN levels predict disease recurrence in individuals with steroid hormone receptor-positive breast cancers who are undergoing treatment with the estrogen receptor antagonist tamoxifen [49]. Progranulin may promote tumor growth in many ways. Its expression in vitro increases cancer cell proliferation [14, 37, 50–54], invasion [37, 51, 53], and resistance to apoptotic stimuli [51, 54] including some anticancer drugs [55–58]. In mouse models, the overexpression of progranulin by weakly tumor forming cells greatly enhances tumor formation [47, 50, 56], while the targeted depletion of progranulin mRNA expression in highly tumorigenic cells reduces tumor formation [59]. Monoclonal antibodies directed against progranulin inhibit the growth of hepatocellular carcinomas in mice [60], suggesting that progranulin is a viable target for the development of novel agents to combat some forms of cancer. Other attributes of tumorigenesis that are influenced by progranulin include roles in cancer stem cells [58, 61], with stroma formation [62, 63] and with immune surveillance of developing tumors [64]. Given that progranulin exerts complex and multifaceted actions on tumor development, it offers many possible avenues for further investigation and, potentially, for the development of novel therapies (Fig. 3).

2.7 Progranulin and the Brain

The first indications of that there are roles for progranulin in the brain came from histological studies [26] and from the observation that progranulin contributes to the hormonally regulated induction of a male phenotype in the developing hypothalamus [65, 66]. The neurobiology of progranulin, however, remained relatively unappreciated until the dramatic discovery that mutation of a single copy of *GRN*, the progranulin gene, causes frontotemporal dementia (FTD) in human patients [67, 68]. Haploinsufficiency of progranulin due to nonsense-mediated decay of the mutated progranulin mRNA results in a slow, progressive destruction of the frontal and temporal lobes and, ulti-

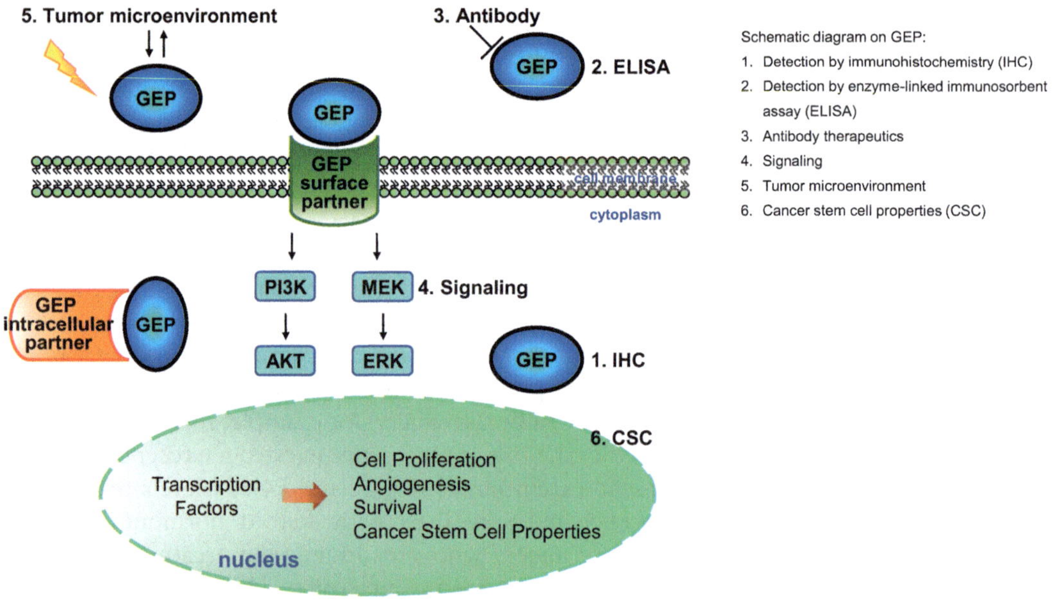

Fig. 3 Progranulin/GEP and cancer. Progranulin levels are highly elevated in a range of cancers, and its level often correlates with a poor prognosis. Later chapters in the book describe methods to characterize progranulin levels in tumors and in circulation using immunohistochemical and enzyme-linked immunoabsorbent assay techniques. Progranulin influences many of the well-established hallmarks of carcinogenesis and regulates, among other parameters, cancer cell proliferation, survival, stem cell properties, and angiogenesis. In later chapters, methods are discussed to investigate how antibodies directed against progranulin inhibit tumor growth, how progranulin stimulates oncogenic signaling pathways, how progranulin regulates events in the tumor microenvironment, and the role of progranulin in the regulation of cancer stem cell properties

mately, death. *GRN*-dependent FTD is characterized by TDP43 proteinopathy, that is, the accumulation of ubiquitinated cellular inclusion of a protein called TAR-DNA-binding protein (TARDBP43 or TDP43). Why progranulin deficiency causes FTD is not fully understood, with evidence that it supports neuronal survival and neurite outgrowth [18, 25], modulates neuroinflammation [33, 69], and influences neuronal lysosome function [70, 71]. A link between progranulin and lysosome function, which was somewhat unexpected, was established in progranulin knockout mice [72, 73] and in rare patients who have mutations of both *GRN* alleles [74]. In both instances, the complete loss of progranulin manifests in the brain with the appearance of age-dependent neuronal ceroid lipofuscinosis which is a form of lysosome storage disease. Given that the loss of progranulin is so harmful to the brain, it was of interest to ask whether the opposite might apply, namely, whether supplementing progranulin levels in a diseased brain would be beneficial. This has been proven to be the case, at least in mice. The targeted delivery of progranulin to disease-related brain areas using lentiviral vectors was found to greatly reduce the severity of Parkinson's disease- [75] and Alzheimer's disease-like [76] phenotypes in experimental mouse models. Most neurons of the brain cannot be replaced, and mechanisms must, therefore, exist to protect them from premature death. Given that partial loss of progranulin results in a neuronal demise associated with FTD while *GRN* gene delivery impedes neurodegeneration, it is reasonable to suggest that progranulin has a critical role in preventing neuronal cell death. A better appreciation of how progranulin supports neuronal longevity in the brain may ultimately suggest new therapeutic strategies to improve the treatment of a variety of neurodegenerative conditions.

3 Conclusion

Progranulin influences many physiological and pathological processes (Fig. 4). These include protective and regenerative processes where the immunomodulatory and growth factor-like properties of progranulin act in conjunction to prevent excessive tissue injury

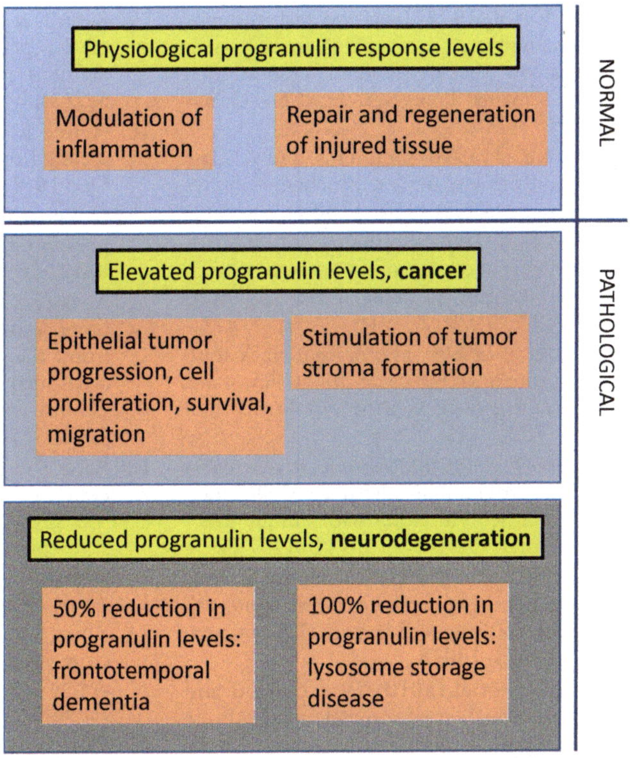

Fig. 4 Normal and pathological actions of progranulin depend upon progranulin levels. At physiological levels, progranulin has protective and regenerative roles in adult tissues. Progranulin levels are often elevated in cancers and contribute to tumor progression. Low levels of progranulin result in neurological disorders. Thus, mutation of one copy of *GRN* results in frontotemporal dementia, and mutation of both copies of the GRN gene results in neuronal ceroid lipofuscinosis, a lysosome storage disorder

and to promote tissue recuperation. Progranulin is, therefore, highly relevant to inflammatory disease research and regenerative medicine. Progranulin is overexpressed in several cancer types and contributes to many aspects of cancer progression. Progranulin cancer research will continue to provide insight into tumor biology, and, importantly, it may provide novel strategies to improve cancer diagnosis and treatment. Research into the neurobiology of progranulin can be expected to continue to reveal a wealth of knowledge impinging on the long-term maintenance of neuronal viability and, by extension, suggest new strategies to slow or stop the progression of a number of neurodegenerative diseases. The success of these research programs depends upon the use of suitable methods and experimental models to probe the biological and pathological properties of progranulin. The development and application of many of these techniques are discussed in this publication.

References

1. Bhandari V et al (1992) Isolation and sequence of the granulin precursor cDNA from human bone marrow reveals tandem cysteine-rich granulin domains. Proc Natl Acad Sci U S A 89(5):1715–1719
2. Bateman A et al (1990) Granulins, a novel class of peptide from leukocytes. Biochem Biophys Res Commun 173(3):1161–1168
3. Shoyab M et al (1990) Epithelins 1 and 2: isolation and characterization of two cysteine-rich growth-modulating proteins. Proc Natl Acad Sci U S A 87(20):7912–7916
4. Hrabal R et al (1996) The hairpin stack fold, a novel protein architecture for a new family of protein growth factors. Nat Struct Biol 3(9):747–752
5. Tolkatchev D et al (2008) Structure dissection of human progranulin identifies well-folded granulin/epithelin modules with unique functional activities. Protein Sci 17(4):711–724
6. Plowman GD et al (1992) The epithelin precursor encodes two proteins with opposing activities on epithelial cell growth. J Biol Chem 267(18):13073–13078
7. Songsrirote K et al (2010) Development and application of mass spectrometric methods for the analysis of progranulin *N*-glycosylation. J Proteomics 73(8):1479–1490
8. Bhandari V, Bateman A (1992) Structure and chromosomal location of the human granulin gene. Biochem Biophys Res Commun 188(1):57–63
9. Baba T et al (1993) Exon/intron organization of the gene encoding the mouse epithelin/granulin precursor (acrogranin). FEBS Lett 322(2):89–94
10. Palfree RG et al (2015) The evolution of the secreted regulatory protein Progranulin. PLoS One 10(8):e0133749
11. Cadieux B et al (2005) The zebrafish progranulin gene family and antisense transcripts. BMC Genomics 6:156
12. Anakwe OO, Gerton GL (1990) Acrosome biogenesis begins during meiosis: evidence from the synthesis and distribution of an acrosomal glycoprotein, acrogranin, during Guinea pig spermatogenesis. Biol Reprod 42(2):317–328
13. Baba T et al (1993) Acrogranin, an acrosomal cysteine-rich glycoprotein, is the precursor of the growth-modulating peptides, granulins, and epithelins, and is expressed in somatic as well as male germ cells. Mol Reprod Dev 34(3):233–243
14. Zhou J et al (1993) Purification of an autocrine growth factor homologous with mouse epithelin precursor from a highly tumorigenic cell line. J Biol Chem 268(15):10863–10869
15. Parnell PG et al (1992) Transforming growth factor e: amino acid analysis and partial amino acid sequence. Growth Factors 7(1):65–72
16. Parnell PG et al (1990) Purification of transforming growth factor type e. J Cell Biochem 42(2):111–116

17. Halper J, Moses HL (1987) Purification and characterization of a novel transforming growth factor. Cancer Res 47(17):4552–4559
18. Van Damme P et al (2008) Progranulin functions as a neurotrophic factor to regulate neurite outgrowth and enhance neuronal survival. J Cell Biol 181(1):37–41
19. De Muynck L et al (2013) The neurotrophic properties of progranulin depend on the granulin E domain but do not require sortilin binding. Neurobiol Aging 34(11):2541–2547
20. Zhu J et al (2002) Conversion of proepithelin to epithelins: roles of SLPI and elastase in host defense and wound repair. Cell 111(6):867–878
21. Kessenbrock K et al (2008) Proteinase 3 and neutrophil elastase enhance inflammation in mice by inactivating antiinflammatory progranulin. J Clin Invest 118(7):2438–2447
22. Tang W et al (2011) The growth factor progranulin binds to TNF receptors and is therapeutic against inflammatory arthritis in mice. Science 332(6028):478–484
23. Suh HS et al (2012) Regulation of progranulin expression in human microglia and proteolysis of progranulin by matrix metalloproteinase-12 (MMP-12). PLoS One 7(4):e35115
24. Ong CH et al (2006) Regulation of progranulin expression in myeloid cells. Am J Physiol Regul Integr Comp Physiol 291(6):R1602–R1612
25. Ryan CL et al (2009) Progranulin is expressed within motor neurons and promotes neuronal cell survival. BMC Neurosci 10:130
26. Daniel R et al (2000) Cellular localization of gene expression for progranulin. J Histochem Cytochem 48(7):999–1009
27. He Z et al (2003) Progranulin is a mediator of the wound response. Nat Med 9(2):225–229
28. Diaz-Cueto L et al (2000) Modulation of mouse preimplantation embryo development by acrogranin (epithelin/granulin precursor). Dev Biol 217(2):406–418
29. Qin J et al (2005) Effects of progranulin on blastocyst hatching and subsequent adhesion and outgrowth in the mouse. Biol Reprod 73(3):434–442
30. Desmarais JA et al (2008) Spatiotemporal expression pattern of progranulin in embryo implantation and placenta formation suggests a role in cell proliferation, remodeling, and angiogenesis. Reproduction 136(2):247–257
31. Toh H et al (2013) Expression of the growth factor progranulin in endothelial cells influences growth and development of blood vessels: a novel mouse model. PLoS One 8(5):e64989
32. Zhao YP et al (2013) The promotion of bone healing by progranulin, a downstream molecule of BMP-2, through interacting with TNF/TNFR signaling. Biomaterials 34(27):6412–6421
33. Yin F et al (2010) Exaggerated inflammation, impaired host defense, and neuropathology in progranulin-deficient mice. J Exp Med 207(1):117–128
34. Kayasuga Y et al (2007) Alteration of behavioural phenotype in mice by targeted disruption of the progranulin gene. Behav Brain Res 185(2):110–118
35. Kawase R et al (2013) Deletion of progranulin exacerbates atherosclerosis in ApoE knockout mice. Cardiovasc Res 100(1):125–133
36. Zhang Y, Bateman A (2011) The glycoprotein growth factor progranulin promotes carcinogenesis and has potential value in anti-cancer therapy. J Carcinog Mutagen 2:001
37. Cheung ST et al (2004) Granulin-epithelin precursor overexpression promotes growth and invasion of hepatocellular carcinoma. Clin Cancer Res 10(22):7629–7636
38. Serrero G, Ioffe OB (2003) Expression of PC-cell-derived growth factor in benign and malignant human breast epithelium. Hum Pathol 34(11):1148–1154
39. Jones MB et al (2003) The granulin-epithelin precursor/PC-cell-derived growth factor is a growth factor for epithelial ovarian cancer. Clin Cancer Res 9(1):44–51
40. Cuevas-Antonio R et al (2010) Expression of progranulin (Acrogranin/PCDGF/Granulin-Epithelin precursor) in benign and malignant ovarian tumors and activation of MAPK signaling in ovarian cancer cell line. Cancer Invest 28(5):452–458
41. Jones MB et al (2006) The granulin-epithelin precursor is a steroid-regulated growth factor in endometrial cancer. J Soc Gynecol Investig 13(4):304–311
42. Pan CX et al (2004) PC cell-derived growth factor expression in prostatic intraepithelial neoplasia and prostatic adenocarcinoma. Clin Cancer Res 10(4):1333–1337
43. Donald CD et al (2001) Expression of progranulin and the epithelin/granulin precursor acrogranin correlates with neoplastic state in renal epithelium. Anticancer Res 21(6a):3739–3742
44. Lovat F et al (2009) Proepithelin is an autocrine growth factor for bladder cancer. Carcinogenesis 30(5):861–868
45. Liau LM et al (2000) Identification of a human glioma-associated growth factor gene, granulin, using differential immuno-absorption. Cancer Res 60(5):1353–1360
46. Wang W et al (2003) PC cell-derived growth factor (granulin precursor) expression and

action in human multiple myeloma. Clin Cancer Res 9(6):2221–2228
47. Matsumura N et al (2006) Oncogenic property of acrogranin in human uterine leiomyosarcoma: direct evidence of genetic contribution in in vivo tumorigenesis. Clin Cancer Res 12(5):1402–1411
48. Serrero G et al (2012) Progranulin (GP88) tumor tissue expression is associated with increased risk of recurrence in breast cancer patients diagnosed with estrogen receptor positive invasive ductal carcinoma. Breast Cancer Res 14(1):R26
49. Koo DH et al (2012) Progranulin as a prognostic biomarker for breast cancer recurrence in patients who had hormone receptor-positive tumors: a cohort study. PLoS One 7(6):e39880
50. He Z, Bateman A (1999) Progranulin gene expression regulates epithelial cell growth and promotes tumor growth in vivo. Cancer Res 59(13):3222–3229
51. He Z et al (2002) Progranulin (PC-cell-derived growth factor/acrogranin) regulates invasion and cell survival. Cancer Res 62(19):5590–5596
52. Lu R, Serrero G (2001) Mediation of estrogen mitogenic effect in human breast cancer MCF-7 cells by PC-cell-derived growth factor (PCDGF/granulin precursor). Proc Natl Acad Sci U S A 98(1):142–147
53. Monami G et al (2006) Proepithelin promotes migration and invasion of 5637 bladder cancer cells through the activation of ERK1/2 and the formation of a paxillin/FAK/ERK complex. Cancer Res 66(14):7103–7110
54. Kong WJ et al (2007) PC cell-derived growth factor overexpression promotes proliferation and survival of laryngeal carcinoma. Anticancer Drugs 18(1):29–40
55. Pizarro GO et al (2007) Prosurvival function of the granulin-epithelin precursor is important in tumor progression and chemoresponse. Int J Cancer 120(11):2339–2343
56. Tangkeangsirisin W et al (2004) PC cell-derived growth factor mediates tamoxifen resistance and promotes tumor growth of human breast cancer cells. Cancer Res 64(5):1737–1743
57. Wong NC et al (2014) Antibody against granulin-epithelin precursor sensitizes hepatocellular carcinoma to chemotherapeutic agents. Mol Cancer Ther 13(12):3001–3012
58. Bandey I et al (2015) Progranulin promotes Temozolomide resistance of glioblastoma by orchestrating DNA repair and tumor stemness. Oncogene 34(14):1853–1864
59. Lu R, Serrero G (2000) Inhibition of PC cell-derived growth factor (PCDGF, epithelin/granulin precursor) expression by antisense PCDGF cDNA transfection inhibits tumorigenicity of the human breast carcinoma cell line MDA-MB-468. Proc Natl Acad Sci U S A 97(8):3993–3998
60. Ho JC et al (2008) Granulin-epithelin precursor as a therapeutic target for hepatocellular carcinoma. Hepatology 47(5):1524–1532
61. Cheung PF et al (2011) Granulin-epithelin precursor is an oncofetal protein defining hepatic cancer stem cells. PLoS One 6(12):e28246
62. Elkabets M et al (2011) Human tumors instigate granulin-expressing hematopoietic cells that promote malignancy by activating stromal fibroblasts in mice. J Clin Invest 121(2):784–799
63. Dong T et al (2016) PGRN promotes migration and invasion of epithelial ovarian cancer cells through an epithelial mesenchymal transition program and the activation of cancer associated fibroblasts. Exp Mol Pathol 100(1):17–25
64. Cheung PF et al (2014) Granulin-epithelin precursor renders hepatocellular carcinoma cells resistant to natural killer cytotoxicity. Cancer Immunol Res 2(12):1209–1219
65. Suzuki M et al (2000) Suppression of copulatory behavior by intracerebroventricular infusion of antisense oligodeoxynucleotide of granulin in neonatal male rats. Physiol Behav 68(5):707–713
66. Suzuki M et al (1998) Identification of a sex steroid-inducible gene in the neonatal rat hypothalamus. Neurosci Lett 242(3):127–130
67. Baker M et al (2006) Mutations in progranulin cause tau-negative frontotemporal dementia linked to chromosome 17. Nature 442(7105):916–919
68. Cruts M et al (2006) Null mutations in progranulin cause ubiquitin-positive frontotemporal dementia linked to chromosome 17q21. Nature 442(7105):920–924
69. Martens LH et al (2012) Progranulin deficiency promotes neuroinflammation and neuron loss following toxin-induced injury. J Clin Invest 122(11):3955–3959
70. Hu F et al (2010) Sortilin-mediated endocytosis determines levels of the frontotemporal dementia protein, progranulin. Neuron 68(4):654–667
71. Tanaka Y et al (2013) Increased lysosomal biogenesis in activated microglia and exacerbated neuronal damage after traumatic brain injury in progranulin-deficient mice. Neuroscience 250:8–19

72. Ahmed Z et al (2010) Accelerated lipofuscinosis and ubiquitination in granulin knockout mice suggest a role for progranulin in successful aging. Am J Pathol 177(1):311–324
73. Kleinberger G et al (2010) Increased caspase activation and decreased TDP-43 solubility in progranulin knockout cortical cultures. J Neurochem 115(3):735–747
74. Smith KR et al (2012) Strikingly different clinicopathological phenotypes determined by progranulin-mutation dosage. Am J Hum Genet 90(6):1102–1107
75. Van Kampen JM et al (2014) Progranulin gene delivery protects dopaminergic neurons in a mouse model of Parkinson's disease. PLoS One 9(5):e97032
76. Minami SS et al (2014) Progranulin protects against amyloid beta deposition and toxicity in Alzheimer's disease mouse models. Nat Med 20(10):1157–1164

Part II

Analytical and In Vitro Methods for the Study of Progranulin Biology

Chapter 2

Chromatographic Methods for the Purification of Granulin Peptides

Andrew Bateman, Babykumari P. Chitramuthu, and Hugh P. J. Bennett

Abstract

Progranulin is composed of seven repeating cysteine-rich granulin domains. In some cells and tissues, the progranulin is fragmented by proteolysis to generate the granulin modules as individual peptides, which are collectively referred to as granulins. These peptides are often biologically active, but the activity need not be identical to that of the parental progranulin from which they are derived. Thus, some granulin peptides stimulate cell proliferation, as does progranulin itself, while other granulin peptides suppress proliferation. Similarly, some granulin peptides promote inflammation even though progranulin itself suppresses inflammation. Investigating the structural and biological properties of granulin peptides is challenging. Here we discuss methods that employ reversed-phase high-performance liquid chromatography (RP-HPLC) and in some instances size-exclusion high-performance liquid chromatography (SE-HPLC) to isolate granulin peptides from tissues, in particular those that are rich in inflammatory cells such as neutrophils, bone marrow, or hematopoietic organs of teleost fish.

Key words Granulin, Peptide extraction, High-performance liquid chromatography, Neutrophils, Carp spleen

1 Introduction

Mammalian progranulin consists of seven granulin modules called granulin A to granulin G. Granulin modules possess a structural motif of 12 cysteines that form 6 internal disulfide bridges, the only exception being granulin G which has 10 cysteines [1]. Some of the granulin modules exist as discrete granulin peptides [2], sometimes also called epithelin peptides [3], which have been generated from progranulin by proteolysis. Granulin peptides have been isolated, for example, from human neutrophils [2] and urine [4], rodent bone marrow [2] and kidneys [3], and fish hematopoietic tissue [5]. Mammalian tissues contain relatively low levels of granulin peptides whereas the fish hematopoietic tissues are remarkably rich in granulins [5]. This abundance of granulins in fish tissue, together with methods to isolate them to a high degree of purity,

Andrew Bateman et al. (eds.), *Progranulin: Methods and Protocols*, Methods in Molecular Biology, vol. 1806,
https://doi.org/10.1007/978-1-4939-8559-3_2,

greatly facilitated the three-dimensional structural analysis of the granulin fold [6] by solution nuclear magnetic resonance (NMR). It has been known for some time that the granulin peptides have biological activities, but often their activity is not the same as that of the parental progranulin. For example, progranulin stimulates cell proliferation [3, 7], as do some granulin peptides, while other granulin peptides inhibit proliferation [3, 7]. Progranulin generally suppresses inflammation [8] while some granulin peptides stimulate the secretion of pro-inflammatory cytokines [8]. Unraveling this apparent dichotomy between progranulin action and the sometimes opposing activity of its granulin peptide fragments has proven difficult, not least because the peptides are not easily available for study. Here we will discuss methods to purify granulin peptides. The methods presented focus on the purification of the peptides from tissues, but they can readily be modified to purify granulin peptides generated from recombinant expression systems. Three tissues, namely, human neutrophils, rat bone marrow, and fish hematopoietic tissue, will be discussed using three different but related purification methods. This should enable the reader to adapt the methodology to develop customized strategies to suit their individual purification needs.

Granulin peptides can be purified to homogeneity using RP-HPLC by using modifications of methods first designed for the isolation of peptide hormones [9, 10]. In RP-HPLC the sample is passed at high pressure over columns that are packed with a stationary phase of silica beads that have been chemically coated with alkyl chains [11]. The samples are injected onto the HPLC column in an aqueous buffer, the mobile phase, that contains an acidic counterion such as trifluoroacetic acid [12]. Upon injection, the peptides are retained by association with the alkyl coating of the silica beads and become immobilized at the head of RP-HPLC column. The hydrophobicity of the mobile fluid phase is then increased in a gradient fashion using a solvent such as acetonitrile while the concentration of the counterion is kept constant throughout. The increasing hydrophobicity of the mobile phase displaces the peptides from their association with the coated beads, and the peptides elute from the column according to their relative hydrophobicity with the less-hydrophobic peptides eluting before the more-hydrophobic peptides. The purification of granulin peptides employs octadecyl carbon chain (C_{18})-bonded silica RP-HPLC columns. The original granulin purifications reported in the literature were performed using Waters μBondapak C_{18} columns; however other C_{18} columns work just as well (e.g. the Vydac series). A high-salt acidic tissue extraction medium is employed as this favors the extraction of peptides, such as granulins. The peptides are readily soluble in the extraction medium whereas the majority of the larger proteins precipitate out of solution. To further concentrate the peptide component of the extract before its final purification by

RP-HPLC, the extract is subjected to reversed-phase enrichment through the use of Sep-Pak C_{18} cartridges. The Sep-Pak cartridges contain loosely packed C_{18}-coated silica beads at low pressure. They are useful for batch extraction but do not provide chromatographic resolution to achieve purification on their own. The Sep-Paks serve to desalt the peptide extract and concentrate it, as well as further eliminate any remaining higher molecular weight proteins as these are, in general, not well retained on the Sep-Pak cartridges.

The systems devised to purify granulins and other small peptides use volatile solvents throughout. By avoiding salt-based buffer solutions, chromatography samples can be dried completely and subsequently analyzed chemically and biologically without recourse to desalting procedures that can reduce yields especially when dealing with low-abundance peptides. Purification is achieved by using a number of HPLC strategies in series all with the common feature that organic solvents, usually acetonitrile, are used. This includes size-exclusion (SE) chromatography. The column packing of the high-performance SE columns is relatively inert to the nature of the running buffer and can be operated using organic solvents. Other variables include the use of trifluoroacetic acid (TFA) and heptafluorobutyric acid (HFBA). TFA and HFBA are fully dissociated organic acids that retain their hydrophobic character and act as counterions by associating with the basic charges within peptide sequences at low pH (i.e., arginine, lysine, histidine, and the amino-terminal amine group). Employing different counterions and pH changes the separation selectivity of reversed-phase columns and facilitates purifications by employing multiple chromatographic runs using different solvent systems but the same reversed-phase column [13].

Granulin peptides are most easily isolated from isolated neutrophil or tissues rich in neutrophils. Thus, human granulin peptides were originally isolated from the peritoneal exudate of patients with peritonitis. Although this is a very rich source of neutrophils, it is difficult to obtain. Human granulin peptides can instead be isolated from blood neutrophils with the method of purification being mostly identical. Precautions must be taken whenever handling human material to prevent the risk of contamination. Rat granulin peptides can be isolated from bone marrow. Fish granulin peptides can be isolated from head kidney and spleen. The head kidney of teleost fish has similar functions as the bone marrow in mammals [14, 15]. As with mammals, the fish spleen is enriched in blood cells. The method for fish granulin peptide isolation was developed for use with the carp, *Cyprinus carpio* [6].

Three related strategies are presented for the purification of granulin peptides (Fig. 1). In the first strategy, purification is achieved by three sequential chromatographic steps employing first a C_{18} RP-HPLC gradient fractionation using 0.1% TFA as counterion and acetonitrile as the hydrophobic solvent, followed by SE-HPLC using

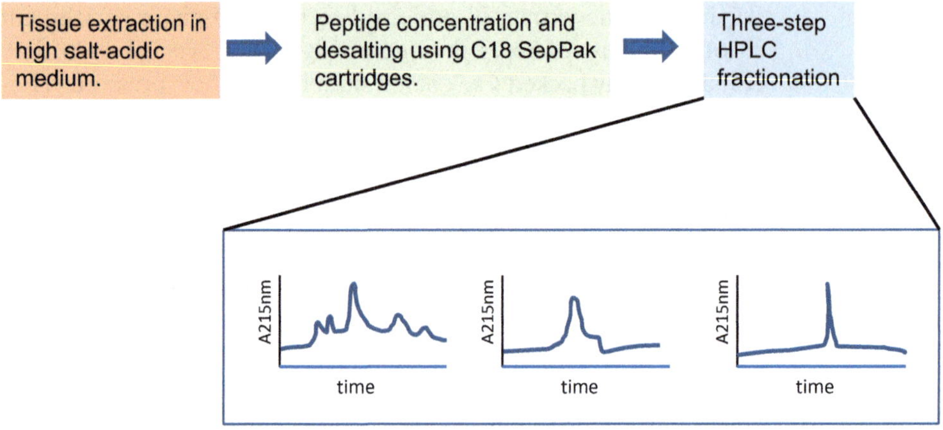

Example	First fractionation	Second fractionation	Third fractionation
Human granulins.	RP-HPLC, 0.1% TFA, Acetonitrile solvent.	Size exclusion HPLC.	RP-HPLC, 0.1% TFA, Acetonitrile solvent.
Rat bone marrow granulins.	RP-HPLC, 0.1% TFA, Acetonitrile solvent.	RP-HPLC, 0.13% HFBA, Acetonitrile solvent.	Size exclusion HPLC.
Fish spleen, head kidney granulins	RP-HPLC, 0.1% TFA, Acetonitrile solvent.	RP-HPLC, 0.13% HFBA, Acetonitrile solvent.	RP-HPLC, 0.1% TFA, Acetonitrile solvent.

Fig. 1 Strategies for the isolation of granulin peptides. Tissues are first extracted in a high-salt/acidic extraction medium. Peptides in the extract are concentrated and desalted using C_{18} Sep-Pak cartridges. This is followed by three-step HPLC fractionation. Three variations of the chromatographic strategy are summarized. The fractionation columns outline the type of HPLC employed (reversed phase, RP, or size exclusion), the counterion used (trifluoroacetic acid, TFA, or heptafluorobutyric acid, HFBA), and the volatile solvent, acetonitrile. For details please consult the text

Protein-Pak 125 columns, and a final C_{18} RP-HPLC separation employing again a 0.1% TFA counterion and acetonitrile gradient mobile phase. This will be exemplified using human granulins as the example [2]. The second strategy, exemplified by the purification of granulin peptides from rat bone marrow [2], uses sequential RP-HPLC fractionations employing two different counterions, namely, 0.1% TFA in the first fractionation and 0.13% HFBA in the second fractionation step. The final step of purification employs size-exclusion HPLC. The third variation of the method, exemplified here by extraction of fish granulins, also employs three sequential chromatographic steps: first a C_{18} RP-HPLC gradient fractionation with a 0.1% TFA as the counterion and acetonitrile as the solvent, followed by RP-HPLC fractionation with 0.13% HFBA as the counterion, and a final RP-HPLC purification employing 0.1% TFA and acetonitrile gradient fractionation.

The initial reports on the isolation of the granulin peptides illustrated the use of amino acid analysis of fraction aliquots to assay for the characteristic 12 cysteine signature of granulins. Very few laboratories have access to amino acid analysis, and instead we recommend the use of mass spectrometric screening of fractions to identify the elution position of the granulin peptides. This works well for non-glycosylated granulins but is complicated when the granulin peptide is glycosylated, as in the case of human granulin C, or if the proteolysis to the granulin peptide is incomplete.

2 Reagents and Materials

2.1 Tissue Extraction

1. 1% Dextran T500 (Pharmacia, Uppsala, Sweden) is prepared by dissolving 10 g of dextran T500 in 1000 ml of 0.9% NaCl (*see* **Note 1**).
2. 0.15 M Ammonium chloride (NH_4CL). NH_4Cl (8.26 g/l), potassium bicarbonate (1.0 g/l)), and EDTA, tetrasodium ethylene-diaminetetraacetic acid (0.037 g/l) in water (pH 7.3).
3. Buffered salt solution. Glucose 5.5×10^{-3} M, 1.0 g/liter; $CaCl_2 \cdot 2H_2O$, 5.0×10^{-5} M 7.4 mg/l; $MgCl_2 \cdot 6H_2O$ 9.8×10^{-4} M, 0.19 g/l; KCl 5.4×10^{-3} M, 0.40 g/l; TRIS 0.145 M, 17.6 g/l. Dissolve in 950 ml distilled water and adjust the pH to 7.6 using concentrated HCl, then bring to 1000 ml final volume.
4. Ficoll-Paque PLUS density gradient centrifugation medium (GE Healthcare Life Sciences).
5. Trypan blue, 0.4% solution (Gibco).
6. Microscope slides.
7. Wright's stain (Sigma-Aldrich).
8. Tissue homogenizer. "A Cole-Parmer Ultrasonicator" 4710 series 100 watts max mode, sonic probe length 1.8 cm (7 inches), tip diameter 3 mm), or a Polytron tissue homogenizer (Brinkmann, Mississauga, Ontario, Canada) are suitable.
9. Extraction medium, 1 M hydrochloric acid, 5% formic acid (v/v), 1% trifluoroacetic acid (v/v), and 1 g sodium chloride (w/v) per 1000 ml. This can be stored for several months.
10. Sep-Pak C_{18} Classic Cartridge, 360 mg Sorbent per Cartridge, 55–105 μm Particle Size (Waters Corporation, Milford, MA, USA).
11. 15 ml polypropylene conical centrifugation tubes.

12. 50 ml polypropylene conical centrifugation tubes.
13. SpeedVac lyophilizer.
14. Low-speed centrifuge.
15. High-speed centrifuge.
16. Ice.

2.2 Chromatography

1. Milli-Q Ultrapure filtered and deionized water (EMD Millipore, Billerica, Massachusetts).
2. Trifluoroacetic acid (TFA), HPLC grade, ≥99.0% pure.
3. Heptafluorobutyric acid (HFBA); HPLC grade.
4. Acetonitrile, HPLC grade.
5. Solution A. 0.1% TFA in Milli-Q purified water. Add 1 ml of anhydrous TFA in 999 ml water.
6. Solution B. 0.1% TFA in 80% acetonitrile. Add 1 ml of anhydrous TFA in 199 ml Milli-Q water, add 800 ml acetonitrile.
7. Solution C. 0.13% HFBA. Add 1.3 ml HFBA to 998.7 ml Milli-Q water.
8. Solution D. 80% acetonitrile in 0.13% HFBA. Add 0.13 ml HFBA to 198.7 ml Milli-Q water. Make up to 1000 ml with 800 ml acetonitrile.
9. 1.5 ml Eppendorf tubes.
10. Microfuge.
11. μBondapak C_{18} 15–20 μm, 7.8 mm × 300 mm (Waters Corporation, Milford, MA, USA).
12. μBondapak C_{18} 15–20 μm, 3.9 mm × 300 mm (Waters Corporation, Milford, MA, USA).
13. Two Protein-Pak 125 Columns, 10 μm, 19 × 300 mm (Waters Corporation, Milford, MA, USA).
14. Fraction collector (capable of collecting 1 min fractions).
15. 3 ml Polypropylene collection tubes, preferably with conical bottoms.
16. Hamilton liquid syringe, 500 μl with a blunt-tipped needle.
17. High-performance liquid chromatography instrument; many instruments are suitable, but must be equipped with two pumps both capable of pumping solvents at 1.5 ml/min that can be programmed to deliver gradient flows, large capacity sample loop (500 μl to 2 ml), and UV detection monitors operating at 215 nm and 280 nm (the 280 nm detection is optional).

2.3 Animals

1. Live carp, approximate weight of 2.2–3.6 kg (5 to 7 pounds) obtained at a commercial fish market.
2. Adult Sprague-Dawley rats.

3 Methods

3.1 Tissue Preparation and Extraction

3.1.1 Human Neutrophils

1. Separate fresh human venous blood collected with citrate as anticoagulant into red blood cell and leukocyte fractions by gravity sedimentation in 1% dextran T500 at room temperature. Gently mix blood 1:1 with the dextran T500 solution and allow to stand for approximately 30 min at room temperature.
2. Remove the residual erythrocytes trapped in the leukocyte sediment by lysis in 0.15 M ammonium chloride. Mix cells in the ammonium chloride by gentle up and down pipetting with approximately 10 to 1 volume-to-volume ratio of ammonium chloride to pelleted cells. Stand for 5 min at room temperature.
3. Spin at 400 × *g* for 10 min at room temperature. Collect the pellet and wash in balanced salt solution. If the pellet appears to contain residual contamination with red blood cells, repeat the procedure.
4. Suspend the washed pellet in 10 ml of balanced salt solution and layer over 3 ml of Ficoll-Paque (Pharmacia) in a 15 ml conical sterile centrifuge tube at room temperature.
5. Spin in a centrifuge at 100 × *g* for 15 min. Neutrophils pellet at the bottom of the tube while lymphocytes, monocytes, and platelets form a band at the upper face of the Ficoll-Paque layer. Collect the pellet (*see* **Note 2**).
6. Resuspend the neutrophil pellet in 5–10 ml balanced salt solution.
7. To assess cell number and integrity, take an aliquot, typically 50 μl, and dilute 1:1 with trypan blue. Count viable cell in a hemocytometer (*see* **Note 3**). Cells should be greater than 95% viable.
8. To confirm that the cells are predominantly neutrophils, spread a second aliquot of cells on a clean microscope slide and stain using Wright's stain. Examine under a microscope and evaluate purity of the preparation by counting cells with neutrophil nucleus morphology versus those without typical neutrophil nucleus morphology.

3.1.2 Isolation of Crude Granule Fraction

This is an optional step. Its purpose is to remove most of the cytosolic contamination and concentrate the granule fraction of the cells. The major contaminating peptide that fractionates close to the granulin peptides is a cytosolic polypeptide called thymosin-β4. The granulin peptides are located within the granule fraction, and therefore this step eliminates much of the thymosin-β4 contamination and simplifies the later chromatographic steps.

1. Lyse the cells in ice cold 20 ml HBSS using 30 s bursts of an ultrasonic homogenizer. Inspect an aliquot under the microscope to ensure complete cell lysis.
2. Remove the cellular debris by centrifugation at 500 × *g* for 10 min.
3. Inspect the supernatant under the microscope to confirm the complete removal of cell debris. Repeat the centrifugation step if the supernatant remains contaminated with cell debris.
4. Centrifuge the supernatant at 5000 × *g* for 20 min at 4 °C and save the pellet which contains the crude granule fraction.

3.1.3 Rat Bone Marrow Extraction

Bone marrow is obtained from the femurs of Sprague-Dawley rats. A typical purification employs 100 femurs.

1. Snip the femurs at one end using fine scissors and aspirate the marrow using a 10 ml syringe containing ice cold extraction medium. Extract the aspirate immediately (go to Subheading 3.2).

3.1.4 Carp Spleen and Head Kidney

1. Dissect the tissues and mince using fine scissors.
2. Disperse the tissue in phosphate-buffered saline, pH 7.2, by up and down aspiration through an 18-gauge needle attached to a 10 ml plastic syringe. Allow large aggregates to settle under gravity and discard.
3. Centrifuge at 1800 × *g* for 15 min.
4. Repeat the dispersal of cells and matrix three times, and discard the supernatant at each step.
5. Suspend an aliquot of the final pellet in trypan blue and examine under the microscope. The preparation should be of cells attached to fibrous matrix fragments and show 95% viability.

3.2 Peptide Extraction, Human and Rat Tissues

1. Suspend either whole cells, crude granule fraction, or bone marrow aspirate in 10–20 ml ice cold extraction medium in a 50 ml polypropylene centrifuge tube.
2. Place the suspension on ice and homogenize at high power using 30 s pulses of ultrasonic homogenizer for 5 min.
3. Spin the extract in a centrifuge at 2000 × *g* for 10 min at 4 °C in polypropylene centrifuge tubes. Save the supernatant.
4. If extracting from whole cells, suspend an aliquot of the pellet in trypan blue and inspect under the microscope to ensure that most cells have been lysed. If a significant number of intact cells remain, repeat the sonication step.

3.3 Peptide Extraction, Carp Tissue

1. Suspend the cell pellet in 35 ml of ice-cold acidic extraction medium.
2. Place the suspension on ice and extract using a Polytron tissue homogenizer (Brinkmann, Mississauga, Ontario, Canada) at high speed.

3. Centrifuge at 1800 × *g* for 15 min. Save the supernatant and keep on ice.
4. Re-extract the pellets with the acidic extraction medium and pool the supernatant with that obtained in the first extraction.

3.4 Desalting and Reversed-Phase Enrichment of Extracted Peptides

This step in the protocol removes contaminating salts and concentrates the peptides in a form suitable for fractionation by HPLC.

1. Attach at least two C_{18} Sep-Pak cartridges to 10 ml syringes.
2. Equilibrate the Sep-Pak cartridges by passing through 5 ml of 80% acetonitrile containing 0.1% TFA.
3. Flush with 10 ml of 0.1% TFA to remove excess acetonitrile.
4. Using the syringe, pass the peptide extracts through the Sep-Pak cartridge.
5. Collect the flow through and pass it through the second equilibrated Sep-Pak cartridge.
6. Wash the Sep-Paks with 10 ml of 0.1% TFA.
7. Elute the peptides by passing through 3 ml of 80% acetonitrile in 0.1% TFA. Collect the eluate and dry in a SpeedVac.

3.5 HPLC of Human Granulin Peptides (See Fig. 2)

1. Equilibrate a μBondapak C_{18} 15–20 μm, 7.8 mm × 300 mm RP-HPLC column (or an equivalent column from other manufacturer) by passing solution B at 1.5 ml/min through the column for 15 min, followed by solution A for 20–30 min at 1.5 ml/min.
2. Pump solution A continuously through the RP-HPLC column at 1.5 ml/min until the gradient elution is initiated.
3. Vortex the lyophilized Sep-Pak eluates in solution A until dissolved. The amount of solution A to use depends on the quantity of lyophilized extract. If the extract does not dissolve fully, add concentrated TFA dropwise and vortex.
4. Transfer to 1.5 ml Eppendorf tubes and spin in a Microfuge at high speed for 5 min.
5. Remove and save the supernatant (*see* **Note 4**).
6. Inject the supernatant onto the RP-HPLC column (*see* **Note 5**). Run a linear RP-HPLC gradient from 100% solution A at 1.5 ml/min to 100% solution B at 1.5 ml/min over 180 min. Collect 1 min fractions in 3 ml polypropylene tubes (*see* **Note 6**). Monitor the UV absorption profile at wavelengths of 210 nm and 280 nm. *See* Fig. 1a, b for a representative chromatogram (*see* **Note 7**).
7. To identify the position of granulin peptides, monitor aliquots (10–50 μl) of UV-absorbing fractions that elute ahead of the defensin peak (*see* Fig. 1a) using mass spectrometry (*see* **Note 8**). The anticipated molecular weights for human granulins are given in Table 1.

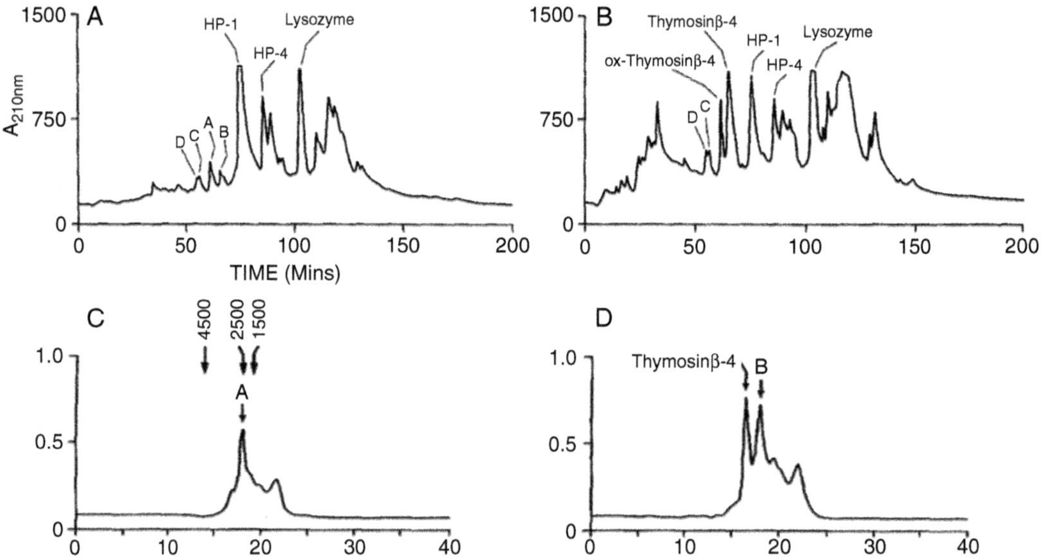

Fig. 2 RP-HPLC fractionation of human granulins. (**a**) The first RP-HPLC fractionation from a crude granule extract obtained from inflammatory exudate cells, and (**b**) from a whole cell extract. In both fractionations the peptide extracts were loaded onto a C_{18} μBondapak column and eluted from 0% to 100% acetonitrile in 0.1% trifluoroacetic acid throughout at 1.5 ml/min for 3 h. The positions of the granulins are marked by A, B, C, and D. HP1 is human defensin 1 and is an abundant peptide in neutrophil granules. Thymosin-β4 is a cytoplasmic marker peptide. (**c**) Size-exclusion HPLC fractionation of granulin A. (**d**) Size-exclusion chromatography of granulin B

8. Lyophilize the fractions in a SpeedVac. Store at −20 °C.
9. Chromatographic **step 2**; high-performance size-exclusion chromatography (Fig. 1c). Connect two Protein-Pak 125 high-performance size-exclusion columns in series. Equilibrate the columns in 40% acetonitrile in 0.1% TFA at a flow rate of 1 ml/min for at least 30 min.
10. Dissolve the fractions of interest from the first RP-HPLC step in a total of 100 μl of a 40% acetonitrile in 0.1% TFA solution. Inject onto the HPLC and begin collection immediately (*see* **Note 9**). Collect 1 min fractions.
11. Remove aliquots (10–50 μl) to screen fractions for peptides of interest by mass spectrometry.
12. Lyophilize fractions and store at −20 °C.
13. Final RP-HPLC chromatographic step. Equilibrate a μBondapak C_{18} 15–20 μm, 3.9 mm × 300 mm (or equivalent column) as described above. Run isocratically at 10% solution B until ready to inject the sample.
14. Dissolve the peptide of interest in solution A.
15. Inject onto the RP-HPLC column.

Table 1
The molecular weight of human, rat, and carp granulin (epithelin) peptides. The molecular weight given is the average molecular weight over each monoisotopic peak, with full disulfide bridges. Values in italics are peptides that have not been isolated and are the predicted values based on the sequence of progranulin. The proteolysis of progranulin can result in ragged termini, extended by or reduced by one amino acid, in which case the molecular weight can readily be calculated from the full progranulin sequence. For example, human granulin A from urine has two alternate amino termini [4], the major form has an amino terminus of DVKC…, the minor form has an amino terminus of VKC

Human granulin	Position in progranulin	Amino terminus	Carboxy terminus	Molecular weight
Paragranulin	18–47	TRCP	PLLD	3142.62
G (module 1)	*58–113*	*GGPC*	*RSCF*	*5651.29*
F (module 2)	123–179	AIQC	TRCI	6232.18
B (module 3)	206–261	VMPC	SKCL	5985.95
A (module 4)	281–336	DVKC	GTCE	6045.86
C (module 5)	364–417	VPCD	GQCQ	5659.30
D (module 6)	442–496	IGCD	RSCE	5806.62
E (module 7)	*518–573*	*DVEC*	*TKCL*	*6257.09*
Rat granulin				
Granulin 2	204–259	VVCP	SKCI	5995.97
Granulin 1	278–334	EVKC	TCEL	6178.03
Carp granulin				
Granulin 1		VIHC	HCLR	6277.21
Granulin 2		VVYC	LCLR	6325.62
Granulin 3		VVFC	LCLR	6303.34

16. Elute the column using a linear gradient of 10 to 40% solution B at a flow rate of 1.5 ml/min for 90 min. Monitor the column eluate by UV absorption at wavelengths of 215 nm and (optionally) at 280 nm.
17. Collect 1 min fractions and remove aliquots of between 10 and 50 μl for mass spectrometry.
18. Lyophilize the fractions in a SpeedVac and store the dried, sealed tubes at −20 °C or colder.

3.6 HPLC of Granulin Peptides from Rat Bone Marrow Extracts (See Fig. 3)

1. Equilibrate μBondapak C_{18} 15–20 μm, 7.8 mm × 300 mm (or equivalent) column in solution B and wash in solution A as above.
2. Inject the sample and elute using a linear gradient of 4 to 48% solution B over 1 h at 1.5 ml/min. Monitor the UV

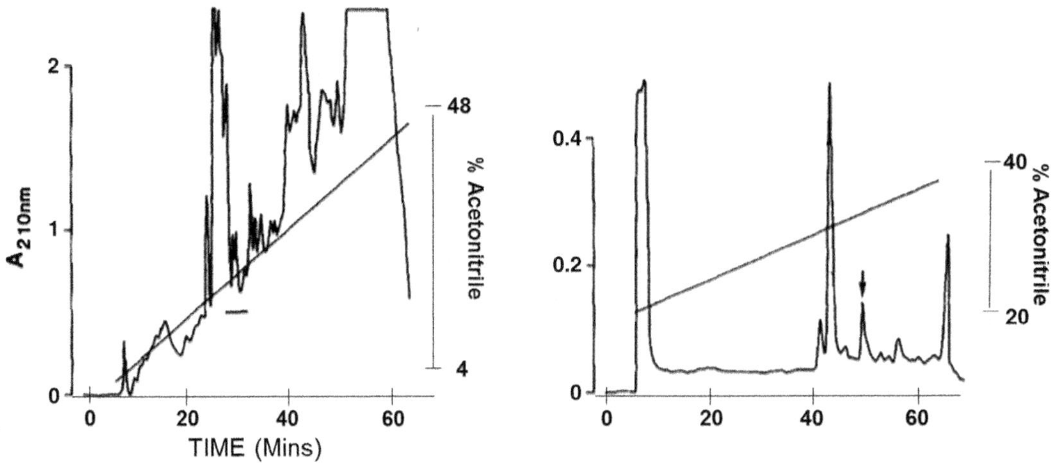

Fig. 3 RP-HPLC purification of rat granulin from bone marrow. In left hand panel the first HPLC fractionation employed acetonitrile/trifluoroacetic acid. The right hand panel the second step of purification employed acetonitrile/HFBA. The bar in the left hand panel indicates where the rat granulin elutes, and in the right hand panel the arrow indicates its elution position

absorption of the eluate at wavelengths of 210 nm and 280 nm. Collect 1 ml fractions. Lyophilize the fraction in a SpeedVac.

3. Equilibrate a μBondapak C_{18} 15–20 μm, 3.9 mm × 300 mm (or equivalent) column in solution D, and wash in solution C. Dissolve the peptides of interest from the initial HPLC fractionation in solution C. Inject onto the RP-HPLC column and elute using a linear gradient from 20% to 40% solution D over 1 h at a flow rate of 1.5 ml/min. Collect 1 min fractions.
4. Lyophilize fraction in a SpeedVac.
5. Perform size-exclusion HPLC to achieve final purification of the peptides as described above.

3.7 HPLC of Carp Granulin Peptides (See Fig. 4)

1. Equilibrate a μBondapak C_{18} 15–20 μm, 7.8 mm × 300 mm (or equivalent) column in solution B, and wash in solution A as above.
2. Continue to pass 100% solution A through the column at 1.5 ml/min until the sample is injected and the elution gradient is initiated.
3. Dissolve and centrifuge the lyophilized Sep-Pak eluate as above.
4. Inject on to the RP-HPLC column and elute using a 1 h linear gradient from 0–50% solution B.
5. Collect 1 ml fractions and lyophilize. The positions of granulin peptides can be determined by mass spectrometry (*see* Fig. 5).

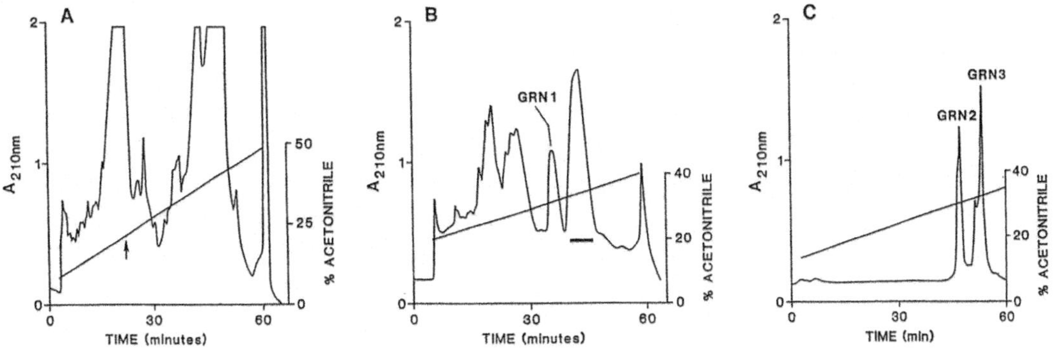

Fig. 4 RP-HPLC purification of carp granulins 1, 2, and 3. (**a**) The peptide-enriched extract from three head kidneys was loaded onto a C_{18} μBondapak column which was then eluted over 1 h at 1.5 ml/min with a linear gradient of 0–50% acetonitrile containing 0.1% trifluoroacetic acid throughout. Fractions 20–26 (indicated by an arrow) were found to contain granulin peptides. (**b**) The material of interest was loaded onto the same C_{18} μBondapak HPLC column which was then eluted over 1 h at 1.5 ml/min with a linear gradient of 20–40% acetonitrile containing 0.13% HFBA throughout. Granulin-1 (GRN1) elutes in fractions 36–40. Fractions 41–48 contain granulin-2 and granulin-3 (indicated by the bar). (**c**) RP-HPLC purification of granulin-2 and granulin-3 (GRN2 and GRN3) was achieved using a gradient of 15–35% acetonitrile over 1 h containing 0.1% trifluoroacetic acid. GRN1 was purified separately using a linear gradient of 12–28% acetonitrile containing 0.1% trifluoroacetic acid at 1.5 ml/min for 1 h

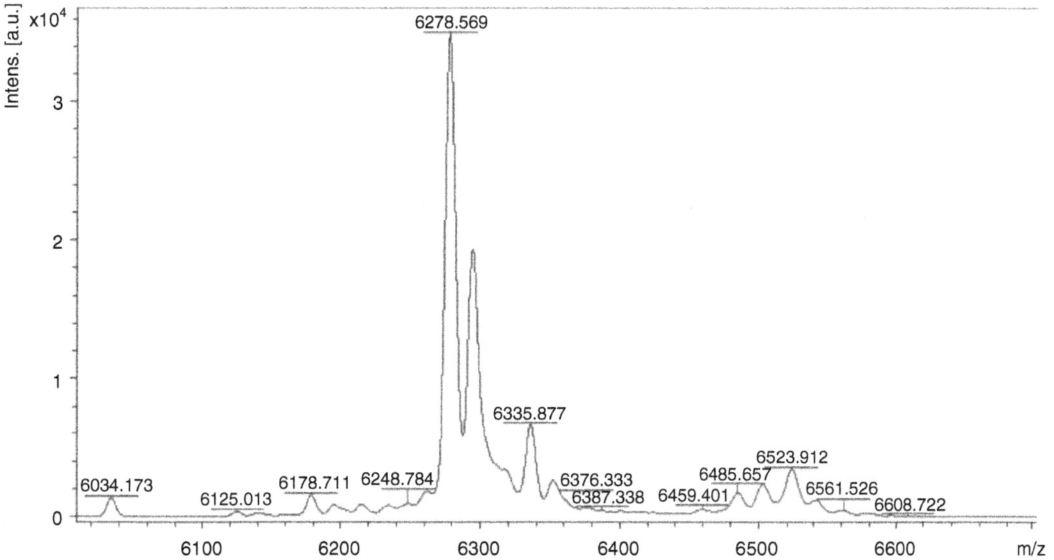

Fig. 5 Mass spectrometric identification of carp GRN1. The molecular weight for GRN1 is 6277.21 (Table 1), giving a predicted quasimolecular ion $[M + H]^+$ of 6278.21. A one-thirtieth aliquot of the chromatography faction from the first trifluoroacetic acid: acetonitrile gradient of the three-step RP-HPLC sequence (Subheading 3.6) was used for matrix-assisted laser desorption ionization time-of-flight mass spectrometry (MALDI-TOF-MS) analysis. The matrix solution was dihydroxybenzoic acid and was prepared at saturation in a solvent mixture (water with 0.1% trifluoroacetic acid: acetonitrile, 1:1 v/v). For the sample-matrix crystallization procedure, the aliquot of sample was then mixed with the saturated matrix in a 1:1 ratio (v/v), and 1 μl of this mixture was directly spotted onto the MALDI target plate. MALDI spectra were acquired on an ultrafleXtreme time-of-flight spectrometer operated in the positive ion, linear mode equipped with a 2 kHz repetition rate smartbeam-II laser (Bruker Daltonics, Proteomics Platform, Centre for Translational Bioolgy, Rersearch Institute of the the McGill Universiy Health Centre)

6. Equilibrate the RP-HPLC column in solution D and wash in solution C. Continue to pass solution C through the column isocratically until the sample is injected and the elution gradient is initiated. Resuspend the fractions of interest in solution C. Elute the column using a linear gradient of 20–40% solution D at 1.5 ml/min. Collect 1 min fractions and lyophilize.
7. Equilibrate a μBondapak C_{18} column in solution B and wash in solution A as above. Inject the fractions of interest and elute at 12–28% solution B for carp granulin 1 or at 15–35% solution B for carp granulins 2 and 3. Collect 1 min fractions and lyophilize.

3.8 Care of the HPLC Columns (See Note 10)

1. When not in use flush columns with at least one and a half column volumes in 100% acetonitrile run through both pumps at 1 ml/min for each pump.
2. Before reusing the column, wash it in at least two column volumes of 0.1% TFA or 0.13% HFBA to remove the acetonitrile then equilibrate the column as described above.

4 Notes

1. The solution can be stored for 4 weeks at 4 °C.
2. To increase the capacity of the separation step, use multiple centrifuge tubes as necessary.
3. Trypan blue stains dead cells blue, living cells remain clear.
4. Do not disturb the pellet as it will, if injected onto the HPLC column, eventually block the filters at the top of the column.
5. In preparative RP-HPLC it is important that the sample loading loop of the injector has sufficient capacity. A 2 ml sample loop is preferable; however if only smaller loops are available, use multiple injections to load the sample. For example, if a 500 μl sample loop is employed, load1.5 ml in three separate 500 μl injections. Leave sufficient time between each injection to allow the sample from the previous injection to exit the loop before injecting the next aliquot; for example, at a 1.5 ml/min flow rate, a 500 μl sample loop is fully emptied in 20 s. The peptides from each injection remain concentrated and stationary at the head of the RP-HPLC column until the elution gradient (the "run") commences.
6. Polypropylene tubes should be used as other plastics can be "attacked" by the acetonitrile and may leach chemicals into the extract. Do not use glass tubes as granulin peptides tend to "stick" to glass.

7. This is a long slow gradient that is useful for scouting the elution position of the peptides of interest, in this case granulins. It can be readily modified in subsequent cycles of purification to terminate earlier once the position of granulin elution has been securely determined for any given column.
8. Matrix-assisted laser desorption mass spectrometry or electrospray mass spectrometry is suitable; the choice of technique and details will depend on the institutional resources available. The anticipated molecular weights for human granulins are given in Table 1. In most cases this analysis needs to be done once only. When the elution positions have been established, the UV profile of the chromatogram will normally be sufficient to locate the granulin peptides.
9. Unlike RP-HPLC columns, multiple injections cannot be performed on size-exclusion columns, and the injection volume must be kept small.
10. To protect the HPLC columns from degradation, they should never be left containing aqueous acidic solutions for prolonged periods. When HPLC columns are not in use, for example, at the end of each day, they must be stored in a nonaqueous organic solvent such as 100% acetonitrile.

References

1. Bhandari V, Palfree RG, Bateman A (1992) Isolation and sequence of the granulin precursor cDNA from human bone marrow reveals tandem cysteine-rich granulin domains. Proc Natl Acad Sci U S A 89:1715–1719
2. Bateman A, Belcourt D, Bennett H, Lazure C, Solomon S (1990) Granulins, a novel class of peptide from leukocytes. Biochem Biophys Res Commun 173:1161–1168
3. Shoyab M, McDonald VL, Byles C, Todaro GJ, Plowman GD (1990) Epithelins 1 and 2: isolation and characterization of two cysteine-rich growth-modulating proteins. Proc Natl Acad Sci U S A 79:12–16
4. Sparro G, Galdenzi G, Eleuteri AM, Angeletti M, Schroeder W, Fioretti E (1997) Isolation and N-terminal sequence of multiple forms of granulins in human urine. Protein Expr Purif 10:169–174
5. Belcourt DR, Lazure C, Bennett HP (1993) Isolation and primary structure of the three major forms of granulin-like peptides from hematopoietic tissues of a teleost fish (*Cyprinus carpio*). J Biol Chem 268:9230–9237
6. Hrabal R, Chen Z, James S, Bennett HP, Ni F (1996) The hairpin stack fold, a novel protein architecture for a new family of protein growth factors. Nat Struct Biol 3:747–752
7. Tolkatchev D, Malik S, Vinogradova A, Wang P, Chen Z, Xu P, Bennett HP, Bateman A, Ni F (2008) Structure dissection of human progranulin identifies well-folded granulin/epithelin modules with unique functional activities. Protein Sci 17:711–724
8. Zhu J, Nathan C, Jin W, Sim D, Ashcroft GS, Wahl SM, Lacomis L, Erdjument-Bromage H, Tempst P, Wright CD, Ding A (2002) Conversion of proepithelin to epithelins: roles of SLPI and elastase in host defense and wound repair. Cell 111:867–878
9. Bennett HP, Browne CA, Solomon S (1981) Purification of the two major forms of rat pituitary corticotropin using only reversed-phase liquid chromatography. Biochemistry 20:4530–4538
10. Browne CA, Bennett HP, Solomon S (1981) Isolation and characterization of corticotropin- and melanotropin-related peptides from the neurointermediary lobe of the rat pituitary by reversed-phase liquid chromatography. Biochemistry 20:4538–4546

11. Regnier FE, Gooding KM (1980) High-performance liquid chromatography of proteins. Anal Biochem 103:1–25
12. Bennett HPJ, Browne CA, Solomon S (1980) The use of perfluorinated carboxylic acids in the reversed-phase HPLC of peptides. J Liq Chromatogr 3:1353–1365
13. Bennett HP (1983) Expansion of the resolving power of reversed-phase columns by manipulating pH and the nature of the ion-pairing reagent. J Chromatogr 266:501–510
14. Havixbeck JJ, Barreda DR (2015) Neutrophil development, migration, and function in teleost fish. Biology 4:715–734
15. Cadieux B, Chitramuthu BP, Baranowski D, Bennett HP (2005) The zebrafish progranulin gene family and antisense transcripts. BMC Genomics 6:156

Chapter 3

Methods for Expression and Purification of Biologically Active Recombinant Progranulin

Chun Yu Zhao and Andrew Bateman

Abstract

Purifying progranulin may be useful in a variety of situations, for example, after it has been mutated or otherwise modified or when working with a species for which commercially produced progranulin is unavailable. A method to express and purify human progranulin is presented. Progranulin is transiently expressed in mammalian cells and isolated from their conditioned medium before purification by reversed-phase high-performance liquid chromatography (RP-HPLC). Human progranulin is used as an example, but the protocol can be applied to any other progranulin protein. Modifications of the expression-purification strategy for metabolic labeling of progranulin and analytical systems based on heparin-affinity chromatography are presented.

Key words Progranulin, pcDNA 3.1, Expression, Purification, COS-7 cells, Transient transfection, Reversed-phase HPLC, Western blot, Heparin-affinity chromatography

1 Introduction

Progranulin is a secreted protein that has growth factor, immunomodulatory, and neurotrophic activities. It is widely expressed in epithelial cells, especially those that cycle rapidly, in immune cells, and the nervous system both during development and in adulthood [1]. Here we will discuss methods to express and purify progranulin. The purification is based on reversed-phase high-performance liquid chromatography (RP-HPLC). Alternative methods to purify recombinant progranulin that employ a C-terminal hexahistidine affinity tag and nickel-affinity column chromatography have been reported, but in our experience, they often give impure preparations without further chromatography. Progranulin binds the membrane protein sortilin [2, 3], and the last three amino acids of progranulin are required for this interaction to occur [4]. Therefore, there is a risk that adding hexahistidine tags at the C-terminus of progranulin will interfere with the sortilin interaction.

Andrew Bateman et al. (eds.), *Progranulin: Methods and Protocols*, Methods in Molecular Biology, vol. 1806,
https://doi.org/10.1007/978-1-4939-8559-3_3,

Progranulin expression vectors can be readily generated using commercially available *PGRN* cDNAs cloned into any standard mammalian expression vector. Progranulin is highly post-translationally modified, and it is therefore important to mammalian cell expression systems for its production. It is essential to retain the functional secretory signaling peptide as this directs progranulin into the endoplasmic reticulum for correct folding, disulfide bridge formation, and glycosylation to occur. Moreover, the secretion of progranulin into the cell culture medium facilitates its purification. The method presented for expressing recombinant progranulin uses COS-7 cells [5, 6], which is a fibroblast-like cell line that was obtained by immortalizing CV-1 cells from monkey kidneys with SV40 large T antigen [7]. Other mammalian expression cell lines such as HEK293 or CHO cells can be used instead of COS-7 cells. Cells are transfected using a lipofection reagent which contains lipid subunits that can form liposomes in an aqueous environment, entrap the transfection DNA, and carry it across the cell membrane [8, 9]. Biologically active recombinant PGRN protein is secreted, allowing it to be purified from medium conditioned by the *PGRN* transfected cells by reversed-phase high-performance liquid chromatography (RP-HPLC) [5, 6]. The elution position of PGRN is confirmed by Western blotting. The resulting protein has high purity and is biologically active, and is suitable for in vitro and in vivo applications [5, 6]. Additional expression procedures will also be presented that employ radiolabeling of the expressed progranulin. This has potential application in the analysis of progranulin biosynthesis and for purification of progranulin from species for which antibodies are not available or for the purification of modified forms of progranulin that do not cross-react well with the available progranulin antibodies. Radiolabeled progranulin could be used as a tracer in, for example, binding studies. Metabolic labeling has the advantage that no chemical alterations are incorporated into the protein that may diminish or abolish its biological activity. A method for analyzing progranulin binding to heparin that uses metabolically labeled recombinant progranulin and an alternative purification strategy based on a heparin-affinity HPLC strategy [10] is discussed.

2 Materials

2.1 Materials for Expression and HPLC Purification of Recombinant Progranulin

2.1.1 Cell Culture

1. Cell: COS-7 cells from ATCC; Catalog number: CRL1651.
2. Sterile Falcon T75 75 cm^2 surface area tissue culture flask with vented caps.
3. 15 cm culture dishes: Nunc® petri dishes from Sigma-Aldrich.
4. Inverted microscope (any model provided it is capable of inspecting cells in culture dishes).

5. Dulbecco's Minimum Eagle's Medium (DMEM).
6. FBS: Fetal bovine serum.
7. Cell culture trypsin (Gibco): 0.25% trypsin in 0.53 mM EDTA.
8. Disposable sterile 10 ml plastic pipettes.
9. Motorized pipette filler.
10. Aspiration unit.
11. Sterile 2 ml aspiration pipettes or autoclaved sterile glass Pasteur pipettes.
12. Class II laminar flow hood.
13. CO_2 cell culture incubator.
14. Hemocytometer and cover slides.
15. Trypan blue (0.4%) (Sigma-Aldrich).

2.1.2 Transfection

1. *PGRN* plasmid (*see* Fig. 1).
2. Lipofectamine 2000: from Life Technologies.
3. Platform rocker; any model is acceptable provided it can accommodate cell culture flasks and Petri dishes.
4. PBS: 500 ml PBS from Wisent Bioproducts.
5. DMEM.

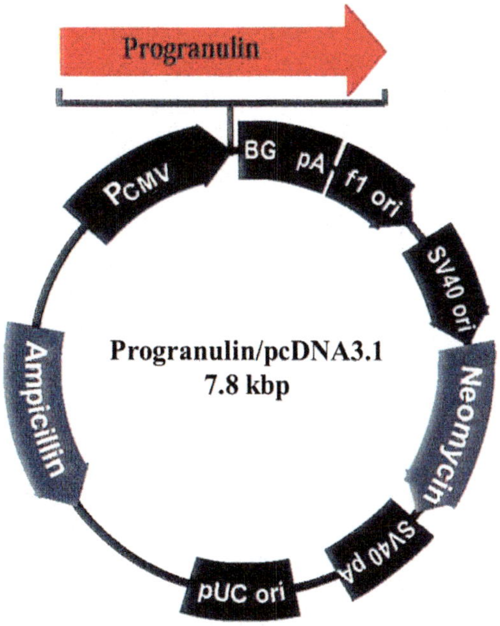

Fig. 1 The construct used for expressing progranulin. A schematic map of the pcDNA3.1 progranulin expression vector. BG pA, bovine growth hormone polyadenylation signal; P_{CMV}, cytomegalovirus promoter; Ampicillin, bacterial resistance marker; SV40 Ori, SV40 promoter and origin; Neomycin, selectable marker for mammalian cells; SV40 pA, SV40 polyadenylation signal; pUC Ori, pUC origin; f1 ori, f1 origin

2.1.3 Purification

1. Amicon Ultra-15 centrifugal filter unit (cutoff size MW 30 kDa) 15 ml volume, Millipore.
2. Trifluoroacetic acid (TFA) from Sigma-Aldrich.
3. Acetonitrile (ACN) from Sigma-Aldrich.
4. HPLC grade water from Sigma-Aldrich.
5. A Vydac 214TP C4 reversed-phase HPLC column.
6. HPLC system. Any HPLC system is acceptable that has two pumps, is programmable, and can monitor the column effluent at 215 nm and, optionally, at 280 nm. For ease of injection an injection loop of at least 500 μl capacity is recommended.
7. Fraction collector (which must be capable of collecting 1 min fractions).
8. 3.5 ml polypropylene collection tubes with conical bases.

2.1.4 Western Blot

1. Bio-Rad mini-Protean gel apparatus mini cell for 1-D vertical gel electrophoresis and electrophoresis power supply.
2. Reagents to pour slabs for SDS polyacrylamide gel electrophoresis gels; acrylamide, Tris–HCL (pH 6.8), Sodium dodecyl sulfate (SDS), ammonium persulfate, *N,N,N′,N′*-tetramethylethylenediamine (TEMED), dd water.
3. 2× Laemmli Sample Buffer from Bio-Rad.
4. β-mercaptoethanol.
5. Water bath capable of heating to 100 °C.
6. Pre-stained protein ladder, low-molecular-weight range (10–180 kDa).
7. 10× Electrophoresis running buffer (288 g glycine, 60.4 g Tris base, 20 g SDS, 1.80 l deionized water).
8. Transfer buffer; 25 mM Tris, 190 mM glycine, 20% methanol.
9. Nitrocellulose membrane, 0.45 μm pore size, from Amersham Bioscience.
10. 3% Milk solution, use nonfat powdered milk in Tris-buffered saline pH 7.6 plus 0.05% Tween 20 detergent.
11. Anti-progranulin polyclonal antibodies (R&D antihuman progranulin antibody AF2420).
12. Anti-rabbit IgG horseradish peroxidase (anti-rabbit IgG HRP) from GE Healthcare, UK.
13. ECL Western blotting detection reagents kit from GE Healthcare.
14. Amersham ECL Hyperfilm from GE Healthcare.

2.2 Materials for Modified Expression and Purification Protocols

2.2.1 Radiolabeled Progranulin

1. ^{35}S-L-Cysteine (1000 Ci/mmol).
2. DMEM.
3. 0.1 M Acetic acid.
4. Dialysis tubing (12,000 kDa cutoff).
5. Waters C18 Sep-Pak cartridge.
6. 80% Acetonitrile in 0.1% trifluoroacetic acid (TFA).
7. 0.1% TFA.
8. 5% Acetonitrile in 0.1% trifluoroacetic acid.
9. Plastic 6 ml scintillation vials with caps (Sigma-Aldrich).
10. Opti-Fluor scintillation fluid.
11. Liquid scintillation beta counter.
12. Gel drier.
13. Kodak BioMax™ XAR™ X-ray film (Fisher).
14. Intensifier screen.
15. Autoradiography cassette.

2.2.2 Heparin-Affinity Chromatography of Progranulin

1. Shodex heparin-affinity HPLC column.
2. 0.05 M Tris–Cl (pH 7.1) for preconditioning of the column.
3. 0.01 M NaCl in 0.05 M Tris–Cl (pH 7.1).
4. 1.0 M NaCl in 0.05 M Tris–Cl (pH 7.1).
5. HPLC system (see above).
6. Milli-Q grade deionized distilled water for washing the column and pump system before storage.

3 Methods

3.1 Expression and HPLC Purification of Recombinant Progranulin

The method for expressing recombinant progranulin takes advantage of its secretion into the culture medium to simplify the purification.

3.1.1 COS-7 Cell Culturing

All steps involving the manipulation of cell cultures must be performed in a Class II laminar flow hood. All equipment must be sterile and sterile technique must be observed throughout. Incubation steps must be performed in a 37 °C incubator with a 5% CO_2 atmosphere.

1. Incubate the COS-7 cell culture in DMEM medium with 10% FBS in a T75 culture flask at 37 °C in a suitable incubator with 5% CO_2 in air atmosphere until cells reach 80% confluence.

3.1.2 Subculturing COS-7 Cells

(Volumes are given for 75 cm^2 culture dishes.) Remove and discard the culture medium by aspiration either using sterile 2 ml aspiration pipettes or autoclaved sterile Pasteur pipettes attached to an aspiration unit.

1. Remove and discard the culture medium by aspiration using either sterile 2 ml aspiration pipettes or autoclaved sterile Pasteur pipettes attached to an aspiration unit.
2. Briefly rinse the cell layer with 1 ml of 0.25% (w/v) trypsin in 0.53 mM EDTA solution. Remove the trypsin by aspiration to remove all traces of serum that contains trypsin inhibitor.
3. Add 2.0 ml of trypsin-EDTA solution to cells at about 80% confluence in a T75 flask and incubate for 5 min 37 °C in a 5% CO_2 atmosphere. Observe cells under an inverted microscope until cell layer is dispersed (usually within 5 to 10 min; *see* **Note 1**).
4. Remove the trypsin-EDTA cell suspension using a 2 ml pipette and place in a sterile 15 ml centrifuge tube and spin at 200 × *g* for 5 to 10 min. A cell pellet should be visible after the suspension has been centrifuged. Carefully remove the trypsin solution without disturbing the pellet.
5. Add 7.0–8.0 ml of DMEM with 10% FBS and suspend the cells by gently pipetting up and down.
6. Mix a 10 μl aliquot of cell suspension and with an equal volume of trypan blue solution. Count the cells using a hemocytometer to determine total cell number.
7. Add appropriate aliquots of the cell suspension to new T75 tissue culture flasks to achieve a sub-cultivation ratio of 1:4 to 1:8. Bring the final volume of DMEM with 10% FBS up to 15–20 ml in each flask.
8. Incubate cultures at 37 °C in a 5% CO_2 atmosphere. Change the culture medium two to three times per week.

3.1.3 Transient Transfection

1. Trypsinize COS-7 cells as above (Subheading 3.1.2) and plate 2×10^6 cells into sterile 15 cm diameter cell culture Petri dishes. Incubate at 37 °C in a 5% CO_2 atmosphere until 80% confluent.
2. One day before performing the transfection, replace the medium with serum-free medium. The medium must not contain antibiotics.
3. For one 15 cm petri dish, prepare two 1.5 ml Eppendorf tubes:
   (a) In the first Eppendorf tube, mix 600 μl of serum-free DMEM with 45 μl of Lipofectamine 2000.
   (b) In a second Eppendorf tube, mix 600 μl of antibiotic-free DMEM with 35 μg *progranulin* plasmid DNA (e.g., 7 μl of *progranulin* plasmid DNA at 5 μg/μl].

4. Wait 2–5 min. Then combine the contents of both tubes into one.
5. Incubate on a platform rocker at room temperature for 45 min in the dark at low speed.
6. Wash the cells three times with sterilized PBS.
7. Add 13.8 ml of antibiotic-free DMEM to the 1.2 ml of the Lipofectamine 2000, *progranulin* plasmid DNA mixture from the **step 5**.
8. Aspirate the PBS from the COS-7 cell Petri dish and add the DMEM/lipofectamine/plasmid mixture dropwise to the cells. Incubate the cells at 37 °C in a 5% CO_2 atmosphere.
9. Change media after 6–12 h (can be up to 24 h) with DMEM.
10. Incubate the transfected cells in serum-free DMEM for 48 h at 37 °C in a 5% CO_2 atmosphere. During this time the cells will secrete progranulin protein into the DMEM media.
11. After 48 h, collect the medium by centrifuging at 1000 × *g* for 5 min to get rid of the cells. Collect the supernatant which contains the secreted progranulin.

3.1.4 Filtration

1. Add up to 15 ml of the centrifuged conditioned medium obtained from transient transfection (*see* Subheading 3.1.3) containing progranulin to an Amicon Ultra-15 centrifugal filter unit. (Add only 12 ml if using a fixed angle rotor.)
2. Place the capped filter unit into a centrifuge rotor and counterbalance with a similar unit.

   When using a fixed angle rotor, place the filter unit with the membrane panel facing up and spin at 5000 × *g* maximum for 30 min. If using a swinging bucket, rotor spin at 4000 × *g* for 30 min.
3. Inspect the filter unit to ensure complete filtration has been achieved.
4. To recover the concentrate, withdraw the sample from the bottom of the filter chamber using a side-to-side sweeping motion with a pipette tip to ensure total recovery. For optimal recovery, remove the concentrated sample immediately after centrifugation and place in a polypropylene collection tube (tube A).
5. Rinse the filter chamber three times with 200 μl of the filtrate collected at the bottom of the filter column and transfer to tube A.

3.1.5 HPLC Purification

The HPLC elution conditions presented are those we have found that work well but can be modified by trial and error to suit individual needs. In HPLC chromatography, a gradient of increasing concentrations of an organic solvent, in this case acetonitrile, is created in an ionic solvent, in this case 0.1% TFA. At a critical concentration of acetonitrile, progranulin is displaced from the top of the HPLC column and is eluted along the column to the collector. Purification is achieved as different proteins elute at different concentrations of the organic solvent.

1. Prepare fresh solution of 500 ml of 0.1% TFA in HPLC grade water. This is run through the HPLC pump A, and will be referred to as solution A.
2. Prepare fresh elution solution of 500 ml of 80% acetonitrile in 0.1% TFA in HPLC grade water. This is run through the HPLC pump B, and will be referred to as solution B.
3. Start the UV detector; choose UV absorbance at 215 nm and 280 nm.
4. Use a C4 reversed-phase HPLC column to purify progranulin. Before loading a sample onto the HPLC column, it must be conditioned by washing in solution B and then in elution solution A as follows:

   1st conditioning wash: run the HPLC at 100% solution B for 15 min at 1.5 ml/min.

   2nd conditioning wash: run the HPLC at 100% solution A for 25 min at 1.5 ml/min.
5. In the meantime, prepare the sample for loading. In an Eppendorf tube (tube B), pipette 990 μl of Milli-Q water and add 10 μl of TFA.
6. Pipette 100 μl of the 1:100 diluted TFA from tube B, into tube A (*see* Subheading 3.1.4 above) which contains the concentrate of the conditioned medium after ultrafiltration (*see* **Notes 2–4**).
7. After acidification some protein may precipitate out of solution. It is vital to remove the precipitated protein so as not to damage the column. Spin in a 1.5 ml microcentrifuge tube at top speed to pellet any precipitate. The supernatant is now ready to be injected. Use 400 μl for each loading injection onto the HPLC column.
8. Continue to run the HPLC at 100% solution A and a flow rate of 1.5 ml/min. Inject 400 μl of the acidified medium containing progranulin through the injector port. If it is necessary to inject more than 400 μl, allow the HPLC to run for 2.5 min with 100% elution solution A at 1.5 ml/min, but do not start the elution program. Perform a second injection again using up to 400 μl. Repeat as often as needed to load the full sample (*see* **Note 5**). The volume of 400 μl is recommended for a 500 μl injection loop. If the injector loop is larger or smaller than 500 μl, adjust the injection volume accordingly.
9. Once loading is complete, run the HPLC using a linear elution gradient of 100% solution A to 100% solution B (i.e., from 0% to 80% acetonitrile in 0.1% TFA) over 60 min at a flow rate of

1.5 ml/min. Collect fractions every minute in 3.5 ml polypropylene tubes with conical bottoms (*see* **Note 6**). Progranulin will elute at around the 35th min (*see* Fig. 1). This will vary if other columns are used or if the column is old, and so must be determined by experiment (*see* **Note 7**).

10. After the elution is complete, to store the column safely it should be washed with 100% acetonitrile by putting the elution buffer delivery tubes for both pump A and pump B tubes into a bottle of 500 ml acetonitrile and washing the HPLC column for 30 min at a flow rate of 1.5 ml/min. This ensures that no residual water or acid remains on the column or in the HPLC pumps during storage (*see* **Note 8**).
11. Turn off the pump, the detector, and the computer.
12. Keep the collected fractions at −80 °C for long-term storage.

3.1.6 Western Blot Characterization (see Note 9)

1. Prepare an SDS-polyacrylamide gel for the Western blot using a Mini-Protean slab gel TM apparatus (Bio-Rad). Use 0.75- to 1-mm-thick polyacrylamide gels with a resolving phase of 10% acrylamide and stacking phase of 4.5% acrylamide prepared using acrylamide, Tris–HCL (pH 6.8), SDS, ammonium persulfate, TEMED, and dd water.
2. Select fractions to be analyzed based on UV profile of the HPLC chromatogram (*see* Fig. 2). Lyophilize 30 μl of each fraction using a SpeedVac concentrator (which takes 60–90 min) and resuspend in 1× Laemmli buffer with 10% β-mercaptoethanol to bring the volume back to 30 μl. Heat for 5 min at 100 °C to denature the sample.
3. Load 2 μl of the sample into the SDS-polyacrylamide gel. Load protein standards into a track of the gel.
4. Electrophorese at 10–20 V/cm with constant volts for 2–4 h.
5. To transfer proteins from the SDS-polyacrylamide gel to a 0.45 μM nitrocellulose membrane, stabilize the gel in the transfer buffer, 25 mM Tris, 190 mM glycine, 20% methanol, for 30 min. Electro-transfer proteins to the membrane at 200 mA for 90 min.
6. Block the nitrocellulose membrane in 3% milk solution at 4 °C for 1 h.
7. Wash three times with PBS.
8. Add 10 μl of anti-PGRN polyclonal antibodies (1:1500) in 15 ml 3% milk solution and keep at 4 °C overnight.
9. Wash three times with PBS.
10. Add 2 μl of anti-rabbit IgG HRP (1:7500) in 15 ml 3% milk solution and leave at 4 °C for 1 h.
11. Wash three times with PBS.
12. Add 2 ml ECL mixture for 1 min, then expose on film and develop the film.

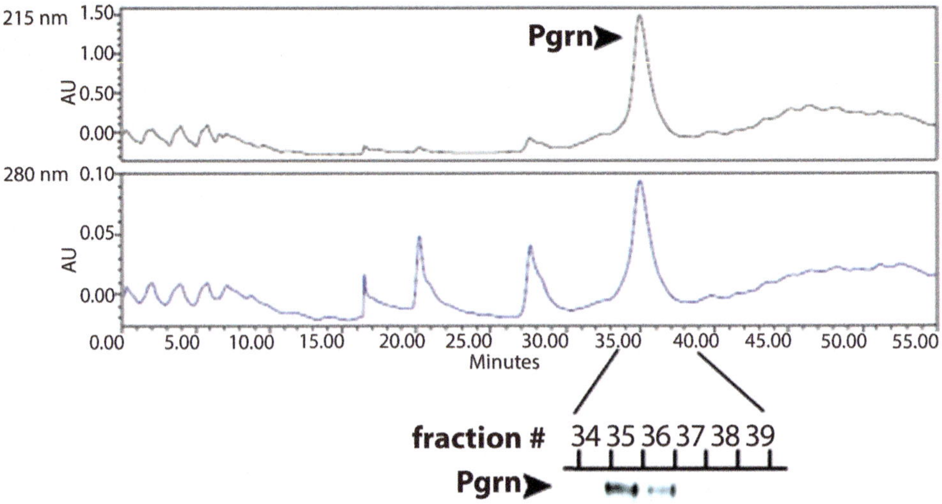

Fig. 2 PRGN HPLC elution graph. A C4 reversed-phase HPLC column was conditioned using an isocratic phase of 0.1% trifluoroacetic acid for 10 min at a 1.5 ml/min flow rate, followed by a linear gradient from 0% acetonitrile to 80% acetonitrile in 0.1% trifluoroacetic acid throughout over 60 min at a 1.5 ml/min flow rate. The column elution flow was monitored by UV absorbance at both 215 nm and 280 nm. Each of the 1 min fractions was collected individually. Progranulin is eluted at around the 35th min as shown by the Western blot (lower panel). Progranulin absorbs UV at both 215 nm and 280 nm. A number of components that absorb UV light only at 280 nm are present in the earlier part of the chromatogram and are due to the presence of residual contaminants such as phenol red from the cell culture medium

3.1.7 Storage of Progranulin

1. Progranulin is a highly disulfide-rich protein which raises problems with storage, as disulfide bonds could oxidize and progranulin loses its biological activities upon storage. We store the protein as lyophilized aliquots kept frozen at −80 °C which will be resuspended only just before use (*see* **Note 8**). Progranulin tends to self-aggregate when concentrated and some loss may occur at this step.
2. Progranulin aliquots can be resuspended in water, PBS buffers, and cell culture media (with or without serum).

3.2 Modifications of the Basic Expression/Purification Protocol

3.2.1 Radiolabeling Progranulin

Progranulin can be radiolabeled by addition of ^{35}S-labeled cysteine into the cell culture medium. All steps must be performed in a dedicated radiation hot lab. For analytical purposes, where purified protein is not required, the radiolabeled protein can be analyzed by autoradiography. With suitable precautions for the handling of radioactive materials, radiolabeled progranulin can be purified by reversed-phase HPLC as above.

1. Transfect cells as described in Subheading 3.1.3.
2. After 24 h wash cells with serum-free DMEM. Incubate the cells in serum-free DMEM supplemented with 15 μl/ml ^{35}S-L-Cysteine (1000 Ci/mmol) for 20 h.

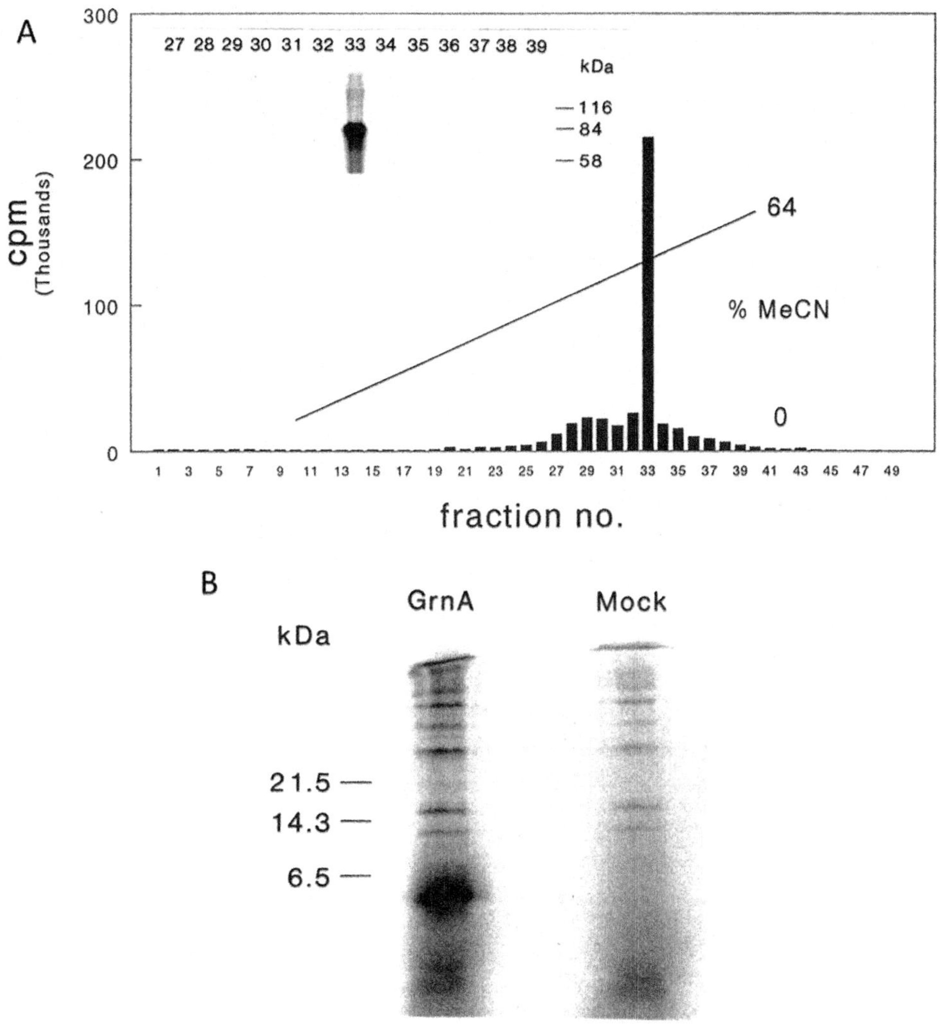

Fig. 3 (**a**) ^{35}S-L-Cysteine-labeled progranulin purified by reversed-phase HPLC showing radiolabel incorporation and autoradiography of the labeled protein. (**b**) SDS-PAGE of ^{35}S-L-Cysteine-labeled granulin A (GrnA), one of the small peptides that can be generated from progranulin. Mock transfected cells were transfected with an empty vector and then treated to the same ^{35}S-L-Cysteine metabolic labeling procedure as the GrnA cells

3. Collect the supernatant after centrifuging at 900 × *g* (or top speed in a benchtop centrifuge) for 5 min to get rid of the cells.
4. Dialyze overnight against 0.1 M acetic acid to remove the unincorporated ^{35}S-L-Cysteine. Dispose of all radioactive liquids and contaminated labware in an appropriate radioactive waste disposal container.
5. If required, purify the ^{35}S-L-Cysteine-labeled progranulin by reversed-phase HPLC as in Subheadings 3.1.4 and 3.1.5 (Fig. 3a).
6. To determine the elution position of the ^{35}S-L-Cysteine-labeled progranulin, measure the β-emission of the labeled products by

scintillation counting. Transfer 50 μl aliquots of each HPLC fraction into 3 ml of scintillation fluid in scintillation vials. Place the vials into a liquid scintillation beta counter and count the radioactive incorporation.

7. To confirm the identity of the labeled progranulin, determine the size of the ^{35}S-L-Cysteine-labeled progranulin by autoradiography. Perform SDS-PAGE on an aliquot of the labeled progranulin using a 15% polyacrylamide gel (*see* Subheading 3.1.6).
8. Dry the gel under vacuum in a gel drier at 55 °C for 2 to 12 h and autoradiograph using Kodak BioMax™ XAR™ X-ray film in an autoradiography cassette with an intensifier screen.

   This method has been adopted to express and label individual granulin modules. For example, cDNA encoding a granulin A module was ligated to a Ly6C leader sequence and cloned into the eukaryotic expression vector pcEXV-3. The transfection and radiolabeling is performed as described above. The labeled peptide is then concentrated using silica-based reversed-phase cartridges rather than dialysis.
9. Unincorporated ^{35}S-L-Cysteine is removed by passing the supernatant through a Waters C18 Sep-Pak cartridge. First, to precondition the Sep-Pak cartridge attach it to a 10 ml syringe, then pass through 10 ml of 80% acetonitrile in 0.1% trifluoroacetic acid followed by 50 ml of 0.1% trifluoroacetic acid.
10. Acidify the conditioned medium before loading onto the cartridge by adding trifluoroacetic acid until the phenol red indicator dye in the cell culture medium turns yellow. Load the conditioned medium onto the Sep-Pak cartridge. Unincorporated ^{35}S-cytseine passes through the cartridge.
11. Wash the Sep-Pak cartridge using a solution of 10 ml 5% acetonitrile in 0.1% trifluoroacetic acid. This removes any residual unincorporated ^{35}S-cytseine remaining in the cartridge. Elute the peptide using 1 ml of 80% acetonitrile in 0.1% TFA. Dispose of all radioactive waste in appropriate radioactive waste containers.
12. The β-emission of the labeled products can be measured by scintillation counting, SDS-PAGE.

3.2.2 Heparin-Affinity Chromatography of Progranulin

One use of radiolabeled progranulin is in chromatography applications that use high-salt elution that interferes with antibody recognition. For example, we investigated the binding of progranulin to heparin using heparin-affinity chromatography using the radiolabeled protein (Fig. 3).

1. ^{35}S-L-Cysteine-labeled progranulin purified by reversed-phase HPLC (above) was applied to a Shodex heparin-affinity HPLC column pre-equilibrated with 0.05 M Tris–Cl pH 7.1.

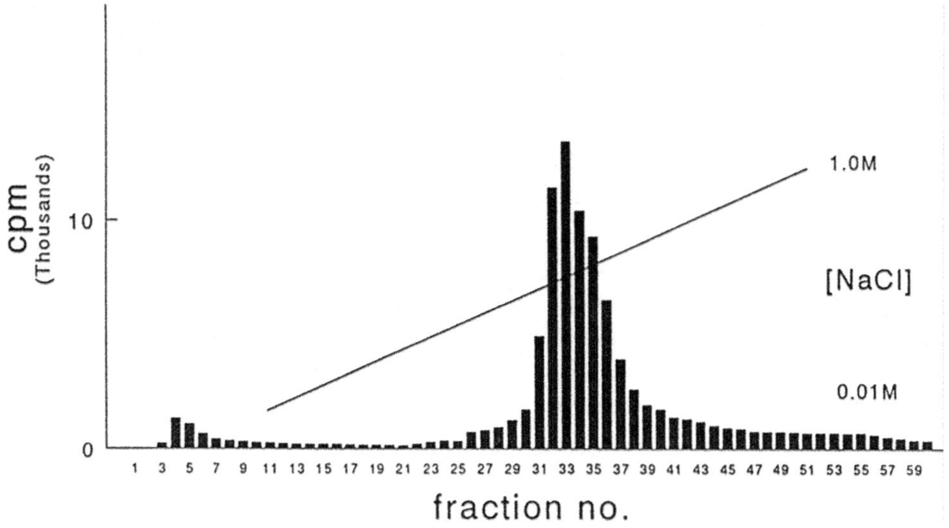

Fig. 4 ^{35}S-L-Cysteine PGRN, purified by HPLC, was eluted from a heparin column by increasing concentrations of NaCl

2. The heparin-bound proteins were then eluted with a linear gradient of NaCl from 0.01 M to 1.0 M NaCl in Tris–Cl pH 7.1 over 30 min at a flow rate of 0.6 ml/min.
3. Fractions were monitored for β-emission by liquid scintillation as in Subheading 3.2.1.
4. The NaCl-containing buffers can be damaging to the metal components of the HPLC apparatus. Thoroughly remove residual buffers by passing HPLC grade distilled water through the pumps and injection system of the HPLC preferably overnight. Do not use organic solvents in the HPLC apparatus after using NaCl-containing solvents without first washing out the injector and pumps in water.

Progranulin elutes between 0.4 M and 0.55 M NaCl. This is a less avid binding than would be expected for heparin-binding growth factors such as those of the fibroblast growth factor family but is above physiological salt concentrations and suggests that progranulin may bind to heparin-like glycans in vivo. The radiolabeled granulin A peptide is eluted at 0.25 M NaCl, indicating a weak affinity for heparin (Fig. 4).

4 Notes

1. To avoid clumping of the cells, do not agitate the cells by hitting or shaking the culture flask while waiting for the cells to detach. Cells that are difficult to detach may be placed at 37 °C to facilitate dispersal.

2. Acidification is necessary for the HPLC purification. Residual phenol red from the culture medium will turn yellow upon acidification.
3. The diluted TFA can be added in greater or smaller amounts; if residual phenol red pH indicator from the original medium is present it will become pale straw colored as the preparation is acidified.
4. This step is to acidify the protein as this enables it to bind to the HPLC column. The hydrophobic column will retain the protein during loading (in 100% buffer A), and then during the elution the protein will be released and eluted due to the increasing concentration of acetonitrile.
5. Provided no acetonitrile is passing through the column, i.e., while pumping 100% solution A and before beginning the elution gradient, the protein remains adsorbed at the head of the column, and therefore the injection step can be repeated multiple times without affecting the quality of the chromatography.
6. Some plastics are damaged by exposure to acetonitrile, but polypropylene tubes do not react with acetonitrile. Glass tubes are not recommended as progranulin sticks to glass. Conical bottom tubes are better than round bottom tubes if the tube contents are lyophilized as the protein tends to smear over the surface of the round-bottomed tubes upon freeze-drying.
7. There is a delay between the peak seen on the UV chromatogram and the time it arrives in the collection tube because of a "dead volume" from when the sample passes from the spectrophotometer to the collection tubes. For a dead volume of 1.5 ml, the delay between the detector and the collector is 1 min. The dead volume, and therefore the delay, will vary between instruments.
8. The HPLC column can be reloaded and repeat HPLC runs performed provided the column is washed in the 80% acetonitrile solution B before each HPLC run to condition the column and then washed in 0.1% TFA solution A for 20–25 min before the next injection. The fractions can also be stored after lyophilization in a SpeedVac, but repeated freeze/thawing, or lyophilization, may result in loss of sample and should be kept to a minimum.
9. Once the progranulin-containing fractions have been confirmed by Western blot analysis, the progranulin-containing fractions should be pooled, lyophilized, and resuspended in water and the concentration of protein determined by any of several commssercially available protein determination kits. The progranulin can then be aliquoted into the desired concentrations for storage or immediate use as required.

References

1. Cenik B, Sephton CF, Cenik BK et al (2012) Progranulin: a proteolytically processed protein at the crossroads of inflammation and neurodegeneration. J Biol Chem 287:32298–32306
2. Hu F, Padukkavidana T, Vægter CB, Brady OA, Zheng Y, Mackenzie IR, Feldman HH, Nykjaer A, Strittmatter SM (2010) Sortilin-mediated endocytosis determines levels of the frontotemporal dementia protein, progranulin. Neuron 68:654–667
3. Tanimoto R, Morcavallo A, Terracciano M, Xu SQ, Stefanello M, Buraschi S, Lu KG, Bagley DH, Gomella LG, Scotlandi K, Belfiore A, Iozzo RV, Morrione A (2015) Sortilin regulates progranulin action in castration-resistant prostate cancer cells. Endocrinology 156:58–70
4. Zheng Y, Brady OA, Meng PS, Mao Y, Hu F (2011) C-terminus of progranulin interacts with the beta-propeller region of sortilin to regulate progranulin trafficking. PLoS One 6:e21023
5. He Z, Bateman A (1999) Progranulin gene expression regulates epithelial cell growth and promotes tumor growth in vivo. Cancer Res 59:3222–3229
6. He Z, Ismail A, Kriazhev L, Sadvakassova G et al (2002) Progranulin (PC-cell-derived growth factor/acrogranin) regulates invasion and cell survival. Cancer Res 62:5590–5596
7. Gluzman Y (1981) SV40-transformed simian cells support the replication of early SV40 mutants. Cell 23:175–182
8. Ohki EC, Tilkins ML, Ciccarone VC et al (2001) Improving the transfection efficiency of post-mitotic neurons. J Neurosci Methods 112:95–99
9. Dalby B, Cates S, Harris A et al (2004) Advanced transfection with Lipofectamine 2000 reagent: primary neurons, siRNA, and high-throughput applications. Methods 33:95–103
10. James S, Gibbs BF, Toney K, Bennett HP (1994) Purification of antimicrobial peptides from an extract of the skin of *Xenopus laevis* using heparin-affinity HPLC: characterization by ion-spray mass spectrometry. Anal Biochem 217:84–90

Chapter 4

Large-Scale Generation of Recombinant Granulin Peptides in *E. coli*

Dmitri Tolkatchev

Abstract

Generating milligram quantities of correctly folded granulin molecules with properly formed disulfide bonds and biologically relevant activities may represent a considerable challenge. Here I describe a protocol for obtaining well-folded human granulins A, C, and F by expressing them as thioredoxin fusion proteins in Origami (DE3) *Escherichia coli* cells promoting disulfide bond formation in the cytoplasm environment. The thioredoxin tag is removed by proteolytic cleavage with enterokinase and granulins which are purified by reversed-phase HPLC. Well-folded disulfide species display lower retention time than misfolded species and therefore can be readily purified.

Key words Granulin, Progranulin, Large-scale bacterial expression, Disulfide bond, Purification, HPLC

1 Introduction

Despite fast-growing interest in progranulin/granulin (PGRN/GRN) biology [1, 2] and GRN-based therapeutics, the progress in understanding its functions is hampered by difficulties in high-yield production of properly folded, active PGRN/GRN proteins. One major obstacle is the large number of disulfide bonds per GRN module. Cysteines contribute more than 20% to the amino acid composition of each GRN repeat. If oxidized spontaneously in denaturing conditions, each 12-cysteine GRN motif can theoretically form 10,395 disulfide-bonding patterns, while the entire PRGN, with its 88 cysteines, can form an astronomic number of disulfide isomers [3]. Depending on a specific biological or biochemical assay, incorrectly formed disulfide isomers may display either a full activity or a partial/complete loss of functionality.

When PGRN is expressed naturally in mammalian cells, in vivo oxidative folding pathways supported by the cellular protein folding machinery [4, 5] ensure proper maturation of the protein. However, it is not clear how consistent and effective this

Andrew Bateman et al. (eds.), *Progranulin: Methods and Protocols*, Methods in Molecular Biology, vol. 1806,
https://doi.org/10.1007/978-1-4939-8559-3_4,

mechanism is when PGRN and especially its fragments are produced by recombinant overexpression. The question about proper folding of recombinant PGRN/GRN is particularly important when they are produced in evolutionary distant host systems. Ideally, recombinant PGRN and its fragments obtained by a specific protocol in a specific expression system must undergo rigorous physicochemical characterization to establish if they are properly oxidized and folded. In practice, such comprehensive characterization may require several milligrams of a chemically homogeneous PGRN/GRN protein.

We have developed a protocol for high-yield *Escherichia coli* expression and purification of recombinant human GRN domains [6]. Out of seven purified GRN domains, three granulins—GRN A, GRN C, and GRN F—were isolated as chemically homogeneous HPLC fractions with defined three-dimensional structures. The correct disulfide-bonding pattern of GRN A was confirmed using chemical methods and NMR. We have determined the three-dimensional structure of GRNs A, C, and F and found that for all of them, the N-terminal half-domain is better structured while the C-terminal half-domain is somewhat more disordered. GRNs A and F demonstrated distinct biological activities when tested in proliferation assays with the MDA-MB-468 cancer cell line, i.e., GRN A was a potent inhibitor of cell growth, whereas GRN F was stimulatory.

In this chapter I will describe in detail the subcloning, expression, and purification of the recombinant human GRNs in an *E. coli* expression system allowing for cytoplasmic disulfide bond formation. Our initial attempts to express GRN A in the *E. coli* periplasm with an expectation to induce disulfide bond formation [7] were unsuccessful, because our recombinant construct turned out to be very toxic to *E. coli* cells (Vinogradova and Tolkatchev, unpublished). We have succeeded in producing milligram quantities of the GRN peptides expressed as thioredoxin fusion proteins in the Origami (DE3) (Novagen) *E. coli* strain. The Origami strain carries glutathione reductase and thioredoxin reductase mutations which promote formation of disulfide bonds in the *E. coli* cytoplasm [8, 9]. Upon proteolytic digestion separating GRN peptides from the thioredoxin tag, well-structured isomers were purified by reversed-phase HPLC chromatography. The well-folded GRN species typically displayed lower HPLC retention time than misfolded disulfide species that were always present in the preparation.

2 Materials and Equipment

2.1 GRN Subcloning

1. A PGRN/GRN DNA template (*see* **Note 1**).
2. Forward and reverse cloning primers annealing to the DNA template (Table 1) (*see* **Note 2**).

Table 1
Examples of oligonucleotide primers for *NcoI/Hin*dIII subcloning of GRN A, GRN C and GRN F. Underlined sequences represent *NcoI* and *Hind*III recognition sites in the forward and reverse primers, respectively

Forward primers:
Granulin A: 5′-CTACAGCCATGGATGTGAAATGTGACATGGAGGTGAG-3′
Granulin C: 5′-TGAAGACCATGGTCCCCTGTGATAATGTCAGCAG-3′
Granulin F: 5′-ACTCCGCCATGGCCATCCAGTGCCCTGAT-3′
Reverse primers:
Granulin A: 5′-CTGGTGAAGCTTCTGTTCACAGGTACCCTTCTGC-3′
Granulin C: 5′-CTCGCTAAGCTTCTGACACTGCCCCTCAGC-3′
Granulin F: 5′-GCCCGTAAGCTTGATGCAGCGGGTGTGAAC-3′

3. T7 terminator primer 5′-GCT AGT TAT TGC TCA GCG GTG G-3′.
4. pET-32b vector (Novagen/EMD Millipore, MA).
5. Subcloning efficiency DH5α competent cells (Life Technologies/Thermo Fisher Scientific, MA).
6. *Nco*I-HF and *Hind*III-HF restriction endonucleases (New England Biolabs, MA).
7. One*Taq* DNA polymerase (New England Biolabs, MA) (*see* **Note 3**).
8. dNTP solution mix (New England Biolabs, MA).
9. T4 DNA ligase (New England Biolabs, MA).
10. Ultrapure H_2O (e.g., Milli-Q): autoclave.
11. Sterile 50% (v/v) glycerol: mix 63 g glycerol with 50 mL distilled water, autoclave.
12. 3 M sodium acetate, pH 5.0: autoclave.
13. 0.5 M EDTA, pH 8.0: stir 186.1 g disodium EDTA·$2H_2O$ into 800 mL of distilled water, add 18–20 g of NaOH pellets until EDTA is completely dissolved, adjust pH to 8.0, and adjust volume to 1 L.
14. 50× TAE buffer: dissolve 141.1 g of Tris base in 300 mL of ddH_2O, add 28.6 mL of glacial acetic acid, add 50 mL of 0.5 M EDTA (pH 8.0), and bring the final volume to 0.5 L.
15. Low EEO/multipurpose agarose (Fisher).
16. QIAquick PCR Purification Kit (Qiagen, Netherlands).
17. QIAprep Spin Miniprep Kit (Qiagen, Netherlands).
18. 1000× Ampicillin solution: 100 mg/mL filter-sterilized solution of ampicillin in water.

19. SOC medium (Life Technologies/Thermo Fisher Scientific, MA).
20. LB medium: dissolve 10 g tryptone, 5 g yeast extract, and 10 g NaCl in 1 L deionized water. Aliquot and autoclave.
21. LBA (LB plus ampicillin) agar plates: dissolve the components of LB medium in deionized water as in **step 20** (scale down to 0.5 L); add 7.5 g agar powder into a dry 1 L Erlenmeyer flask, add 0.5 L of LB medium, cover with foil, autoclave; when the medium is cooled to approximately 55–60 °C, add aseptically 0.5 mL of 100 mg/mL ampicillin, swirl thoroughly, and pour the medium into petri dishes. Let the plates cool until the agar is solid. Store upside down for no longer than 2 weeks at 4 °C.
22. Temperature controlled water bath.
23. PCR thermal cycler.
24. UV-Vis spectrophotometer.
25. Horizontal gel electrophoresis system (Bio-Rad, CA).
26. Shaking incubator for cell culturing.
27. Micro-centrifuge.

2.2 Expression and Purification of the GRN Fusion Protein

1. LB medium: *see* **step 20** of Subheading 2.1.
2. Sterile 50% (v/v) glycerol: **step 11** of Subheading 2.1.
3. 100 mg/mL (1000×) filter-sterilized ampicillin solution in water.
4. Origami (DE3) competent cells (Novagen/EMD Millipore, MA) (*see* **Note 4**).
5. LBA (LB plus ampicillin) agar plates: *see* **step 21** of Subheading 2.1 for additional information.
6. A pET-32b-GRN recombinant expression plasmid encoding a GRN module as a thioredoxin fusion protein.
7. 100 mM (1000×) isopropyl β-D-1-thiogalactopyranoside (IPTG), filter-sterilized.
8. Pierce protease inhibitor tablets (Thermo Scientific).
9. Phenylmethylsulfonyl fluoride (PMSF), 100 mM: prepare 100 mL solution in isopropanol; store at −20 °C for up to 6 months.
10. Urea.
11. 5 M NaOH: carefully dissolve 20.0 g of NaOH in 80 mL water by slowly adding NaOH pellets to water in a glass beaker with stirring. Wear gloves and safety goggles. Adjust to 100 mL after the solution cools down, and store in a polyethylene container.
12. 6 M HCl: to 100 mL water slowly add, with stirring, 100 mL of concentrated hydrochloric acid. Wear gloves and safety goggles, and work in a chemical fume hood.

13. 25/16× Lysis buffer: prepare 1 L of 78 mM sodium phosphate buffer, 469 mM NaCl, 15.6 mM imidazole, pH 8.0 by dissolving 10.8 g $NaH_2PO_4 \cdot H_2O$ (MW 137.99 g/mol), 27.4 g NaCl (MW 58.44 g/mol), 1.06 g imidazole (MW 68.08 g/mol) in approximately 0.8 L water. Adjust pH to 8.0, and adjust the volume to 1 L. Store at 4 °C.
14. Buffer A: prepare 100 mL of 100 mM sodium phosphate buffer, pH 8.0, 10 mM Tris-Cl, 8 M urea. Dissolve in approximately 50 mL of water 1.38 g $NaH_2PO_4 \cdot H_2O$, 0.12 g Tris base (MW 121.1 g/mol), 48.05 g urea (MW 60.06 g/mol), adjust pH to 8.0, and adjust volume to 100 mL.
15. Buffer B: prepare 100 mL of 100 mM sodium phosphate buffer, pH 6.3, 10 mM Tris-Cl, 8 M urea.
16. Buffer C: prepare 100 mL of 100 mM sodium phosphate buffer, pH 5.9, 10 mM Tris-Cl, 8 M urea.
17. Buffer D: prepare 100 mL of 100 mM sodium phosphate buffer, pH 4.5, 10 mM Tris-Cl, 8 M urea.
18. Ni-NTA resin (QIAGEN), washed with water before use.
19. High-speed, large volume centrifuge (e.g., Beckman Coulter J2, J6, or Avanti J series).
20. Ultrasonic cell homogenizer/disruptor (e.g., Branson Sonifier).
21. Protein 1-D electrophoresis system (Bio-Rad, CA).
22. Molecular weight protein markers/protein standards for SDS-PAGE analysis.

2.3 Proteolytic Cleavage of the GRN Fusion Protein and Purification of the GRN Module

1. Buffer E: prepare 200 mL of 50 mM Tris-Cl, pH 7.4, 50 mM NaCl, 8 M urea.
2. Recombinant enterokinase (rEK) reaction buffer: prepare 200 mL of 50 mM Tris-Cl, pH 7.4, 50 mM NaCl.
3. $CaCl_2$, 1 M: prepare 10 mL in water, and store at room temperature.
4. 6 M HCl.
5. Urea.
6. 0.1% Trifluoroacetic acid (TFA) in water (HPLC grade).
7. 0.1% TFA in acetonitrile (HPLC grade).
8. 0.1% TFA in 45%/55% acetonitrile/water mixture.
9. Recombinant enterokinase (rEK) (*see* **Note 5**).
10. Phenylmethylsulfonyl fluoride (PMSF), 100 mM in isopropanol.
11. Centriprep Centrifugal Filter Units, 10 kDa MW cutoff (EMD Millipore).

12. Dialysis tubing, 10–15 kDa MW cutoff.
13. Sep-Pak C_{18} cartridges (Waters).
14. Polyethersulfone (PES) filters (0.22 or 0.45 micron pore size).
15. HPLC instrument.
16. Vydac 218TP54 C18 reversed-phase HPLC column.
17. Freeze-dryer equipped with an −84 °C condenser.

3 Methods

3.1 GRN Subcloning (See Note 6)

The purpose of performing experiments described in this section is to incorporate the GRN gene into the pET-32b vector. The pET-32b contains a thioredoxin protein tag facilitating disulfide bond reshuffling and therefore promoting correct disulfide pairing. The construct also contains two His-tags. Constructs with GRN inserts must undergo sequencing before expression to confirm the absence of mutations.

3.1.1 Day 1: Preparation of the Digested Vector and Insert

1. Set up the following PCR reaction (*see* **Note 7**):
   - 20–50 pg DNA template (per~6 kbp plasmid).
   - 200 μM each dNTP.
   - 0.2 μM each of the two subcloning primer (forward and reverse).
   - 10 μL 5× OneTaq Standard Reaction Buffer (supplied with the enzyme).
   - H_2O added to a final reaction volume of 50 μL.
   - 0.6 units One*Taq* DNA polymerase.
2. Use the following PCR program:
   - PCR1: Initial denaturation for 5 min at 94 °C.
   - PCR2: Denature for 1 min at 94 °C.
   - PCR3: Anneal primers for 30 s at 55 °C (5 °C below primers' melting temperature Tm).
   - PCR4: Extend DNA for 20 s (1 min per kbp) at 68 °C.
   - PCR5: Repeat steps PCR2 through PCR4 29 times (for a total of 30 cycles).
   - PCR6: Final extension for 5 min at 68 °C.
3. Use 3–5 μL of the PCR product to verify its size by agarose gel electrophoresis on a 2% gel in 1× TAE buffer.
4. Purify the PCR product using the QIAquick PCR Purification Kit (*see* **Note 8**).

5. Use 5 μL of the purified PCR reaction to determine the DNA concentration by measuring the absorbance at 260 nm. Calculate the DNA concentration using the approximate formula: 1 OD_{260} unit = 50 ng/μL.
6. Set up double digestion reactions for the pET-32b vector and the purified PCR-generated insert as follows:
   - DNA (1500 ng of the vector or 500 ng of the insert).
   - 5 μL of 10× CutSmart Buffer (supplied with the enzymes).
   - H_2O added to a final reaction volume of 50 μL.
   - 15 U of each restriction endonuclease (0.75 μL of *Nco*I-HF (20 U/μL) plus 0.75 μL of *Hind*III-HF (20 U/μL)).
7. Incubate the *Nco*I/*Hind*III double digestion reactions for 2 h at 37 °C.
8. Inactivate the restriction endonucleases at 80 °C, for 20 min. Freeze the inactivated mixtures until next day.

3.1.2 Day 2: Ligation and Transformation

9. Purify and quantify the digests by repeating **steps 4** and **5**.
10. Set up a ligation reaction by mixing on ice:
    - 10 ng digested and purified pET-32b vector.
    - 10–15 ng digested and purified GRN insert.
    - 1 μL of 10× T4 DNA Ligase Reaction Buffer (supplied with the enzyme).
    - H_2O added to a final reaction volume of 10 μL.
    - 0.5 μL of T4 DNA Ligase.
11. Incubate the ligation reaction mixture at room temperature (22–25 °C) for 1 h.
12. Thaw on ice 50 μL of competent DH5α cells, add 2 μL of the ligation mixture. Incubate the mixture on ice for 30 min.
13. Perform a heat shock by placing the tubes with the cells and DNA into a 42 °C water bath for 30 s.
14. Put the tubes back on ice for 2 min.
15. Add 1 mL of SOC medium (no antibiotic added) on ice, and incubate in a 37 °C shaking incubator at 250 rpm for 45 min.
16. Centrifuge the cells at 2000 × *g* for 5 min, aspirate approximately 0.9 mL of supernatant, and resuspend the cells in the remaining 0.1 mL medium solution. Plate 15–20 μL on one LBA agar plate and the remaining 80–85 μL on another. Incubate the plates overnight at 37 °C.

3.1.3 Day 3: Colony Screening

17. Use sterile pipetting 200 μL tips to transfer four to six *E. coli* colonies individually into separate labeled tubes containing 50 μL of sterile water and resuspend.

18. Immediately transfer 10 μL of each colony suspension into a sterile mixture of 30 μL LB medium with 20 μL of 50% glycerol. Label and freeze at −80 °C. This will be the DH5α colony mini-glycerol stock for future propagation if the sequence contains the expected recombinant GRN insert.
19. Lyse the remaining colony suspensions in water by incubating at 100 °C for 1–2 min.
20. Repeat **steps 1–3** with the following modifications: scale down PCR reaction volumes to 25 μL; use the T7 terminator primer as the reverse primer in the PCR reaction; use 2 μL of lysed colonies from **step 19** for template DNA; use 50 °C as annealing temperature in the PCR3 step; and run the program for 35 cycles.
21. At the end of the day, thaw on ice mini-glycerol stocks of 3–4 clones of each construct that were shown to contain an insert of appropriate size in **step 20**. Use 10–20 μL of each glycerol stock to set up overnight cultures in 5–10 mL of LB medium supplemented with 100 μg/mL ampicillin.

3.1.4 Day 4: DNA Purification and Storage

22. Prepare glycerol stocks of the DH5α cells transformed with the constructs by withdrawing aseptically 0.4 mL of each overnight culture and mixing with 0.2 mL of 50% sterile glycerol. Label and store frozen at −80 °C.
23. Purify plasmids in the overnight cultures using the QIAprep Spin Miniprep Kit. Determine plasmid concentrations as in **step 5**. An expected yield is approximately 5–10 μg of plasmid in 50 μL of EB solution. Send 5–10 μL (500–1000 ng) of each plasmid for Sanger DNA sequencing, and keep the remaining plasmid solutions frozen at −20 °C or at a lower temperature.

3.2 Expression and Purification of the GRN Fusion Protein

The protocol below describes production of GRNs in rich LB medium. High-yield expression of recombinant GRNs was initially developed for obtaining sufficient (milligram) amount of ^{15}N and/or ^{13}C-labeled proteins for NMR analysis. For isotope labeling, a BioExpress-labeled cell growth medium (Cambridge Isotope Laboratories, Tewksbury, MA) at 0.5× dilution should be used instead of LB.

3.2.1 Day 1

1. Transform 50 μL of competent Origami (DE3) with 5 ng of pET-32-GRN by following the transformation procedure described in **steps 12–16** of Subheading 3.1.

3.2.2 Day 2

2. Pick a single colony from the plate with the pET-32-GRN transformants (*see* **Note 9**), and set up an overnight culture by inoculating 2 mL of LB medium supplemented with 100 μg/mL ampicillin (LBA medium). Incubate in a 37 °C shaking incubator at 250 rpm.

3.2.3 Day 3

3. Prepare glycerol stock of the Origami (DE3) cells transformed with the construct by withdrawing aseptically 0.4 mL of the culture and mixing with 0.2 mL of 50% glycerol. Label and store frozen at −80 °C until future use.
4. Transfer 1 mL of the overnight culture into 25 mL of fresh LBA medium, and let the culture grow at 22 °C, 250 rpm, during the day. Growing the transformed Origami (DE3) cells at 22 °C instead of at 37 °C helps to decrease the proportion of incorrectly folded GRN modules (*see* **Note 10**).
5. At the end of the day, transfer the culture into 1 L fresh LBA medium, and shake overnight at 22 °C, 220 rpm.

3.2.4 Day 4

6. Next day, when the optical density of the culture reaches 0.8–0.9 at 600 nm, induce the culture by adding IPTG to the final concentration of 0.1 mM. Incubate for 6 h at 22 °C, 220 rpm.
7. Pellet cells by centrifugation at 4 °C, 4000 × *g* for 20 min. Freeze the pellet at −20 °C (*see* **Note 11**).

3.2.5 Day 5

8. Thaw the frozen cell pellet on ice, and resuspend in 50 mL of the 25/16× lysis buffer (*see* **Note 12**). Add and dissolve one Pierce protease inhibitor tablet and 0.5 mL 100 mM PMSF.
9. Sonicate the resuspended cells on slush ice for 5 min using pulses of 5–10 s and rest periods of 20 s to prevent overheating.
10. Centrifuge lysed cells at 15,000 × *g*, 4 °C to remove insoluble cell debris. Collect the supernatant.
11. Estimate the fusion protein concentration by running an SDS-PAGE with a molecular weight protein standard (*see* **Note 12**).
12. To each 16 mL of the cleared cell lysate, add 12 g of crystalline urea (*see* **Notes 13–15**). Warm up to room temperature, and shake gently until urea is dissolved.
13. Add Ni-NTA slurry (approximately 1 mL of resin per each 10 mg of the fusion protein), and shake gently for 1 h at room temperature.
14. Prepare buffers A, B, C, and D.
15. Use a column with 1 cm internal diameter and 10–15 cm length per 2–2.5 mL Ni-NTA resin. If using more resin, adjust the column diameter to ensure that the settled bed height does not exceed 3 cm.
16. Load the column with the lysate Ni-NTA resin, collect the flow-through, and save for SDS-PAGE analysis.
17. Wash the resin successively with 10 bed column volumes of buffers A, B, and C.
18. Elute the fusion protein with six bed volumes of buffer D.

3.3 Proteolytic Cleavage of the GRN Fusion Protein and Purification of the GRN Module

3.3.1 Day 1

1. Exchange the buffer D to the buffer E by dialysis or several concentration/dilution cycles using a concentrator.
2. Concentrate the sample to approximately 10–20 mg/mL of protein.

3.3.2 Day 2

3. Using a one-step dilution, dilute the sample with the rEK Reaction Buffer. The final urea concentration should be 0.5–1.0 M (depending on the protein solubility), and the final protein concentration should be 0.5–1.2 mg/mL (*see* **Note 16**).
4. Supplement the protein solution with 2 mM $CaCl_2$, and add an appropriate amount (*see* **Note 17**) of rEK. Typically, one unit of commercial rEK is used to cleave 50 μg of fusion protein.
5. Incubate overnight at room temperature (22–25 °C), or longer, until 75–80% of the fusion protein is cleaved.

3.3.3 Day 3

6. Test the extent of the proteolytic cleavage on SDS-PAGE.
7. Stop the reaction by adding 1 mM PMSF, and incubate for 1 h at room temperature.
8. Acidify the solution by adding 0.1 M HCl. If protein precipitation is observed, add crystalline urea until the solution is clear.
9. Apply the reaction mixture onto a primed Sep-Pak column, collect the flow-through, and save for SDS-PAGE analysis. Use 1 g of Sep-Pak sorbent per 10 mg of protein.
10. Wash the column with six bed volumes of 0.1% TFA in water.
11. Elute GRN stepwise with 0.1% TFA/acetonitrile/water mixtures. Start with 20% acetonitrile, and increase the acetonitrile concentration by 5% at each step until all the protein is eluted (typically, at 50% acetonitrile). Use 4–6 bed volumes of eluent at each step. Test collected fractions for the presence of GRN or thioredoxin on a 20% SDS-PAGE.
12. Freeze the fractions containing GRN on dry ice and lyophilize.

3.3.4 Day 4

13. Reconstitute dried GRN fractions with water and filter on a 0.22–0.45 μM membrane (*see* **Note 18**).
14. Purify by HPLC on a reversed-phase C18 column. Well-folded GRN fractions have lower retention time and can be readily separated from misfolded fractions (Fig. 1).
15. Freeze the fractions of interest on dry ice and lyophilize.

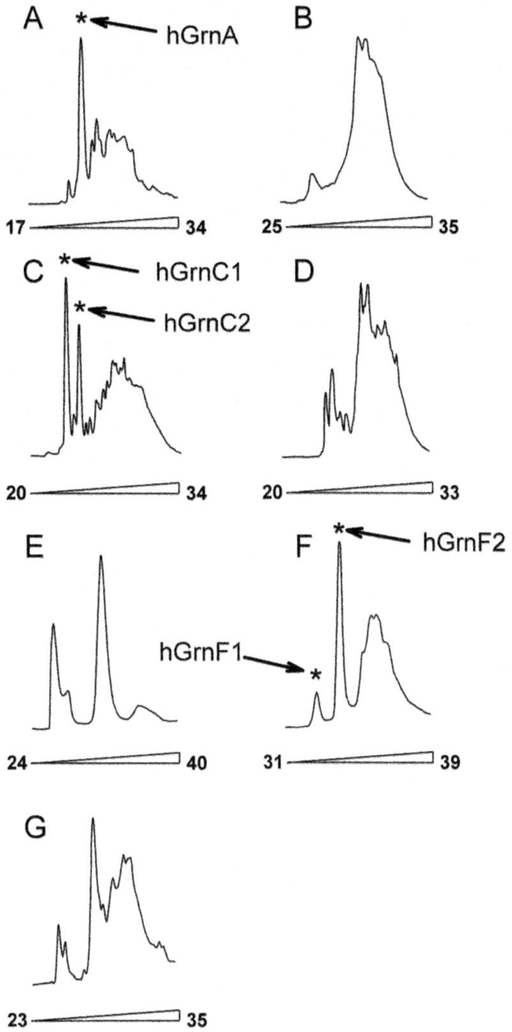

Fig. 1 Reversed-phase HPLC purification of human granulins A–G. Each granulin was purified on a Vydac C_{18} reversed-phase column with a 1%/min acetonitrile gradient in the presence of 0.1% TFA. Numbers at the bottom of each profile indicate the acetonitrile percentage concentration. Asterisks mark well-folded disulfide species, which elute earlier than a mixture of poorly separated misfolded disulfide isomers. The fraction hGrnF1 represents a well-folded N-terminal fragment of the human GRN F. Reproduced from [6] with permission from John Wiley and Sons

4 Notes

1. To improve protein yields, the DNA template can be optimized for *E. coli* expression (e.g., as in [10]) and produced by gene synthesis.

2. Examples of cloning primers annealing to a DNA template coding for the human PGRN (NCBI Reference Sequence: NM_002087.3) are shown in Table 1. The primers are designed to have an approximate melting temperature of 60 °C in the One*Taq* DNA polymerase buffer (http://tmcalculator.neb.com).
3. Typically, the fidelity of the One*Taq* DNA polymerase is sufficient for accurate amplification of DNA inserts smaller than 700–800 bp. The enzyme can be used for both GRN subcloning and PCR screening of clones.
4. Origami (DE3) cells are unable to produce leucine due to the Δ*(ara–leu)7697* deletion and will not grow on a minimal medium unless it is supplied with exogenous leucine [11]. If expression on a minimal medium is desired (e.g., for NMR applications), Origami B (DE3) (Novagen/EMD Millipore, MA) strain may be a proper substitution as an expression host.
5. Commercial recombinant enterokinase (rEK) is available from several sources (e.g., EMD Millipore). A cost-effective way to produce large amounts of pure rEK is described in [12]. The plasmid for the rEK expression (#49048) can be obtained at Addgene, a nonprofit plasmid repository. Another advantage of using the in-house produced rEK is the ability to control its purity and to minimize nonspecific proteolytic cleavage by contaminants.
6. Take particular precautions to avoid sources of nuclease contamination when manipulating DNA. Use disposable examination gloves. Always autoclave ultrapure water and plasticware regardless of their DNase-free certification. For reliable results, keep enzymes and perishable reagents in mini-coolers. Aliquot primers, DNA template, dNTP mixture, and ligase buffer to minimize detrimental effects of freeze–thaw cycles.
7. To minimize nonspecific priming, assemble PCR reactions on ice, and transfer reaction mixtures into a thermocycler preheated to 94 °C, which is the starting temperature for a typical PCR program.
8. When purifying DNA using the QIAquick PCR Purification Kit, before loading the sample onto the column, always add 10 μL of 3 M sodium acetate, pH 5.0 (this step is considered optional in the QIAquick PCR Purification Kit manual). When working with short DNA fragments (100–300 bp), this addition ensures a somewhat higher yield.
9. Typically Origami (DE3) cells grow markedly slower than a more common protein expression *E. coli* host BL21 (DE3). It may take longer than usual for an Origami (DE3) colony to reach the size of an approximately 1 mm in diameter. If colo-

nies are still very small in the morning following the day of transformation, keep them at 37 °C until the end of the day.

10. The schedule of the Origami (DE3) cells culturing may be somewhat adjusted by altering the amount of cells transferred between flasks in the **steps 4** and **5** of Methods. It is not advisable to raise the incubation temperature above 22 °C, as this will increase the proportion of incorrectly folded GRN modules, which will be directed into inclusion bodies.
11. The frozen cell pellet of **step 7** can be stored for several weeks.
12. Collect small aliquots of sample fractions at every step of preparation starting with the initial cell suspension, and analyze them on a SDS-PAGE. That allows keeping track of protein yields at every step and helps to maximize its recovery. Using a commercial protein standard as a control will allow the fusion protein amount and purity to be estimated with accuracy sufficient for each step of purification.
13. Solid crystalline urea should be added to the lysis buffer immediately before the Ni-NTA purification. Also, heating the urea solution above the room temperature to accelerate dissolution should be avoided. In aqueous solutions upon heating and over time, urea produces isocyanate leading to carbamylation of lysine residues and the protein N-terminus [13]. For the same reason, buffers A, B, C, and D should be prepared fresh, or urea should be added at the last moment. Isocyanate can be effectively removed via hydrolysis by incubating a urea solution for 1 h at pH 2.0 [13, 14].
14. Addition of 12 g urea to 16 mL aqueous solution produces approximately 25 mL 8 M urea (account for the change of the buffer concentrations upon urea dissolution). After urea addition, check and adjust pH, if necessary.
15. Urea is added to the cleared cell lysate to prevent the fusion protein from precipitation on the Ni-NTA resin. Adding urea before the cells are lysed is not recommended, as this will increase the amount of misfolded recombinant protein solubilized from the inclusion bodies.
16. Although the optimal pH range for the rEK activity is 7.0–8.0, the protease can work sufficiently well at pH as low as 6.3–6.5. Lowering the pH may increase the fusion protein solubility, reduce the urea concentration needed to keep the fusion protein in solution, and ultimately decrease the extent of nonspecific proteolysis. Performing the digestion at lower pH also reduces the risk of undesirable disulfide bond reshuffling.

17. To establish the amount of rEK needed for the cleavage, test several rEK concentrations in 50 μL aliquots of the protein solution. Use the rEK concentration that will cleave ~70–80% of the fusion protein overnight. Using an excess of rEK will increase the nonspecific proteolysis of the fusion protein.
18. The choice of polyethersulfone membranes to filter protein solution is dictated by their low protein-binding capacity. Other types of low protein-binding membranes can be considered as well.

References

1. Bateman A, Bennett HP (1998) Granulins: the structure and function of an emerging family of growth factors. J Endocrinol 158:145–151
2. Bateman A, Bennett HP (2009) The granulin gene family: from cancer to dementia. BioEssays 31:1245–1254
3. Benham CJ, Saleet Jafri M (1993) Disulfide bonding patterns and protein topologies. Protein Sci 2:41–54
4. Sevier CS, Kaiser CA (2002) Formation and transfer of disulphide bonds in living cells. Nat Rev Mol Cell Biol 3:836–847
5. Sevier CS, Kaiser CA (2008) Ero1 and redox homeostasis in the endoplasmic reticulum. Biochim Biophys Acta 1783:549–556
6. Tolkatchev D, Malik S, Vinogradova A et al (2008) Structure dissection of human progranulin identifies well-folded granulin/epithelin modules with unique functional activities. Protein Sci 17:711–724
7. Nakamoto H, Bardwell JC (2004) Catalysis of disulfide bond formation and isomerization in the *Escherichia coli* periplasm. Biochim Biophys Acta 1694:111–119
8. Prinz WA, Åslund F, Holmgren A et al (1997) The role of the Thioredoxin and Glutaredoxin pathways in reducing protein disulfide bonds in the *Escherichia coli* cytoplasm. J Biol Chem 272:15661–15667
9. Bessette PH, Åslund F, Beckwith J et al (1999) Efficient folding of proteins with multiple disulfide bonds in the *Escherichia coli* cytoplasm. Proc Natl Acad Sci U S A 96: 13703–13708
10. Puigbo P, Guzman E, Romeu A et al (2007) OPTIMIZER: a web server for optimizing the codon usage of DNA sequences. Nucleic Acids Res 35:W126–W131
11. Durfee T, Nelson R, Baldwin S et al (2008) The complete genome sequence of *Escherichia coli* DH10B: insights into the biology of a laboratory workhorse. J Bacteriol 190:2597–2606
12. Skala W, Goettig P, Brandstetter H (2013) Do-it-yourself histidine-tagged bovine enterokinase: a handy member of the protein engineer's toolbox. J Biotechnol 168:421–425
13. Stark GR, Stein WH, Moore S (1960) Reactions of the cyanate present in aqueous urea with amino acids and proteins. J Biol Chem 235:3177–3181
14. Shaw WH, Bordeaux JJ (1955) The decomposition of urea in aqueous media. J Am Chem Soc 77:4729–4733

Chapter 5

Nuclear Magnetic Resonance Spectroscopy in Analysis of Granulin Three-Dimensional Structure and Cysteine Bridging

Dmitri Tolkatchev

Abstract

Granulin (GRN) structural motif represents a ladderlike stack of β-hairpins reinforced with six parallel disulfide bridges. When GRNs are produced in a recombinant protein expression host (e.g., in bacteria) or via chemical synthesis, the formation of disulfide bridges from thiols undergoing uncontrolled oxidation may be random. As a consequence, the resulting protein could be a mixture of a large number of disulfide species. Incorrectly folded GRNs may behave abnormally in bioassays; therefore isolation and identification of properly structured, chemically homogenous GRN peptides is very important for biological relevance of the GRN effects observed in the tests. Protein nuclear magnetic resonance (NMR) spectroscopy is an excellent tool for identification and characterization of well-structured GRN disulfide species produced in an *Escherichia coli* expression system. At first, GRN disulfide species are crudely separated by reversed-phase HPLC chromatography. Obtained fractions are screened by 1D (one-dimensional) proton NMR for the presence of well-folded GRN species. The well-folded GRNs are ^{15}N-labeled and purified, and NMR is used to determine their three-dimensional structure and assign disulfide pairing patterns. Additionally, NMR characterization of model peptides derived from the GRN amino acid sequences can help resolve ambiguities in disulfide bond assignment. This approach was first successfully used to obtain biologically active human GRNs, but it can be easily expanded to GRN peptides from other species and/or generated by other methods.

Key words Granulin, Protein NMR, Hairpin stack, Disulfide species, Three-dimensional structure

1 Introduction

A consensus primary amino acid sequence of a granulin (GRN) structural motif contains 12 cysteines forming 6 disulfide bonds [1, 2]. The disulfide bonds stabilize a unique, ladderlike three-dimensional structure of GRN known as a "hairpin stack" [3, 4]. The first GRN three-dimensional structure was obtained for the highly abundant carp granulin-1 (CG1), which was purified from source [3]. CG1 is an extremely stable protein, with a very well-defined structure allowing unambiguous assignment of disulfide

Andrew Bateman et al. (eds.), *Progranulin: Methods and Protocols*, Methods in Molecular Biology, vol. 1806,
https://doi.org/10.1007/978-1-4939-8559-3_5,

bonds based on NMR data alone. Until more structural information on other GRN motifs is available, CG1 will remain a reference tertiary structural motif defining the consensus GRN protein fold and the pattern of disulfide bond formation.

Since the structure of CG1 was determined, rigorous research efforts have focused on mammalian granulins. Extensive testing of GRNs in various cell and animal systems will greatly benefit from the availability of protocols for large-scale production of correctly folded GRNs. Considering the large number of disulfide bonds in GRNs, characterization of disulfide pairing patterns and the three-dimensional fold of produced GRNs is important to ensure the quality of these protocols, chemical uniformity of generated GRNs, and reproducibility of biological assays.

To obtain milligram quantities of human GRNs required for structural and cellular studies, we expressed recombinant GRN constructs in *Escherichia coli* [4]. One of the major concerns for recombinant expression of cysteine-rich GRN domains is the proper disulfide pairing, particularly when they are produced in an evolutionary distant host. Even though we performed the expression of GRNs in Origami (DE3), which is an *E. coli* strain promoting cytoplasmic cysteine oxidation [5, 6], the absence of a cellular protein folding machinery existing in eukaryotic cells [7, 8] could lead to formation of incorrect disulfide bonding patterns. Out of seven expressed and purified human GRNs, GRN A and GRN F displayed distinct biological activities [4] which implied that they were properly folded. Consistently, the disulfide pairing of GRN A [4], which we determined by Cys–Cys NOE (nuclear Overhauser effect) cross-peak analysis [9] and the cysteine cyanylation/cleavage methodology [10–14], matched the consensus disulfide bonding pattern of CG1 [3]. The biologically active disulfide species of GRN A and GRN F were partially structured and had a well-defined N-terminal half-domain and a somewhat flexible C-terminal half-domain. Four of the recombinantly expressed in *E. coli* human GRN domains—B, D, E, and G—did not produce folded proteins, and they only displayed weak, possibly non-specific activities. The absence of a well-defined structure in the recombinant GRNs B, D, E, and G may have been caused by incorrect cysteine pairing.

These data suggest that at least some extent of structure formation supported by properly formed disulfide bonds is essential for the functionality of GRN structural modules. On the other hand, some degree of GRN flexibility in a form of intrinsically disordered regions [15] is allowed and may be even advantageous, especially if the GRN domain binds to more than one protein partner. In general, establishing biological functions of a particular GRN will involve structural characterization performed in parallel with the assignment of disulfide bridges.

Typical phases of protein structure determination by NMR include expression of the protein of interest in appropriate culture media allowing for protein isotopic labeling, protein purification, optimizing experimental conditions for NMR data collection, recording NMR spectra, spectral assignment, and structure calculations. These general approaches have been reviewed by many authors (e.g., in [16–19]). In this chapter I will focus on NMR analysis specific to the GRN architecture. Additionally, I will describe the use of model (truncated) GRN peptides [20–22] to elucidate structural features and disulfide connectivities of GRN domains.

2 Materials and Equipment

2.1 Screening for Well-Folded GRN Disulfide Species by 1D ^1H NMR

1. Lyophilized expressed recombinant GRN (a mixture of disulfide species).
2. 0.1% trifluoroacetic acid (TFA) in water (HPLC grade).
3. 0.1% TFA in acetonitrile (HPLC grade).
4. Polyethersulfone (PES) filters (0.22–0.45 micron pore size) for protein samples.
5. Nylon membranes (0.22–0.45 micron pore size) for filtering HPLC solvents.
6. Ultrapure water (Milli-Q, "Type 1", or similar): autoclave.
7. Deuterium oxide, >99.9 atom% D in ampules.
8. 1% sodium azide: prepare fresh in 10 mL of ultrapure autoclaved water, aliquot, store frozen at −20 °C.
9. 0.5 M EDTA, pH 8.0: to 18.6 g disodium EDTA·$2H_2O$ in 80 mL of ultrapure water, slowly add 1.8–2.0 g of NaOH pellets with stirring until EDTA is completely dissolved, adjust pH to 8.0, adjust volume to 100 mL, autoclave.
10. 20 mM EDTA, pH 8.0: add 40 μL 0.5 M EDTA, pH 8.0, to 960 μL ultrapure autoclaved water, store frozen at −20 °C.
11. 0.5 M disodium phosphate: dissolve 0.71 g anhydrous Na_2HPO_4 (MW 141.96 g/mol) in 8 mL ultrapure autoclaved water, adjust volume to 10 mL, aliquot, store frozen at −20 °C.
12. 1 M NaCl: dissolve 0.58 g sodium chloride (MW 58.44 g/mol) in 8 mL ultrapure autoclaved water, adjust volume to 10 mL, aliquot, store frozen at −20 °C.
13. 6 M HCl: to 50 mL ultrapure water slowly add, with stirring, 50 mL of concentrated hydrochloric acid. Wear gloves and safety goggles; work in a chemical fume hood.

14. 1 M and 100 mM HCl: add 200 or 20 μL of 6 M HCl to 1 mL or 1.18 mL ultrapure water, respectively.
15. 5 M NaOH: carefully dissolve 10.0 g of NaOH in 40 mL ultrapure water by slowly adding NaOH pellets to water in a glass beaker with stirring. Wear gloves and safety goggles. Adjust to 50 mL after the solution cools down, and store in a polyethylene container.
16. 1 M and 100 mM NaOH: add 200 or 20 μL of 5 M NaOH to 800 μL or 980 μL ultrapure water, respectively.
17. 20 mL glass scintillation vials.
18. Parafilm.
19. 5 mm NMR tubes for 500 MHz frequency or higher.
20. pH microelectrode for NMR samples (e.g., Mettler Toledo extra-long pH microelectrode or similar).
21. HPLC instrument.
22. Freeze-dryer equipped with a −84 °C condenser.
23. UV-Vis spectrophotometer.
24. Vydac 218TP54 C18 or Waters XBridge C18 reversed-phase HPLC column (4.6 × 250 mm).
25. ExPASy ProtParam tool: used to estimate the protein concentration using the OD_{280} reading of the protein solution.

2.2 NMR Characterization of Well-Folded GRN Disulfide Species

1. ExPASy ProtParam tool: used to determine the theoretical pI value (isoelectric point) of proteins in order to determine the carrier solvent for HPLC purification.
2. Lyophilized partially purified well-folded HPLC fractions of recombinant ^{15}N-labeled GRNs.
3. 1% ammonium bicarbonate in water (HPLC grade): dissolve 10 g of ammonium bicarbonate in 900 mL water (HPLC grade), add water to 1 L, filter through a 0.22 μm nylon membrane (prior to preparing, *see* Subheading 3.2, **steps 1** and **2**).
4. 100 mM sodium phosphate buffer, pH 11.2 (HPLC grade): dissolve 14.2 g anhydrous Na_2HPO_4 (MW 141.96 g/mol) in 900 mL water (HPLC grade), adjust pH to 11.2, adjust volume to 1 L, filter through a 0.22 mm nylon membrane (prior to preparing, *see* Subheading 3.2, **steps 1** and **2**).
5. 0.1% trifluoroacetic acid (TFA) in water (HPLC grade).
6. Acetonitrile (HPLC grade).
7. 0.1% TFA in acetonitrile (HPLC grade).
8. Waters XBridge C18 reversed-phase HPLC column, 4.6 × 250 mm (*see* **Note 1**).
9. Mass spectrometry (MS) for confirmation of the identity of the peptides.

2.3 Identification of Disulfide Connectivities through NMR Characterization of Model Peptides

1. Synthesized purified model peptide.
2. Reducing Buffer: 0.1 M Tris–HCl, pH 8.3, 1 mM EDTA, 100 mM DTT. Prepare 100 mL by dissolving 1.21 g Tris (MW 121.14 g/mol) in 80 mL ultrapure water, add 0.2 mL 0.5 M EDTA, adjust pH to 8.3 with 6 M HCl, adjust volume to 100 mL. Add 1.54 g dithiothreitol powder (DTT, MW 154.2 g/mol) per 100 mL of buffer immediately before use (*see* **Note 2**).
3. 6 M HCl.
4. 0.1% TFA.
5. 60% acetonitrile in 0.1% TFA.
6. 1% ammonium acetate buffer, pH 8.5, 1 L.
7. 0.5 M citric acid: dissolve 5.25 g citric acid monohydrate (MW 210.14) in 40 mL hot water, cool down. Adjust volume to 50 mL.
8. 0.5 M tris-acetate buffer, pH 8.0, 2 mM EDTA: prepare 100 mL by dissolving 6.06 g Tris (MW 121.14 g/mol) in 60 mL ultrapure water, add 0.4 mL 0.5 M EDTA, adjust pH to 8.0 with glacial acetic acid under chemical fume hood, adjust volume to 100 mL.
9. 20 mM tris-(2-carboxyethyl)-phosphine (TCEP) in 0.17 M citrate buffer, pH 3.0: to 0.17 M citric acid monohydrate (MW 210.14), and add 0.17 M trisodium citrate dihydrate (MW 294.12) until pH is equal to 3.0. Dissolve 50 mg TCEP (MW 250.19 g/mol) in 10 mL citrate buffer immediately before use.
10. Anisole.
11. Silver trifluoromethanesulfonate.
12. Dry ether.
13. 0.5 M HCl/50% dimethyl sulfoxide (DMSO).
14. Supersaturated 2.2 M iodoacetamide solution: heat up a hypodermic needle on an open flame and pierce two holes in the cap of a black microcentrifuge tube. Weigh 100 mg of iodoacetamide directly into the tube; add 200 μL of 0.5 M Tris-acetate buffer, pH 8.0, 2 mM EDTA. Close the cap, and heat up the tube to 65 °C until crystals dissolve. Carefully cool to room temperature avoiding any disturbances, e.g., shaking or tipping the tube. Use within 5–10 min of preparation.
15. A flask with a nitrogen purge setup.
16. Hypodermic needles (12–14 gauge) with a syringe.
17. Black disposable 1.5 mL microcentrifuge tubes with a snap cap.
18. Small source of open flame (Bunsen burner or similar).

19. Vortex mixer.
20. C18 reversed-phase HPLC column
21. Sep-Pak C_{18} Cartridges (Waters).
22. 65 °C water bath.

3 Methods

3.1 Screening for Well-Folded GRN Disulfide Species by 1D ^1H NMR

The purpose of the following NMR analysis is to identify GRN disulfide species that form a well-defined three-dimensional structure. These species are assumed to also have the correct, native disulfide connectivities.

1. Dissolve a mixture of GRN disulfide species—approximately 1–5 mg of total protein—in 0.1% TFA (*see* **Note 3**).
2. Filter the solution through a 0.22–0.45 μm PES membrane filter, and load the filtered solution onto a C18 reversed-phase HPLC column.
3. Separate the GRN disulfide species using a 1%/min acetonitrile gradient in the presence of 0.1% TFA.
4. Collect fractions with distinct HPLC peaks.
5. Lyophilize the fractions in 20 mL glass scintillation vials. Store in vials sealed with Parafilm at −20 °C until use (*see* **Note 4**).
6. Dissolve each fraction in 0.46 mL of ultrapure water, add 12 μL of 0.5 M disodium phosphate, 90 μL of 1 M NaCl, 6 μL of 20 mM EDTA, and 6 μL of 1% sodium azide (*see* **Notes 5** and **6**).
7. Estimate the protein concentration by OD_{280} using the ExPASy ProtParam tool [23]. A typical concentration needed to record a one-dimensional (1D) ^1H NMR spectrum is approximately 0.1 mM or higher.
8. Adjust pH to 6.5 by adding small (0.5–5 μL) portions of 1 M or 100 mM HCl or NaOH.
9. Add 40 μL deuterium oxide (D_2O) (*see* **Note 4**).
10. Transfer the solution into a 5 mm NMR tube.
11. Record 1D ^1H NMR spectra using an appropriate water suppression scheme (e.g., WATERGATE [24] or double PFG spin echo [25]).
12. Visually inspect the spectra to identify fractions containing well-folded protein species (*see* **Note 7**).

3.2 NMR Characterization of Well-Folded GRN Disulfide Species (Disulfide Bond Assignment and Structure Calculations)

1. Determine the theoretical pI value of the entire expressed protein construct using the ExPASy ProtParam tool [23].
2. If the pI value is below 7, prepare and use 1% ammonium bicarbonate solution as a carrier solvent for HPLC purification. If the pI value is between 7 and 10, prepare and use 100 mM sodium phosphate buffer, pH 11.2.
3. Dissolve a well-folded GRN fraction in 0.1% TFA and load onto a reversed-phase HPLC column.
4. Separate the GRN disulfide species using a 1%/min acetonitrile gradient in 1% ammonium bicarbonate or 100 mM sodium phosphate buffer, pH 11.2.
5. Acidify fractions of interest to pH below 6.5 by adding 5–10 μL 6 M HCl immediately after collection (*see* **Note 8**).
6. Freeze-dry and reconstitute purified GRN fractions in 0.1% TFA, apply onto a reversed-phase HPLC column using 100% of 0.1% TFA as a carrier solvent.
7. Desalt the GRN peptide by running 100% of 0.1% TFA through the column with the absorbed peptide (*see* **Note 9**).
8. Elute the peptide using a 1%/min acetonitrile gradient in 0.1% TFA. Freeze-dry the collected peptide fraction.
9. Use mass spectrometry (MS) to confirm the identity of the peptide (*see* **Notes 10–12**).
10. Use 2.5–3 mg of GRN to prepare an NMR sample as outlined in **steps 6–10** of Subheading 3.1.
11. Record [^{15}N,^{1}H]-HSQC spectra of ^{15}N-labeled GRN using an appropriate water suppression scheme to identify the number of observed cross-peaks and establish the resonance chemical shift dispersion (*see* **Notes 13** and **14**). An example of the spectrum is shown in Fig. 2.
12. Record the following 2D spectra: ^{15}N-decoupled TOCSY, ^{15}N-decoupled NOESY, and MJ-[^{15}N,^{1}H]-HSQC [26] (*see* **Note 15**).
13. Record the following 3D spectra: ^{15}N-edited TOCSY-HSQC and ^{15}N-edited NOESY-HSQC [16].
14. Lyophilize the sample and dissolve in deuterium oxide.
15. Immediately record a series of [^{15}N,^{1}H]-HSQC spectra with 1–2 h intervals (*see* **Note 16**).
16. Assign proton and ^{15}N chemical shifts via standard spin system identification followed with sequential assignment through NOE connectivities.
17. Perform the NOE cross-peak assignment.

18. Determine $^{3}J_{NH\text{-}H}{}^{\alpha}$ coupling constants from MJ-[^{15}N,^{1}H]-HSQC.
19. Assign disulfide bonds via assignment of NOE connectivities between CβH–CβH and CαH–CβH atoms of cysteine pairs (*see* **Note 17**).
20. Calculate the 3D structure based on distance and dihedral angle constraints provided by the NOE peaks, $^{3}J_{NH\text{-}H}{}^{\alpha}$ coupling constants, data on slowly exchanging protons, disulfide connectivities, and chemical shift index.

3.3 Identification of Disulfide Connectivities Through NMR Characterization of Model Peptides

To identify remaining disulfide bridges, complementary approaches, such as NMR analysis of model peptides, should be used. GRN domain can be viewed as consisting of two disulfide linked domains—the N-terminal subdomain and the C-terminal subdomain [3, 4]. Each subdomain represents a pair of hairpins stabilized by two intra-subdomain disulfide bonds. The N-terminal and C-terminal subdomains are held together by two inter-subdomain disulfide bonds. To generate model peptides, the entire GRN domain is "dissected" into two designed model peptides, the N-terminal and the C-terminal. Depending on the specific disulfide bond to be assigned, cysteine residues forming inter-subdomain disulfide bonds are unaltered, blocked, or substituted with serines. These model peptides are chemically synthesized, and disulfide bonds are formed via a combination of thiol air oxidation and directed approaches [27]. The disulfide bonds in oxidized model peptides are assigned as described by Gray [28, 29]. The NMR spectra of generated peptides serve as a reference for the full-length GRN domain. The comparison between the amide proton chemical shifts of a particular disulfide-bonded isomer of a model peptide and the GRN provides information on the disulfide connectivities. Identical disulfide connectivities will result in similar chemical shift patterns [4, 21]. Additionally, model peptides with a particular forced disulfide bond can be used as a "litmus test" for steric compatibility between the formation of the disulfide bridge and the peptide structural integrity [4]. Incorrect disulfide bonds are not likely to produce well-folded peptides (*see* **Note 18**).

1. Identify a model peptide for chemical synthesis. The synthetic peptide will have a sequence with two or four cysteines out of six substituted with serines and/or cysteines blocked with *S*-acetamidomethyl groups (Acm) (*see* **Note 19**).
2. Incubate 20 mg of peptide in 2 mL of reducing buffer under nitrogen for 2–4 h.
3. Acidify by adding 0.1 mL 6 M HCl.
4. Apply on a Sep-Pak C_{18} Cartridge. Wash with six column volumes of 0.1% TFA. Elute with six column volumes of 60% acetonitrile in 0.1% TFA (*see* **Note 20**).

5. Air-oxidize overnight (with stirring) 10–20 mg of the peptide in 1 L of ammonium acetate buffer (*see* **Note 21**).
6. Adjust pH to 5.0, and apply onto a C18 reversed-phase HPLC column (*see* **Note 22**). Elute with a 1%/min acetonitrile gradient in 0.1% TFA. Freeze-dry.
7. Dissolve the dried peptide (5–10 mg) in 0.5 mL TFA.
8. Deprotect the *S*-acetamidomethyl-cysteines by adding 10 mL anisole and 27 mg of silver trifluoromethanesulfonate and incubate at 4 °C for 1.5 h.
9. Add 1 mL of ice-cold dry ether to the mixture, centrifuge the precipitate, wash the precipitate with the same volume of ice-cold dry ether twice, and collect the peptide by centrifugation.
10. Dissolve the peptide in 0.5 M HCl/50% DMSO and oxidize for 8–16 h at room temperature (*see* **Note 23**).
11. Dilute ten times with HPLC grade water and purify on a C18 reversed-phase HPLC column using a 1%/min acetonitrile gradient in 0.1% TFA (*see* **Note 24**). Freeze-dry.
12. To assign/confirm disulfide connectivities in a model peptide, one volume of 0.2 mg of peptide (HPLC eluent in 0.1% TFA mixed with acetonitrile) mix with two volumes of TCEP solution, incubate for 10 min at 65 °C for complete reduction (*see* **Note 25**).
13. Apply immediately onto a C18 reversed-phase HPLC column; use the HPLC profile as a reference for the HPLC retention time of a completely reduced peptide species.
14. One volume of 0.5–1.0 mg of the peptide (0.2–0.3 mL HPLC eluent in 0.1% TFA mixed with acetonitrile) mix with two volumes of TCEP solution, incubate for 5–10 min at room temperature to produce partially reduced intermediates of the peptide (*see* **Note 25**).
15. Apply immediately onto a C18 reversed-phase HPLC column and purify the intermediates by HPLC.
16. Draw a collected HPLC fraction of a partially reduced peptide intermediate (0.25 mL) into a syringe with a needle. Carefully, without shaking, place a black tube with the supersaturated iodoacetamide solution on top of a vortex mixer. Insert the needle into the pierced cap of the tube and forcibly squirt the peptide into the solution while rigorously stirring the tube.
17. Incubate for 30 s at room temperature.
18. Quench with 0.4 mL 0.5 M citric acid and purify on HPLC.

19. Collect the alkylated intermediate, reduce the remaining disulfide bonds as in **step 12**, and purify on HPLC.
20. Confirm the peptide identity by mass spectrometry (+57 mass changes per each alkylated cysteine).
21. Assign/confirm disulfide bonds by identifying pairs of alkylated cysteines using N-terminal peptide sequencing (Edman degradation).
22. Use 2.5–3 mg of the peptide to prepare an NMR sample as outlined in **steps 6–10** of Subheading 3.1.
23. Record 2D TOCSY and NOESY spectra.
24. Assign proton chemical shifts via spin system identification followed with sequential assignment through NOE connectivities.

4 Notes

1. Many silica-based HPLC columns (e.g., Vydac 218TP series) are not chemically stable at pH > 6. Waters XBridge pH-resistant columns or other pH-resistant alternatives should be used for GRN purification in alkaline conditions.
2. It is important to use DTT before the end of its useful life. After purchase store DTT well sealed at 4 °C or −20 °C. Allow the bottle to warm up to room temperature before opening. Discard after 1 year.
3. Take particular precautions to avoid sources of protease contamination. Use disposable examination gloves. Autoclave plastic ware.
4. Deuterium oxide and lyophilized GRN are hygroscopic, and their exposure to atmosphere should be minimized.
5. For a standard 5 mm NMR tube and a standard Varian probe, a sample volume of 600 μL is required. Bruker 5 mm NMR probes require sample volumes of 450 μL. The use of Shigemi microtubes allows reducing sample volumes by approximately 50%.
6. The buffer composition may strongly affect resonance linewidths of a GRN NMR spectrum. To obtain high-quality spectra with the narrowest lines, the concentration of NaCl can be varied between 0 and 200 mM and pH between 5.0 and 6.8.
7. The difference between well-folded and unfolded/misfolded GRN species is easy to detect without resonance assignment. Unfolded/misfolded species display poor dispersion of amide proton chemical shifts (1.0–1.2 ppm), extensive resonance overlap, and broad resonance lines. In contrast, well-folded

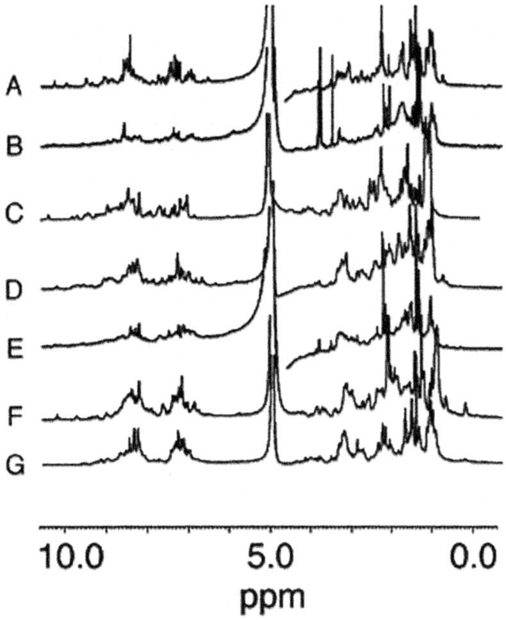

Fig. 1 Proton NMR spectra of human granulins A–G in sodium phosphate buffer, pH 6.5, at 25 °C. Spectra of GRN A, C, D, and F are characteristic of well-folded species, while spectra of GRN B, E, and G represent unfolded/misfolded species

GRN species display sharp resonance signals and visibly larger chemical shift dispersion. An example of 1D ^{1}H NMR spectra for several GRN species is shown in Fig. 1. Spectra of GRN A, C, D, and F are characteristic of well-folded species, while spectra of GRN B, E, and G represent unfolded/misfolded species.

8. Exposure of GRN to alkaline conditions should be minimal to avoid peptide degradation/modification. GRN peptides are generally very stable in the pH range of 5.0 to 6.5.
9. It is particularly important for NMR analysis to eliminate traces of ammonium bicarbonate. Running a 0.1% TFA solution for 1 h at 1 mL/min through a column with an absorbed peptide is sufficient.
10. Preparation of recombinant GRNs involves enterokinase cleavage of expressed GRN fusion proteins. Treatment with enterokinase can lead to non-specific proteolysis of GRN. SDS-PAGE is often used as a criterion for the quality and homogeneity of protein preparations. However, SDS-PAGE may not be a sufficiently accurate method to detect the loss of several N-terminal or C-terminal amino acid residues. If the unwanted cleavage is in the middle of the GRN sequence, it may still be challenging to detect it by SDS-PAGE. Full reduction of disulfide bonds in GRN is difficult to achieve, and the efficiency of

the reaction is dependent on the storage history of the reducing agent (e.g., DTT, commonly used to reduce protein samples for SDS-PAGE analysis). The remaining disulfide bonds may keep two chains together and appear as a single band on the SDS-PAGE with a "correct" molecular weight.

11. MS is an ultimate method to establish if such non-specific cleavage occurred. A peptide bond hydrolysis in the middle of the GRN chain held together by disulfide bonds will manifest as an increase of the molecular mass of GRN by 18 compared to the theoretical value. In addition to backbone integrity, MS may also be able to resolve fully and partially oxidized GRN disulfide species, since a disulfide bond formation results in a −2 Da mass change.
12. In interpreting MS results, it is important to remember that impurities may have different ionization efficiencies and show in MS spectra as prominent peaks even if the impurity level is low. Therefore, to minimize the possibility of obtaining misleading results, MS analysis should be performed on highly pure GRN samples.
13. The number of cross-peaks in an [^{15}N,^{1}H]-HSQC spectrum of a well-folded protein is equal to the sum of the number of backbone amide protons, the number of side chain amide protons of Asn and Gln (two protons per each residue), and the number of indole NH of Trp (Fig. 2). In certain conditions, protons of guanidino groups of Arg can be also detected, if present in a GRN sequence. If many peaks are broadened, heavily overlapped or cannot be accounted for, this indicates the presence of intrinsically disordered regions. Observation of extra peaks/peak satellites is indicative of an impurity.
14. Detailed description of experimental aspects of protein NMR spectroscopy, peak assignment, and structure calculations is beyond the scope of this chapter. The chapter provides information on which set of spectra is sufficient to make the assignments, to determine disulfide connectivities, and to calculate the 3D structure of a GRN molecule. Examples of software programs for NMR data processing are Felix (Felix NMR, San Diego, CA), NMRPipe [30]; for NMR spectra visualization and peak assignment are NMRViewJ [31] (One Moon Scientific), SPARKY (developed by Goddard and Kneller, UCSF); for structure calculations are ARIA-CNS [18, 32], CYANA [33].
15. GRN domains are small β-class proteins. If well structured, an almost complete proton assignment is possible by the use of 2D TOCSY and NOESY spectra via spin system identification followed with sequential assignment through NOE connectivities. ^{15}N-labeling is not required to perform these 2D NMR experiments.

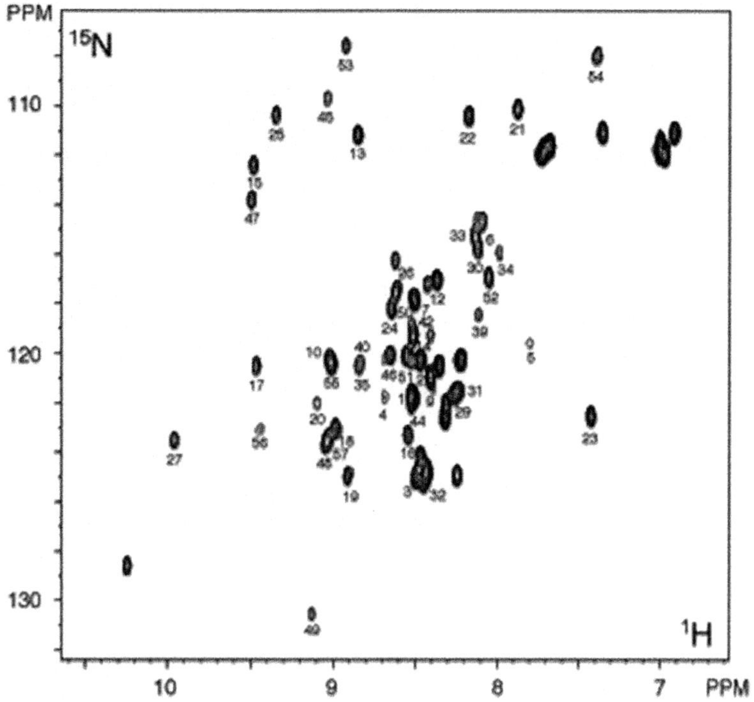

Fig. 2 An [^{15}N,^{1}H]-HSQC spectrum of ^{15}N-labeled human GRN A in 10 mM sodium phosphate, 150 mM NaCl, 0.2 mM EDTA, 10% D_2O at pH 6.8. Residue numbers label HSQC cross-peaks of the corresponding backbone amides. Each resonance peak represents a pair of ^{15}N and ^{1}H atoms separated by one covalent bond. The vertical coordinate axis corresponds to the chemical shift of the ^{15}N atom, while the horizontal axis corresponds to the chemical shift of the ^{1}H atom. The absence of peak satellites indicates that the protein is pure. Differential peak broadening is typical of residues undergoing conformational exchange. Formation of aggregates in solution may also lead to spectral broadening. Reproduced from [4] with permission from John Wiley and Sons

16. In a random coil peptide, the exchange of amide hydrogens to deuterium at pH close to neutral typically is complete before the first NMR spectrum can be recorded [34]. Much slower exchange rates are generally indicative of either hindrance of amide groups from solvent molecules or a formation of intramolecular hydrogen bonds. For a small GRN protein with a ladderlike architecture, the effects of steric hindrance can be neglected. The information on hydrogen bond formation can be included in the structure calculations.
17. Additionally, if an NOE between NH and CαH of two cysteines is observed, these cysteines are unlikely to form a disulfide bridge [9]. This method [9] works well for well-structured regions of GRN. Another NMR-based approach to assign

disulfide bond connectivities has been proposed recently by Poppe et al. [35].

18. Other methods to map disulfide bonds include MS analysis of proteolytic digests and cyanylation/cleavage methodology [10–14]. Proteolytic digestion/MS approach will not work well for GRNs because of the presence of a large number of adjacent cysteines in GRN sequences. Generally speaking, cyanylation/cleavage methodology can be used for complete disulfide bond assignment in GRNs. However, GRN molecules demonstrated a remarkable resistance to disulfide bond reduction, which is the first step reaction before cysteine residues of a particular disulfide bond can be cyanylated. Nevertheless, some of the disulfide connectivities can be identified using this method, particularly if they are located in less structured regions of GRN [4]. Since NMR-based disulfide bond assignment is better suited for well-structured regions, cyanylation/cleavage methodology can serve as an appropriate complementary approach. The technical details of this methodology have been covered extensively [10–14].
19. Model peptides will typically have six cysteines. Depending on the disulfide bonding to test, some pairs will be replaced with serines or *S*-acetamidomethyl-cysteines (Cys(Acm)) to direct a desired disulfide bond formation.
20. Use approximately 1 g of resin per 10 mg of peptide.
21. It is very important to eliminate the possibility of contaminating the solution with proteases. Autoclave solutions, plastic ware and glassware, use examination gloves. Lower concentrations of the peptide during air oxidation reduce the rate of formation of intermolecular disulfide bonds, and therefore favor the formation of intramolecular disulfide bridges.
22. Loading a peptide solution of large volume onto an HPLC system can be performed through one of the solvent lines. After use the solvent line often has to be washed with 70% acetonitrile in 0.1% TFA to remove the absorbed peptide and other contaminants. Alternatively, the solution can be loaded onto a Sep-Pak column, the peptide eluted with 70% acetonitrile in 0.1% TFA, and freeze-dried prior to HPLC purification.
23. The protocol for silver trifluoromethanesulfonate deprotection of Cys(Acm) followed by DMSO oxidation of synthetic peptides was developed by Hunter and Komives to minimize methionine oxidation observed with more commonly used iodine deprotection and oxidation [36].
24. If one-step HPLC purification does not produce a pure enough peptide, use for the second purification either 1%

ammonium bicarbonate or 100 mM sodium phosphate buffer, pH 11.2, as described in Subheading 3.2, **steps 1–8**.

25. Depending on a peptide, the reaction time may have to be adjusted to optimize the yield of reduced species.

References

1. Bateman A, Bennett HP (1998) Granulins: the structure and function of an emerging family of growth factors. J Endocrinol 158:145–151
2. Bateman A, Bennett HP (2009) The granulin gene family: from cancer to dementia. BioEssays 31:1245–1254
3. Hrabal R, Chen Z, James S et al (1996) The hairpin stack fold, a novel protein architecture for a new family of protein growth factors. Nat Struct Biol 3:747–752
4. Tolkatchev D, Malik S, Vinogradova A, Wang P et al (2008) Structure dissection of human progranulin identifies well-folded granulin/epithelin modules with unique functional activities. Protein Sci 17:711–724
5. Prinz WA, Åslund F, Holmgren A et al (1997) The role of the thioredoxin and glutaredoxin pathways in reducing protein disulfide bonds in the *Escherichia coli* cytoplasm. J Biol Chem 272:15661–15667
6. Bessette PH, Åslund F, Beckwith J et al (1999) Efficient folding of proteins with multiple disulfide bonds in the *Escherichia coli* cytoplasm. Proc Natl Acad Sci U S A 96:13703–13708
7. Sevier CS, Kaiser CA (2008) Ero1 and redox homeostasis in the endoplasmic reticulum. Biochim Biophys Acta 1783:549–556
8. Sevier CS, Kaiser CA (2002) Formation and transfer of disulphide bonds in living cells. Nat Rev Mol Cell Biol 3:836–847
9. Klaus W, Broger C, Gerber P et al (1993) Determination of the disulphide bonding pattern in proteins by local and global analysis of nuclear magnetic resonance data: application to flavoridin. J Mol Biol 232:897–906
10. Wu J, Watson JT (1997) A novel methodology for assignment of disulfide bond pairings in proteins. Protein Sci 6:391–398
11. Wu J, Watson JT (1998) Optimization of the cleavage reaction for cyanylated cysteinyl proteins for efficient and simplified mass mapping. Anal Biochem 258:268–276
12. Yang Y, Wu J, Watson JT (1998) Disulfide mass mapping in proteins containing adjacent cysteines is possible with cyanylation/cleavage methodology. J Am Chem Soc 120:5834–5835
13. Qi J, Wu J, Somkuti GA, Watson JT (2001) Determination of the disulfide structure of sillucin, a highly knotted, cysteine-rich peptide, by cyanylation/cleavage mass mapping. Biochemistry 40:4531–4538
14. Wu J (2008) Disulfide bond mapping by Cyanylation-induced cleavage and mass spectrometry. In: Kannicht C (ed) Post-translational modifications of proteins. Humana Press, Totowa, NJ, pp 1–20
15. Uversky VN (2010) Targeting intrinsically disordered proteins in neurodegenerative and protein dysfunction diseases: another illustration of the D2 concept. Expert Rev Proteomics 7:543–564
16. Sattler M, Schleucher J, Griesinger C (1999) Heteronuclear multidimensional NMR experiments for the structure determination of proteins in solution employing pulsed field gradients. Prog Nucl Magn Reson Spectrosc 34:93–158
17. Cavanagh J (ed) (2007) Protein NMR spectroscopy: principles and practice. Academic Press, Amsterdam, Boston
18. Linge JP, Habeck M, Rieping W et al (2003) ARIA: automated NOE assignment and NMR structure calculation. Bioinformatics 19:315–316
19. Downing AK (ed) (2004) Protein NMR techniques. Humana Press, Totowa N.J
20. Tolkatchev D, Xu P, Ni F (2001) A peptide derived from the C-terminal part of a plant cysteine protease folds into a stack of two β-hairpins, a scaffold present in the emerging family of granulin-like growth factors. J Pept Res 57:227–233
21. Tolkatchev D, Ng A, Vranken W et al (2000) Design and solution structure of a well-folded stack of two β-hairpins based on the amino-

terminal fragment of human granulin a. Biochemistry 39:2878–2886
22. Vranken WF, Chen ZG, Xu P et al (1999) A 30-residue fragment of the carp granulin-1 protein folds into a stack of two β-hairpins similar to that found in the native protein. J Pept Res 53:590–597
23. Wilkins MR, Gasteiger E, Bairoch A et al (1999) Protein identification and analysis tools in the ExPASy server. Methods Mol Biol 112:531–552
24. Piotto M, Saudek V, Sklenář V (1992) Gradient-tailored excitation for single-quantum NMR spectroscopy of aqueous solutions. J Biomol NMR 2:661–665
25. Hwang TL, Shaka AJ (1995) Water suppression that works. Excitation sculpting using arbitrary wave-forms and pulsed-field gradients. J Magn Reson A 112:275–279
26. Heikkinen S, Aitio H, Permi P et al (1999) J-multiplied HSQC (MJ-HSQC): a new method for measuring 3 J (H N H α) couplings in 15 N-labeled proteins. J Magn Reson 137:243–246
27. Annis I, Hargittai B, Barany G (1997) [10] disulfide bond formation in peptides. Methods Enzymol 289:198–221
28. Gray WR (1993) Disulfide structures of highly bridged peptides: a new strategy for analysis. Protein Sci 2:1732–1748
29. Gray WR (1993) Echistatin disulfide bridges: selective reduction and linkage assignment. Protein Sci 2:1749–1755
30. Delaglio F, Grzesiek S, Vuister GW et al (1995) NMRPipe: a multidimensional spectral processing system based on UNIX pipes. J Biomol NMR 6:277–293
31. Johnson BA (2004) Using NMRView to visualize and analyze the NMR spectra of macromolecules. In: Protein NMR Techniques. Humana Press, New York, pp 313–352
32. Bardiaux B, Malliavin T, Nilges M (2012) ARIA for solution and solid-state NMR. In: In Protein NMR Techniques. Humana Press, New York, pp 453–483
33. Güntert P, Buchner L (2015) Combined automated NOE assignment and structure calculation with CYANA. J Biomol NMR 62:453–471
34. Bai Y, Milne JS, Mayne L et al (1993) Primary structure effects on peptide group hydrogen exchange. Proteins 17:75–86
35. Poppe L, Hui JO, Ligutti J et al (2011) PADLOC: a powerful tool to assign disulfide bond connectivities in peptides and proteins by NMR spectroscopy. Anal Chem 84:262–266
36. Hunter MJ, Komives EA (1995) Deprotection of *S*-acetamidomethyl cysteine-containing peptides by silver trifluoromethanesulfonate avoids the oxidation of methionines. Anal Biochem 228:173–177

Chapter 6

Data Mining: Applying the AD&FTD Mutation Database to Progranulin

Marc Cruts and Christine Van Broeckhoven

Abstract

The online AD&FTD Mutation Database (http://www.molgen.vib-ua.be/FTDmutations) was conceived to meet the needs of a comprehensive knowledge base of genetic variations in genes associated with monogenic forms of Alzheimer's disease (AD) and frontotemporal dementia (FTD). Today, the AD&FTD Mutation Database provides curated, referenced information of 764 genetic variants in *APP*, *PSEN1*, and *PSEN2* associated with AD and *GRN*, *C9orf72*, *TBK1*, *MAPT*, *VCP*, *CHMP2B*, *TARDBP*, and *FUS* associated with FTD and related diseases. In addition, the database stores demographic and clinicogenetic data of 1646 dementia families associated with these mutations. In FTD, the granulin (*GRN*) gene has the highest number of different mutations (79/231 = 34%) and the second highest number of associated FTD families after *C9orf72*. In addition to the detailed mutation and patient information, summary reports in tabular and graphical formats can be consulted. Further, all variants can be uploaded to the human genome browser for custom-designed analyses.

Key words Alzheimer's disease, Frontotemporal dementia, Database, Genetic variants, In silico analysis

1 Introduction

The AD&FTD Mutation Database provides variant information of the genes associated with monogenic forms of Alzheimer's disease (AD; MIM #104300) and frontotemporal dementia (FTD; MIM #600274) [1]. This locus-specific database (LSDB) was conceived in 1998 [2] in the context of the Mutation Database Initiative (MDI) [3], an initiative originally fostered by the Human Genome Organisation (HUGO, www.hugo-international.org), which has, through the years, evolved to the Human Genome Variation Society (HGVS, www.hgvs.org). HGVS is dedicated to promote and support the collection, documentation, and free distribution of human genomic variation and associated clinical characteristics. The AD&FTD Mutation Database started as a repository of curated genetic, clinical, and biological information of DNA variants in the Mendelian AD genes *APP*, *PSEN1*, and *PSEN2* [2]. Due to

Andrew Bateman et al. (eds.), *Progranulin: Methods and Protocols*, Methods in Molecular Biology, vol. 1806, https://doi.org/10.1007/978-1-4939-8559-3_6,

Table 1
Genes catalogued in the AD&FTD Mutation Database

Primary disease	Symbol[a]	Name[a]	ID[b]	Genomic location	# Mutations[c]
AD	*APP*	Amyloid beta (A4) precursor protein	351	21q21.2	51
	PSEN1	Presenilin 1	5663	14q24.3	219
	PSEN2	Presenilin 2	5664	1q42.13	16
FTD	*C9orf72*	Chromosome 9 open reading frame 72	203,228	9p21.2	1
	CHMP2B	Charged multivesicular body protein 2B	25,978	3p11.2	4
	FUS	FUS RNA binding protein	2521	16p11.2	23
	GRN	Granulin	2896	17q21.32	79
	MAPT	Microtubule-associated protein tau	4137	17q21.1	44
	TARDBP	TAR DNA binding protein	23,435	1p36.22	33
	VCP	Valosin-containing protein	7415	9p13.3	19
	TBK1	TANK binding kinase 1	29110	12q14.2	28

[a]Official gene symbol and name provided by the HUGO Gene Nomenclature Committee (HGNC, www.genenames.org) [9]
[b]Gene ID of the NCBI gene database (www.ncbi.nlm.nih.gov/gene) [7]
[c]Number of variants in the AD&FTD Mutation Database that are marked as pathogenic. The *C9orf72* repeat expansion is counted as one mutation

increasing awareness of genetic and clinical overlaps between AD and FTD, mutations in all known Mendelian FTD genes were added to the AD&FTD Mutation Database from 2004 onward [4, 5]. Today, the Database covers variants in ten genes associated with AD and FTD (Table 1). The AD&FTD Mutation Database is available through a publicly accessible web site at www.molgen.ua.ac.be/ADMutations or www.molgen.ua.ac.be/FTDMutations. The basic genetic information of the mutations stored in the database is at regular intervals shared with the Gen2Phen project (www.gen2phen.org), [6] and the NCBI dbSNP database (www.ncbi.nlm.nih.gov/snp), [7], where they are recorded as clinically significant SNPs with cross-reference to the AD&FTD Mutation Database.

The AD&FTD Mutation Database is of great value in several areas involving molecular genetics of neurodegenerative dementia. In clinical genetic counselling, for example, knowledge of pathological mutations and their genic location assists in developing efficient diagnostic screening protocols based on the region-specific prevalence of disease mutations. Since targeted massive parallel resequencing protocols are becoming mainstream in genetic testing, panels of disease-associated genes are sequenced in parallel, making the development of decision trees in genetic screening unnecessary. Therefore, today the most valuable use of the AD&FTD Mutation Database in clinical genetic settings is to provide a central repository of all genetic variation using unambiguous nomenclature, annotated with various referenced information contributing to the evaluation of its pathogenic nature. Parameters that may assist in making well-founded decisions, like frequencies of occurrence in patients and unaffected individuals, evidence of intrafamilial co-segregation with disease, interspecies codon conservation, genotype-phenotype correlations, in silico predictions, and in vitro and in vivo biochemical consequences, are collected in the Mutation Database. In a research setting, a comprehensive overview of existing information on a particular mutation or gene may facilitate research decisions. Further, aggregate analysis of the genic location and types of all known pathological and neutral variations in a particular gene may reveal valuable indications toward pathomechanistically critical protein domains or motifs.

2 Database Content

2.1 Variant Inclusion Policy

The AD&FTD Mutation Database provides a catalogue of all variants that are either disease-causing, or have an impact on the encoded protein, or both. Because the knowledge of variants with a clinical impact and of benign coding variants is equally important, both are documented. Noncoding neutral variants are excluded because they are not of immediate interest to the Mutation Database user, and because other variant databases like NCBI dbSNP are much better suited to accommodate these variants. AlzGene is an excellent public database of risk and protective genetic variants associated with AD [8].

About 99% of variant information stored in the Mutation Database is extracted from scientific literature, while a small fraction is obtained from direct submission and personal communication. For literature retrieval, the NCBI PubMed biomedical literature database (www.ncbi.nlm.nih.gov/pubmed) is periodically scanned using as query the genes' official symbols and full names as designated by the HUGO Gene Nomenclature Committee (HGNC), [9], supplemented with all gene aliases listed in the NCBI Gene database (www.ncbi.nlm.nih.gov/gene) and unofficial names used in literature. Specifically for *GRN*, the PubMed database is searched for ten terms in all database fields: GRN, GEP, GP88, PEPI, PGRN, PCDGF, granulin, progranulin, proepithelin, and epithelin. Retrieved publications are manually scanned for data on *GRN* variants. Data are then evaluated for experimental quality and unambiguous description. Variant names are checked for consistency with the current HGVS guidelines for mutation nomenclature (www.hgvs.org/varnomen) [10, 11].

Although significantly improved in recent years, genes and variants are not always named according to nomenclature guidelines, and their names may be ambiguous. Consistent variant names following two or more naming systems, e.g., transcript and protein, are considered reliable. If inconsistencies are found or if the nature of the variant remains ambiguous, the authors are contacted for further details. Variants that remain ambiguous despite all efforts are excluded from the Mutation Database to warrant data integrity.

2.1.1 Genetic and Clinicopathological Documentation

Variants are recorded in the Mutation Databases with names at the level of the gene, transcript, and protein, according to the HGVS nomenclature guidelines. In addition, commonly used aliases are also provided. Variant position numbering is relative to stable RefSeq reference sequences [12] for gene (RefSeqGene), RNA, and protein. For historical reasons, genomic numbering of some genes is also provided relative to non-RefSeqGene sequences. For *GRN*, variant position numbers are relative to RefSeq gene sequence NG_007886.1, RefSeq cDNA sequence NM_002087.2, and RefSeq protein sequence NP_002078.1.Mutalyzer [13] is used to assist in generating and verifying variant names. Further, it is indicated whether the variant is experimentally validated in at least one publication or predicted based on indirect evidence, e.g., a predicted change in the protein sequence based on a variant observed in the transcript sequence. All details are documented with references to the original literature reports.

In addition to the mutation descriptors, demographic and clinicopathological information of the mutation carriers is provided. These data include number of independent observations, ethnicity of the mutation carriers, primary clinical or pathological diagnosis, family history, evidence of co-segregation of the mutation with disease within families, and average ages at disease onset and death within the family.

2.2 Database Use

Mutations can be retrieved for one or more genes, primary diagnoses, or publications by clicking the corresponding item in the left menu (Fig. 1). In each case, an overview of all variants meeting the selection criteria is shown in one of three selectable formats: a tabular view, a graphical view, or a view in the UCSC genome browser with the Mutation Database variants displayed as a custom track. In each output format, the variants are clickable to retrieve detailed information of a particular variant.

2.2.1 Tabular View

In the tabular view format, a variant summary table is shown (Fig. 2). Variants are ordered according to their start position in the human reference genome sequence hg19. Clicking on a variant name will open the variant details page (Fig. 3). The table includes nonofficial names by which the variants are easily recognized. These aliases are not in all cases according to HGVS variant nomen-

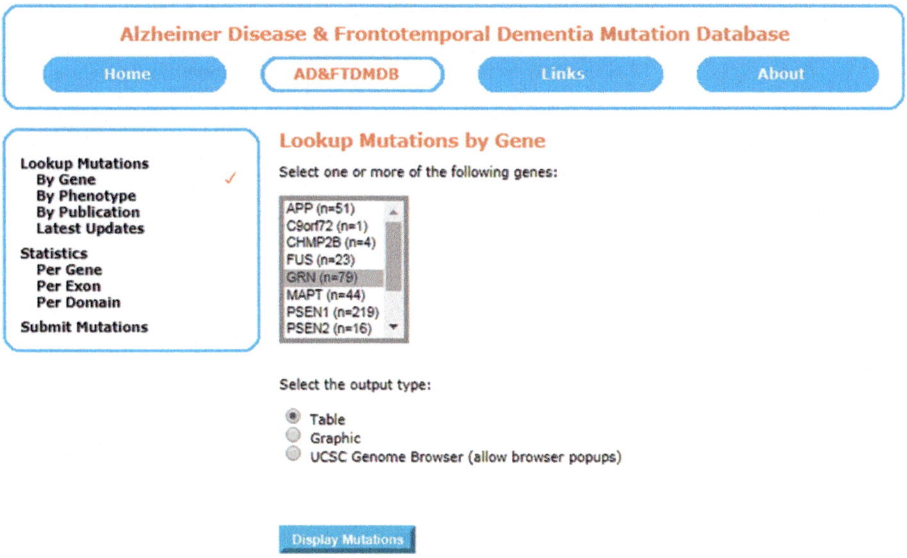

Fig. 1 Variant retrieval by gene name, e.g., *GRN*. Depending on the selected output format, the user is presented with a tabular list (Table) or a graphical presentation (Graphic) of the gene(s) of interest. Alternatively, the list of variants can be uploaded as an hg19 custom track in the UCSC genome browser (https://genome.ucsc.edu)

clature standards [11]. Therefore, the genomic variants names are also shown between brackets. Also included in the table are the genic location, e.g., exon number, the affected protein domain, and literature references. Clicking the variant name opens the variant details page; clicking the references opens an overview of the publications describing a particular variant (*see* **Notes 1** and **2**).

2.2.2 Graphical View

Based on the available knowledge of the protein structure and domain organization, and on which display format is most informative for that particular protein, the graphical view shows the location of the mutations either on a linear, a two-, or three-dimensional diagram of the protein. For *GRN*, a two-dimensional model is chosen (Fig. 4) that illustrates the conserved amino acid positions within each granulin domain, in particular, the fully conserved cysteine residues forming five or six cysteine bonds that control the structure of the granulin peptides [16]. This display format allows quick appreciation of the location of a particular mutation in the GRN peptide structure and its amino acid conservation. Amino acid conservation can provide supportive evidence of the pathogenic nature of a particular *GRN* missense mutation. For example, it can be seen in the figure that the p.C139R mutation affects the third conserved cysteine residue of GranF which predicts a disruptive effect on the protein structure. In accordance with that observation, this mutation was

	Mutation	Exon / Domain	Citations
1	*GRN* delGRN[DR184] ()	Complete gene Complete protein	Gijselinck I, 2008
2	*GRN* IVS1+3A>T (g.-3828A>T)	IVS1 Complete protein	Le Ber I, 2007
3	*GRN* IVS1+5G>C (g.-3826G>C)	IVS1 Complete protein	Cruts M, 2006 Brouwers N, 2007
4	*GRN* delGRN[French] (g.-95_3490del)	IVS1-IVS12 Complete protein	Rovelet-Lecrux A, 2008
5	*GRN* Met1 (g.1A>G)	EX2 Signal peptide	Le Ber I, 2008
6	*GRN* Met1 (g.2T>C)	EX2 Signal peptide	Baker M, 2006 Gass J, 2006
7	*GRN* Met1 (g.3G>A)	EX2 Signal peptide	Cruts M, 2006
8	*GRN* Val5Leu (g.13G>C) Pathogenic nature unclear	EX2 Signal peptide	López de Munain A, 2008
9	*GRN* Trp7Arg (g.19T>C) Pathogenic nature unclear	EX2 Signal peptide	Le Ber I, 2008
10	*GRN* Ala9Asp (g.26C>A)	EX2 Signal peptide	Mukherjee O, 2006 Gass J, 2006 Spina S, 2007 Ghetti B, 2008 Mukherjee O, 2008 Spina S, 2008 Kelley BJ, 2009 Yu CE, 2010
11	*GRN* Leu14 (g.42G>A) Not pathogenic	EX2 Signal peptide	Guerreiro RJ, 2008 Guerreiro RJ, 2010
12	*GRN* Arg19Trp (g.55C>T) Not pathogenic	EX2 ParaGran	Gass J, 2006 Schymick J, 2007 Guerreiro RJ, 2010 Del Bo R, 2011
13	*GRN* Asp22fs (g.63_64insC)	EX2 ParaGran	Gass J, 2006 Pietroboni AM, 2011
14	*GRN* Cys31fs (g.90_91insCTGC)	EX2 ParaGran	Baker M, 2006 Gass J, 2006 Beck J, 2008 Rohrer JD, 2008 Rohrer JD, 2009 Yu CE, 2010 Rohrer JD, 2010

Fig. 2 Extract of tabular view of *GRN* variants. Variants are listed with their aliases and genomic names between brackets. Also the genic location (EX = exon, IVS = intron) and affected protein domain are indicated. For each variant, all publications covered by the AD&FTD Mutation Database are shown

observed in 2 FTD patients and in 1 AD patient, but not in >900 control subjects. Further, it was shown that this mutation results in reduced serum GRN levels which are comparable to that in carriers of established *GRN* loss-of-function mutations [18]. Together, these data suggest that this mutation is likely to be pathogenic.

2.2.3 Genome Browser Track

The AD&FTD Mutation Database offers the option to upload all or selected variants into the UCSC genome browser (genome.ucsc.edu) as a custom track for the hg19 human reference genome sequence in the Browser Extensible Data (BED) format. In the

GRN IVS1+5G>C Mutation Details

(Table Legend)

Gene	*GRN*
Name	g.-3826G>C (relative to Met1 in Reverse Complement of AC003043.1) g.5217G>C (relative to nt1 in NG_007886.1)
Alias	IVS1+5G>C
Description	Point mutation in intron 1 splice donor site causing intron 1 retention resulting in nuclear mRNA degradation
Codon Change	From --- to ---
dbSNP ID	rs63750313

Details	Genomic	cDNA	Protein
Observed	g.-3826G>C	c.0	p.0
Predicted		c.-8_-7ins3823	
Region	IVS1		Complete protein

Pathogenecity	CADD (v1.2) score: phred = 14.67 ; raw = 1.252785 Guerreiro classification:
Phenotype	Frontotemporal Dementia Mean of Mean Onset Ages: 61.4y Mean of Mean Ages at Death: 68.7y
Frequency	11 Families
Function	No Functional data available
Citations	**Cruts M**, *Nature* 442: 920-924, 2006 **Brouwers N**, *Archives of Neurology* 64: 1436-1446, 2007 Citation Details

Added: August 23, 2006 ID: 266
Last Modified: January 31, 2008

Fig. 3 Example of a variant details page. Variant names are shown at level of gene, coding sequence, and protein. If the variant was experimentally demonstrated at the respective molecular level in at least one publication, the variant name is shown in the top row of the Details table labeled "Observed"; if it was predicted based on indirect observations, the name is shown in the second row labeled "Predicted." In the pathogenicity section, scores of variant prediction algorithms are shown. CADD provides a prediction score of the relative pathogenicity of the variant [14]; the Guerreiro classification is a system developed to rank AD mutations according to available evidence of pathogenicity [15]. In the Phenotype section, Mean of Mean ages indicate the average of the mean onset ages within each documented family in which the mutation was observed. Frequency indicates the number of independent observations of the variant. Clicking the number displays the patient and family details page

genome browser, the track is entitled "Alzheimer Disease & Frontotemporal Dementia Mutation Database". This view allows to get an overview of the location of the variants relative to the gene structure and, more importantly, shows the location of the variants relative to all other genome annotations available in other data tracks in the UCSC genome browser (Fig. 5), like sequence conservation, transcription factor binding sites, or a custom track showing variants identified in own gene sequencing projects.

Once uploaded to the genome browser, the AD&FTD Mutation Database track can be customized and manipulated as any data track in the BED data format. Clicking a particular muta-

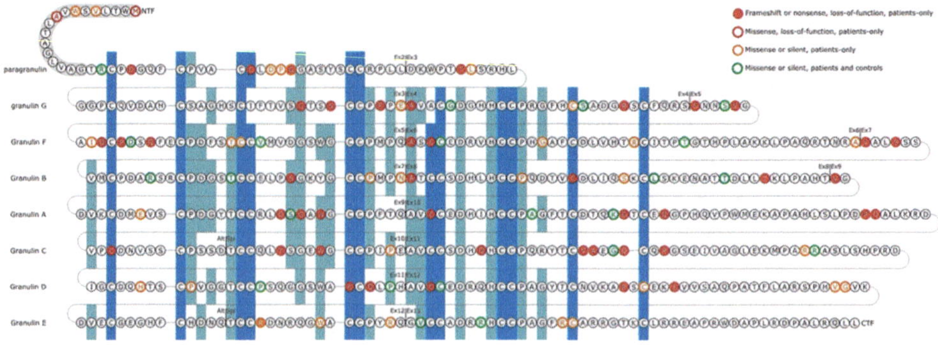

Fig. 4 Interactive graphical presentation of the GRN protein showing the aligned granulin domains A-G and paragranulin. The amino acid positions can be clicked to retrieve a table listing all variants that affect the specific amino acid or are predicted to do so. The cysteine residues forming disulphide bonds essential to maintain the 3D fold of the granulin peptides [16] are highlighted in blue. Other conserved amino acid residues are highlighted in light blue. Exon-exon boundaries indicate the genic localization of the mutations and illustrate the modular structure of the progranulin gene [17]. Frameshift and nonsense mutations, leading to premature termination and predicted to result in loss of transcript due to nonsense-mediated RNA decay, are indicated with filled red circles. Other variants are indicated with open circles, colored depending on the evidence of pathogenicity (*see* text for details)

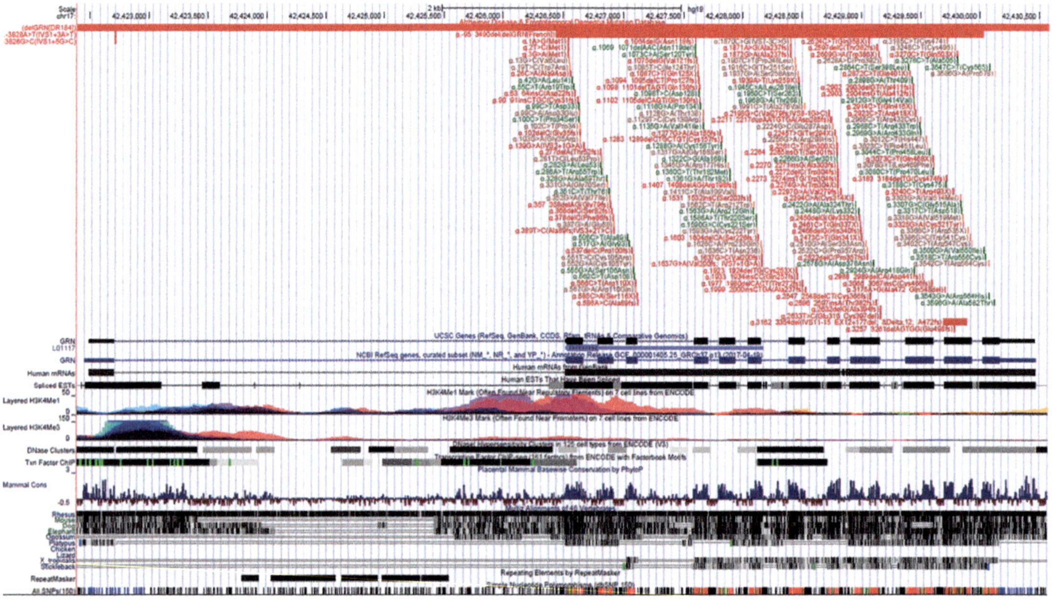

Fig. 5 Screenshot of the UCSC Genome Browser showing the *GRN* variants recorded in the AD&FTD Mutation Database as a custom track on human genome reference sequence hg19. Variants are labeled with their genomic nucleotide change with nucleotide numbering starting at the A of the translation initiation codon, followed by protein change between brackets. Pathogenic variants are colored red, neutral variants green, and variants of unknown pathogenicity orange. To enable this output format, the AD&FTD Mutation Database web site must be allowed to open pop-up windows in the web browser settings, because the UCSC Genome Browser is opened in a separate window

tion inside the track opens a variant details page which contains the Mutation Database ID of the selected variant and a link back to the variant details page in the Mutation Database.

A tab-separated list of all variants can be retrieved using the UCSC Table Browser. This can be done by selecting the Table Browser tool, which can be found in the "Tools" section of the UCSC Genome Browser top menu. The AD&FTDmutDB track can be selected from the "Phenotype and Disease UCSC Genome Browser Associations" group of tracks. The variant data can then be downloaded in any format provided by the Table Browser tool, including BED, GTF, or DNA sequence.

Similarly, a table listing the variants of the AD&FTD Mutation Database annotated with values contained within any other data track available in the Genome Browser can be generated using the Data Integrator tool of the Genome Browser. For example, a table can be generated showing sequence conservation scores of the mutated nucleotide positions or mutation frequencies in the Exome Variant Server (EVS, http://evs.gs.washington.edu/EVS) and Exome Aggregation Consortium (ExAC, exac.broadinstitute.org) databases (Table 2).

Table 2
Example application of the Data Integrator Tool using the AD&FTD Mutation Database custom track in the UCSC Genome Browser

hg19	Variant	Alias	dbSNP	ExAC	EVS
42426534	g.2T>C	Met1	rs63751006	1	
42426587	g.55C>T	Arg19Trp	rs63750723	508	211
42426631	g.99C>A	Asp33Glu	rs63750742	8	
42426635	g.103G>A	Gly35Arg		1	
42427083	g.551T>C	Cys105Arg	rs63750441	1	
42427098	g.566C>T	Arg110X	rs63750411	1	
42427099	g.567G>A	Arg110Gln		11	1
42427605	g.1073C>A	Ser120Tyr	rs63750043	100	2
42427625	g.1094_1095delCT	Pro127fs		1	
42427661	g.1129T>C	Cys139Arg		22	
42427667	g.1135G>A	Val141Ile	rs63749853	5	
42427820	g.1288G>A	Cys158Tyr	rs63750163	1	1
42427849	g.1317G>A	Gly168Ser		1	
42427877	g.1345G>A	Arg177His		5	

(continued)

Table 2 (continued)

hg19	Variant	Alias	dbSNP	ExAC	EVS
42427892	g.1360C>T	Thr182Met	rs63750479	148	42
42428095	g.1563G>A	Arg212Gln	rs63750787	10	5
42428122	g.1590G>C	Cys221Ser		13	
42428158	g.1626C>A	Pro233Gln		2	1
42428169	g.1637G>C	Val200fs		1	
42428169	g.1637G>A	Val200fs	rs63749817	1	1
42428756	g.2224G>C	Glu287Asp	rs63750565	4	1
42428954	g.2422G>A	Ala324Thr	rs63750541	103	18
42429110	g.2578G>A	Asp376Asn	rs143030899	28	12
42429396	g.2864C>T	Ser398Leu	rs148213321	9	1
42429444	g.2912G>T	Gly414Val	rs63750920	1	
42429456	g.2924G>A	Arg418Gln	rs63751100	25	2
42429497	g.2965C>T	Arg432Cys	rs63750130	3	1
42429500	g.2968C>T	Arg433Trp	rs63750412	522	29
42429501	g.2969G>A	Arg433Gln	rs114248177	99	34
42429555	g.3023C>T	Pro451Leu		1	
42429576	g.3044C>T	Pro458Leu	rs63750537	6	
42429610	g.3078G>T	Leu469Phe		1	
42429772	g.3240C>T	Arg493X	rs63751294		1
42429835	g.3303G>A	Val514Met	rs142926942	5	4
42429839	g.3307G>C	Gly515Ala	rs25647	325	132
42429857	g.3325G>A	Cys521Tyr		5	
42429898	g.3366C>T	Arg535X		2	
42430032	g.3500G>A	Val550Ile	rs63750754	27	1
42430074	g.3542C>T	Arg564Cys		1	

Counts of a selection of protein-changing variants in *GRN* recorded in the AD&FTD Mutation Database, in the ExAC and EVS exome sequencing databases

3 Notes

1. The tabular view can be used to retrieve lists of variants for custom handling. However, a more convenient method for downloading variant lists is by uploading selected variants to the UCSC genome browser and using the browser's tools to manipulate custom tracks, as discussed in section 2.2.3.
2. To appreciate the distribution of variants across gene regions like exons or protein domains, summary statistics can be retrieved by clicking the appropriate item in the left menu. From these pages, a list of mutations in a particular genic element, e.g., a specific exon, or a particular protein domain, can be retrieved. Summary statistics are calculated in real time and reflect the actual database content at the time of retrieval.

Acknowledgments

The authors acknowledge the contribution of Eline Wauters PhD and Rita Cacace PhD, who updated the genes and mutations in the AD/FTD Mutation Database and adapted the paper accordingly as well as the figures.

The authors research was funded in part by the Flemish government initiated Impulse Program on Networks for Dementia Research (VIND), the Research Foundation Flanders (FWO) and the University of Antwerp Research Fund, Belgium.

References

1. Cruts M, Theuns J, Van Broeckhoven C (2012) Locus-specific mutation databases for neurodegenerative brain diseases. Hum Mutat. 33(9):1340–1344
2. Cruts M, Van Broeckhoven C (1998) Presenilin mutations in Alzheimer's disease. Hum Mutat. 11(3):183–190
3. Cotton RG, McKusick V, Scriver CR (1998) The HUGO mutation database initiative. Science 279(5347):10–11
4. Rademakers R, Cruts M, Van Broeckhoven C (2004) The role of tau (MAPT) in frontotemporal dementia and related tauopathies. Hum Mutat 24(4):277–295
5. Brouwers N, Sleegers K, Engelborghs S, Maurer-Stroh S, Gijselinck I, van der Zee J et al (2008) Genetic variability in progranulin contributes to risk for clinically diagnosed Alzheimer disease. Neurology 71(9):656–664
6. Webb AJ, Thorisson GA, Brookes AJ, Consortium GP (2011) An informatics project and online "knowledge Centre" supporting modern genotype-to-phenotype research. Hum Mutat. 32(5):543–550
7. Sayers EW, Barrett T, Benson DA, Bolton E, Bryant SH, Canese K et al (2012) Database resources of the National Center for biotechnology information. Nucleic Acids Res 40(Database issue):D13–D25
8. Bertram L, McQueen MB, Mullin K, Blacker D, Tanzi RE (2007) Systematic meta-analyses of Alzheimer disease genetic association studies: the AlzGene database. Nat Genet 39(1):17–23
9. Gray KA, Yates B, Seal RL, Wright MW, Bruford EA (2015) Genenames.org: the HGNC resources in 2015. Nucleic Acids Res 43(Database issue):D1079–D1085
10. den Dunnen JT, Antonarakis SE (2000) Mutation nomenclature extensions and suggestions to describe complex mutations: a discussion. Hum Mutat. 15(1):7–12

11. den Dunnen JT, Antonarakis SE (2003) Mutation nomenclature. Curr Protoc Hum Genet Chapter 7:Unit 7.13
12. Pruitt KD, Tatusova T, Klimke W, Maglott DR (2009) NCBI reference sequences: current status, policy and new initiatives. Nucleic Acids Res 37(Database issue):D32–DD6
13. Wildeman M, van Ophuizen E, den Dunnen JT, Taschner PE (2008) Improving sequence variant descriptions in mutation databases and literature using the Mutalyzer sequence variation nomenclature checker. Hum Mutat. 29(1):6–13
14. Kircher M, Witten DM, Jain P, O'Roak BJ, Cooper GM, Shendure J (2014) A general framework for estimating the relative pathogenicity of human genetic variants. Nat Genet. 46(3):310–315
15. Guerreiro RJ, Baquero M, Blesa R, Boada M, Bras JM, Bullido MJ et al (2010) Genetic screening of Alzheimer's disease genes in Iberian and African samples yields novel mutations in presenilins and APP. Neurobiol Aging 31(5):725–731
16. Bateman A, Bennett HP (1998) Granulins: the structure and function of an emerging family of growth factors. J Endocrinol 158(2):145–151
17. Cruts M, Van Broeckhoven C (2008) Loss of progranulin function in frontotemporal lobar degeneration. Trends Genet 24(4):186–194
18. Sleegers K, Brouwers N, Van Damme P, Engelborghs S, Gijselinck I, van der Zee J et al (2009) Serum biomarker for progranulin-associated frontotemporal lobar degeneration. Ann Neurol 65(5):603–609

Part III

Methods to Investigate the Role of Progranulin in Cancer

Chapter 7

Measurement of Circulating Progranulin (PGRN/GP88/GEP) by Enzyme-Linked Immunosorbent Assay and Application in Human Diseases

Ginette Serrero and David Hicks

Abstract

The enzyme-linked immunosorbent assay (ELISA) is a well-established methodology for detection of analytes in various biological fluids. The assay described herein has been validated for the detection of PGRN/GP88/GEP in blood (serum/ plasma), urine and cerebrospinal fluid (CSF), and synovial fluid and may also be used for breast milk, ductal lavage, nipple aspirates, and saliva. The ability to measure circulating levels of PGRN/GP88/GEP has proven to have clinical utility for several human diseases such as cancer where changes of PGRN/GP88/GEP can be determined as a mean to monitor disease status or response to therapy. In the case of frontotemporal dementia (FTD), the ability to measure PGRN/GP88/GEP levels in plasma and cerebrospinal fluid may be useful in distinguishing PGRN mutation carriers among FTD populations at large. The assay used is a sandwich ELISA where a highly specific antihuman PGRN/GP88/GEP monoclonal antibody is employed as a capture antibody coated on 96-well microplates. After contact with serum (or other bodily fluid), unbound material is washed away before application of another PGRN/GP88/GEP detecting antibody which in turn is detected by a horseradish peroxidase (HRP) conjugated antibody. After further washing to remove all unbound HRP, a substrate (TMB) is added, and after approximately 6 min, a color is developed and can be read as optical density at 620 nm (or 450 nm if using HCL as a stop solution) in a microplate reader. The test described herein is capable of measuring very low levels of PGRN/GP88/GEP such as 0.2 ng/mL as found in CSF of certain FTD patients. Additionally, we have demonstrated the potential clinical utility of measuring the changes of PGRN/GP88/GEP blood levels in cancer patients undergoing therapy.

Key words ELISA, EIA, Immunosorbent, Progranulin, GP88, GEP, PCDGF, Cancer, Blood biomarker, Companion diagnostic, Frontotemporal dementia

1 Introduction

The 88 KDa glycoprotein progranulin (PGRN) also known as GP88, PC cell-derived growth factor (PCDGF), granulin-epithelin precursor (GEP), or acrogranin has been identified by a biological screen as playing a critical role in the biological process of tumorigenesis, cancer cell development, invasiveness, and survival [1]. Included in NIH's Atlas of Genetics and Cytology in Oncology and Hematology,

Andrew Bateman et al. (eds.), *Progranulin: Methods and Protocols*, Methods in Molecular Biology, vol. 1806, https://doi.org/10.1007/978-1-4939-8559-3_7, © Springer Science+Business Media, LLC, part of Springer Nature 2018

PGRN/GP88/GEP has been implicated in a number of cancers and has been shown to be present in aggressive cancers, while it is not normally expressed in corresponding normal tissue or benign disease [2]. In addition, the 88 kDa PGRN/GP88/GEP glycoprotein contains a 17 amino acid signal peptide targeting it for secretion. This compendium of evidence has promoted interest in the development of reagents and methods for the detection and quantification of PGRN/GP88/GEP in tumor tissue and bodily fluids of cancer patients with the specific aim of determining whether measurement of PGRN/GP88/GEP can be used as a disease biomarker for prognosis and monitoring of disease status and/or therapy efficacy.

Enzyme-linked immunosorbent assays (ELISA) and immunohistochemistry (IHC) are well-established laboratory tests to detect proteins in bodily fluids and tissues, respectively. Specific reagent and methods aimed at detecting PGRN/GP88/GEP in bodily fluids by ELISA and its application in pathological studies will be presented in the following sections. The method used to detect and semiquantitatively measure PGRN/GP88/GEP in tissue will be described in a separate chapter.

Enzyme-linked immunosorbent assay (ELISA) is a technique generally performed in microplate formats and is useful for identifying and simultaneously measuring an antigen in large number of samples. ELISAs have three main formats: direct, indirect, and sandwich (Fig. 1). The aim in each is similar in that the matrix captures an analyte (antigen) using a specific antibody (sandwich) or the analyte is captured onto the solid surface of the microplate. The amount of analyte captured is quantified using an enzyme/substrate reaction where the enzyme is conjugated to either a primary or a secondary antibody, and this enzyme reacts with a substrate to provide a measurable colorimetric output. The output can be measured on a standard ELISA reader and reported as the optical density (OD) at a given wavelength. Thus the more analyte captured the greater the colorimetric output. Horseradish peroxidase (HRP) and alkaline phosphatase (AP) are two commonly used enzyme labels for colorimetric systems. Fluorescent

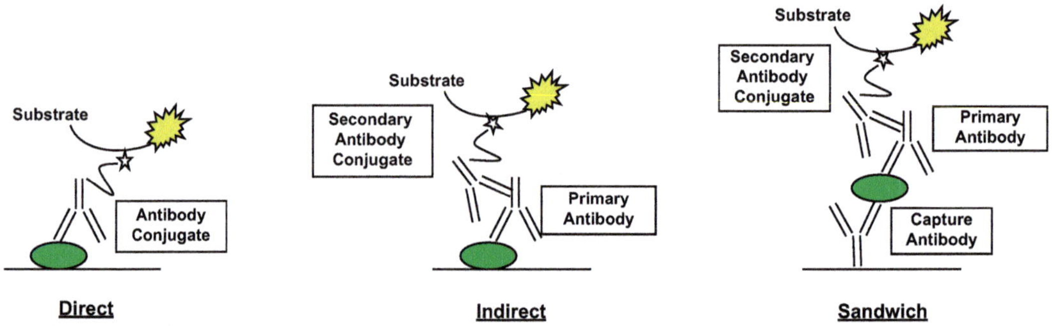

Fig. 1 Various types of ELISA formats

and chemiluminescence-based labels are also being used in clinical chemistry-based assays. No matter what assay format is used, the detection system relies on a highly specific antibody-antigen interaction.

In general, ELISA can be useful to detect analytes in most human biological fluids or tissue extracts providing that there is sufficient analyte present and the detection system is sensitive enough. The assay developed by A&G Pharmaceutical to detect PGRN/GP88/GEP is a sandwich ELISA incorporating a highly specific capture monoclonal anti-GP88 combined with robust and sensitive anti-GP88 antibody-based detection system. This ELISA assay can be used to detect analytes in a variety of human fluids, including but not limited to blood, urine, breast milk, saliva, and cerebrospinal fluid.

2 Materials

ELISA Microplate Reader capable of reading at 620 nm or 450 nm.

ELISA Microplate Washer.

Pipettes and tips.

Tubes for dilutions.

PBS-T Wash buffer (4.3 mM Na_2HPO_4, 1.4 mM KH_2PO_4, 137 mM NaCl, 2.7 mM KCl, 0.05% Tween 20).

96-well microplates—a variety of microplates are available on the market, this protocol uses high binding, round bottom plates.

Calibrators—human serum spiked with known amounts of PGRN/GP88/GEP.

Reagent 1—GP88 standard Curve stock (2 μg/mL).

Reagent 2—coating reagent—highly specific mouse monoclonal antihuman PGRN/GP88/GEP (10 μg/mL).

Reagent 3—detection reagent—rabbit polyclonal anti-PGRN/GP88/GEP (1 μg/mL).

Reagent 4—secondary reagent—goat anti-rabbit IgG conjugated to horseradish peroxidase (1 mg/mL).

Reagent 5—chromogen—3,3′5,5′ Tetramethylbenzidine (TMB) reagent commercially available (store in dark).

Reagent 6—human serum—commercially available used as calibrator and internal standard.

Reagent 7—blocking reagent (4% nonfat dried milk in PBS).

Reagent 8—1% nonfat dried milk in PBS (made from Reagent 7).

Reagent 9—stopping solution—1 N HCl.

As an alternative to the reader developing these reagents, a kit with all reagents needed to perform the tests can be provided by the author via A&G Pharmaceutical Inc. This kit includes:

(a) Ready to use 96-well plate coated with 1 μg/well antihuman PGRN/GP88/GEP antibody and subsequently blocked with 4% milk-PBS.

(b) Detecting anti-PGRN/GP88/GEP rabbit polyclonal antibody.

(c) HRP-conjugated secondary goat anti-rabbit antibody.

(d) Recombinant human PGRN/GP88/GEP/ stock solution 2 μg/mL for standard curve.

(e) Two calibrators (baseline and high) to be used as internal standards to verify assays.

3 Method

In general, ELISA test to measure PGRN/GP88/GEP in bodily fluids uses as reagents a specific antihuman PGRN/GP88/GEP monoclonal antibody as a capture antibody bound to the microplate, a detecting anti-GP88 rabbit polyclonal antibody, an anti-rabbit antibody conjugated to horseradish peroxidase (HRP), and a chromogenic substrate. The capture antibody that binds to human GP88 in a sample of biological fluid (e.g, serum/plasma) is coated onto a microplate. This step is followed by the addition of a rabbit anti-GP88 that binds to any captured GP88, the unbound proteins having been previously washed away. Following a further wash, an anti-rabbit antibody conjugated to HRP binds to the rabbit antibody and after a final wash, the chromogen substrate is added. Any HRP present in the system will convert the chromogen to a color, and after a set time, this reaction is stopped using HCl, and the color is read on a plate reader. Each of these steps has specific incubation times and is interposed with specific washing steps. Results obtained for a sample are interpreted using a standard curve that is formulated using known amounts of the analyte PGRN/GP88/GEP in this assay.

3.1 Specimen Preparation

Our ELISA assay can measure PGRN/GP88/GEP in several human biological fluids. The protocol outlined below describes testing serum and plasma samples. However, the high sensitivity of our ELISA assay has allowed us to measure PGRN/GP88/GEP in urine and in cerebrospinal fluid where the level of PGRN/GP88/GEP is about 30-fold lower than in plasma. Examples of applications with these various biological fluids will be discussed below.

For serum samples, use a serum separator tube (SST) as per manufacturer's instructions, aliquot the serum into labeled tubes. If not using immediately for testing store at ≤20 °C avoiding repeated freeze/thaw cycles (*see* **Note 1**). In the case of plasma, use EDTA or heparin tubes as per manufacturer's instruction, centrifuge, and aliquot plasma into labeled tubes taking care not to disturb or include buffy coat in the removed sample of plasma. If not using immediately for testing, store at ≤20 °C avoiding repeated freeze/thaw cycles.

3.2 Detailed Method for Running a PGRN/GP88/GEP ELISA

3.2.1 Preparation of Coated Plates

1. Coat 1 Costar 96-well (round bottom) high-binding polystyrene plate by adding 100 μL/well of coating antibody reagent (Reagent 2).
2. Cover the plate, and store in a refrigerator (2–8 °C) overnight.

NOTE: For convenience coated plates can be obtained from the authors at A&G Pharmaceutical Inc.

3.2.2 Final Preparation of Coated Plates: Day of Testing

1. Using nonfat dried milk powder, make 100 mL of 4% (w/v) milk in PBS (Reagent 7). Ensure that it is completely dissolved. Using Reagent 7, prepare 100 mL of 1% milk in PBS (Reagent 8). NOTE: Use these solutions freshly, do not store.
2. Wash the coated plate seven times with PBS-T wash buffer (300 μL/well) using an automated washer.
3. "Block" the plate with Reagent 7 (300 μL/well). Incubate at room temperature (RT) for 60 min.
4. Wash the plate seven times with PBS-T using an automated washer as in **step 2**.

3.2.3 Preparation of Standard Curve: Day of Testing

1. Make up a serial dilution of the PGRN/GP88/GEP standard curve stock (Reagent 1) using the suggested dilutions outlined in Table 1. This will provide sufficient standard curve samples for one standard curve in duplicate for one microplate. Keep at 4 °C until ready to dispense into plate.
2. Thaw an aliquot of each previously frozen calibrator sample, and keep at 4 °C until ready to dispense into plate.
3. Dilute test serum samples and calibrator samples in Reagent 8.
   (a) Test samples are generally diluted 1/20 in Reagent 8.
   (b) Calibrators provided with the assays are diluted 1/20 for the baseline calibrator 1 ("baseline") and 1/40 for the "high GP88" calibrator 2.

Ensure at least 200 μL is prepared to enable tests to be run in duplicates. Keep at 4 °C until ready to dispense into plate.

3.2.4 Performing the Assay

1. Wash the plate seven times in PBS-T using an automated washer (*see* **Note 2**).
2. Add standard curve samples, calibrators, and test samples to plate (100 μL/well) as shown in layout (Table 2) (*see* **Note 3**).

   Note: Wells A through H in columns 1 and 2 are the serial dilutions of PGRN/GP88/GEP for the standard curve with concentrations (GP) from 0 to 20 ng/mL (GP0 to GP20) added in duplicate. Wells A3, A4, B3, and B4 are for the calibrators (Cal) 1 and 2. The remaining wells contain 40 unknown samples (S) to be assayed in duplicate (*see* **Note 4**).

Table 1
Dilutions to produce standard curve

Step	Volume of reagent 8	Volume of reagent 1	Final GP88 concentration
1	1980 μL	20 μL of stock (2 ng/μL)	20 ng/mL
2	1000 μL	1000 μL of 20 ng/mL(from step 1)	10 ng/mL
3	1000 μL	1000 μL of 10 ng/mL (from step 2)	5 ng/mL
4	800 μL	200 μL of 20 ng/mL (from step 1)	4 ng/mL
5	1200 μL	800 μL of 5 ng/mL (from step 3)	2 ng/mL
6	1000 μL	1000 μL of 2 ng/mL (from step 5)	1 ng/mL
7	1000 μL	1000 μL of 1 ng/mL (from step 6)	0.5 ng/mL
8	1000 μL	Zero	Blank

Table 2
Plate layout for testing standard curve, calibrators, and samples in duplicate

	1	2	3	4	5	6	7	8	9	10	11	12
A	GP20	GP20	Cal 1	Cal 1	S9	S9	S17	S17	S25	S25	S33	S33
B	GP10	GP10	Cal 2	Cal 2	S10	S10	S18	S18	S26	S26	S34	S34
C	GP5	GP5	S1	S1	S11	S11	S19	S19	S27	S27	S35	S35
D	GP4	GP4	S2	S3	S12	S12	S20	S20	S28	S28	S36	S36
E	GP2	GP2	S4	S5	S13	S13	S21	S21	S29	S29	S37	S37
F	GP1	GP1	S6	S6	S14	S14	S22	S22	S30	S30	S38	S38
G	GP0.5	GP0.5	S7	S7	S15	S15	S23	S23	S31	S31	S39	S39
H	GP0	GP0	S8	S8	S16	S16	S24	S24	S32	S32	S40	S40

After adding the samples, incubate the plate at RT for 120 min.

3. Wash the plate seven times with PBS-T (300 μL/well) using an automated washer.

4. Prepare detecting antibody (Reagent 2) at a final concentration of 1 μg/mL in Reagent 8, and add 100 μL of this solution per well. Incubate the plate at RT for 60 min.
5. Wash the plate seven times with PBS-T using an automated washer as in **step 4**.
6. Prepare HRP goat anti-rabbit IgG secondary antibody (Reagent 3) (stock provided is 1 mg/mL) at 1:3000 in Reagent 8. Add 100 μL/well, and incubate at RT for 60 min.
7. During the incubation time, remove (Reagent 5) TMB from 4 °C and allow to equilibrate to RT.
8. Wash plate seven times in PBS-T using an automated washer as in **step 4**.
9. Add TMB substrate at 100 μL/well, and allow the substrate to develop for ~6 min (*see* **Note 5**). Read absorbance at OD_{620nm} and record results or stop with 1 N HCl, and read absorbance at OD_{450nm} and record these results.

3.2.5 Calculating the Results

1. Average the duplicated optical density (OD) readings for each standard curve point, the calibrators and the samples.
2. Subtract the average blank OD (the readings from wells H1 and H2) from the average OD for each standard curve point, the calibrators and the samples.
3. Create a standard curve using the standard curve data points by plotting the average minus blank OD for each standard on the y-axis and the known concentration on the x-axis. The OD minus blank for each sample and calibrator can then be plotted on the curve, and the concentration of GP88 can be read off the x-axis. Alternatively, this can be performed in a commonly available computer software or statistical packages to generate log/log curve fit to provide automatic sample point generated results. Remember to multiple the reading by the dilution factor for each sample and for the calibrator. Using Microsoft Excel™ a typical set of results can be seen in the following example (Table 3).

Using a standard statistical package, these results can be used to plot the standard GP88 curve (Fig. 2).

On the same plate as the standard curve, we run two calibrators: calibrator 1 consisting of a normal commercial serum and the same serum spiked with known amount of recombinant PGRN/GP88/GEP to provide calibrator 2. These calibrators are run on every plate. As example, the two calibrators and several normal serum samples run together can be seen in Table 4. Calibrator 1 and the normal samples (N) are diluted 1/20, while calibrator 2 is diluted 1/40.

Using the standard curve, the program then interprets the ODs and provides the sample GP88 values as seen in Table 4.

Table 3
Example—original OD readings at 620 nm for standard curve GP88 samples as defined above

Std. (ng/mL)	OD_{620}–1	OD_{620}–2	Ave. of ODs	Ave.-blank	STDEV	CV%
20	2.345	2.352	2.3485	2.172	0.005	0.21
10	1.563	1.598	1.5805	1.404	0.025	1.57
5	0.941	0.934	0.9375	0.761	0.005	0.53
4	0.791	0.803	0.797	0.6205	0.008	1.06
2	0.519	0.525	0.522	0.3455	0.004	0.81
1	0.349	0.355	0.352	0.1755	0.004	1.21
0.5	0.271	0.267	0.269	0.0925	0.003	1.05
0	0.179	0.174	0.1765	0	0.004	2.00

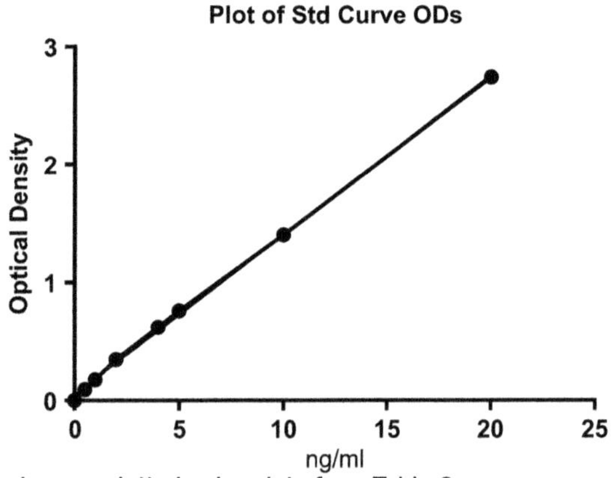

Fig. 2 Standard curve plotted using data from Table 2

Table 4
Example of original OD readings at 620 nm for calibrators and healthy volunteers

Samples	OD_{620}–1	OD_{620}–2	OD Ave	Ave.-blank	STDEV	CV%	GP88 (ng/mL)
Calibrator 1	0.419	0.426	0.4225	0.246	0.005	1.172	29.3
Calibrator 2	0.629	0.632	0.6305	0.454	0.002	0.336	120.0
N1	0.455	0.471	0.463	0.2865	0.011	2.444	35.3
N2	0.438	0.436	0.437	0.2605	0.001	0.324	31.4
N3	0.418	0.429	0.4235	0.247	0.008	1.837	29.4
N4	0.397	0.407	0.402	0.2255	0.007	1.759	26.3
N5	0.472	0.484	0.478	0.3015	0.008	1.775	37.5
N6	0.512	0.503	0.5075	0.331	0.006	1.254	41.8

4 Assay Applications

We have used this assay to determine PGRN/GP88/GEP levels in human serum, plasma, urine, and CSF. The assay has been determined as highly sensitive and accurate over a wide spectrum of PGRN/GP88/GEP levels.

4.1 Increased PGRN/GP88/GEP Circulating Levels in Cancer Patients

Many human cancers have been shown to have elevated blood PGRN/GP88/GEP levels. Using the assay described herein, we have demonstrated elevated levels of GP88 in the blood of patients with lung, breast, prostate, and ovarian cancer (*see* below). These elevated levels appear to correlate with progressive disease in aggressive cancers and may be useful when determining the efficacy of the standard of care agents being used to treat such disease, i.e., a continued rise in serum GP88 may indicate an ineffective treatment strategy. Conversely a decreasing GP88 level may be a sign of treatment efficacy. In particular, application of A&G PGRN/GP88/GEP ELISA test demonstrates that PGRN/GP88/GEP circulating level is elevated in breast [3] and lung cancer [4] patients. Figure 3 exemplifies the difference in circulating PGRN/GP88/GEP in between healthy volunteers/patients with chronic obstructive pulmonary disease (COPD) and lung cancer patients.

4.2 Increased PGRN/GP88/GEP Levels in Serum and Synovial Fluids from Patients with Rheumatoid Arthritis (RA) Compared to Patients with Osteoarthritis (OA)

Application of PGRN/GP88/GEP ELISA assay has allowed us to examine PGRN/GP88/GEP levels in serum and synovial fluids of patients with RA and OA when compared to healthy individuals. Results indicate that PGRN/GP88/GEP serum levels of healthy individuals were significantly lower than patients with arthritis. Moreover, in the serum and synovial fluid of RA patients, PGRN/GP88/GEP levels were significantly elevated and higher than those in OA patients [5].

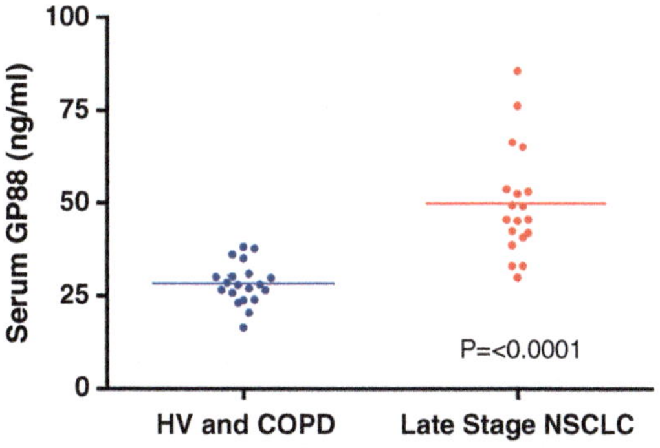

Fig. 3 Comparison of PGRN/GP88/GEP serum concentrations in healthy volunteers and COPD patients versus late-stage non-small cell lung cancer (NSCLC) patients

4.3 PGRN/GP88/GEP ELISA Test Can Measure PGRN in both Plasma and Cerebrospinal Fluid (CSF) of Patients with Frontotemporal Dementia

One form of *Frontotemporal Dementia* (FTD) is recognized as having mutation in the progranulin gene that causes a reduction in its expression and secretion. Our PGRN/GP88/GEP ELISA assay has been used to measure PGRN/GP88/GEP levels in the cerebrospinal fluid and plasma/serum of PGRN mutant FTD patients and non-PGRN/GP88/GEP mutant FTD patients and non-FTD subjects. The advantage of our PGRN/GP88/GEP ELISA test is that the same assay can be used to measure PGRN/GP88/GEP levels in plasma and CSF from the same patients. In these PGRN/GP88 mutant FTD cases, CSF levels of PGRN/GP88/GEP as low as 0.2–0.4 ng/mL can be measured with our ELISA format. In non-PGRN mutant FTD cases, the CSF levels range in the 1–3 ng/mL level.

4.4 Application in Research Related to PGRN/GP88/GEP: Measurement of PGRN/GP88/GEP in Preclinical Studies Such as Human Xenografts Studies in Mice

We have used this ELISA assay in preclinical studies to determine the GP88 levels in the blood of mice bearing human breast and lung cancer xenografts for the development of our anti-GP88 therapy. In this case, measuring circulating GP88 levels secreted by the tumors is used as a surrogate marker for therapy efficiency. In such cases, we have been able to demonstrate that treatment using our therapeutic anti-GP88 monoclonal results in decreasing blood levels of GP88, while at the same time, we see reduced tumor growth rate/tumor regression.

5 Conclusion

An ELISA for measurement of PGRN/GP88/GEP as described herein has a wide application in research and potentially in the clinic. A research ready kit with proprietary antibodies and reagents is available commercially from the authors at A&G Pharmaceutical Inc.

6 Notes

1. Samples of serum/plasma have been tested for recovery of PGRN/GP88/GEP as a result of repeated thaw/freeze cycles and have been found to be stable up to three cycles. However, in general it is advisable to freeze samples in small aliquots so that such freeze/thaw cycles are reduced to a minimum.
2. When using an automated plate washer, it is recommended to add a short "shake" cycle in between each wash cycle to improve the wash function of each well and as such increase the well-to-well reproducibility.
3. To measure the precision of the assay, we suggest that the user obtains 3–5 samples with known PGRN/GP88/GEP levels. These samples can either be human samples or serum spiked

with known amounts of human PGRN/GP88/GEP. These samples can be tested several times on one plate (intra-assay precision) and over several plates (inter-assay precision). Simple statistical techniques can be used to look at standard deviations and coefficient of variance.

4. Note the substrate solution should be colorless prior to adding to the plates. If the solution is colored, then do not use. You should note that the TMB solution once added to the wells will change from the colorless state, progressively becoming bluer with time.
5. When making reagents or aliquoting samples, avoid frothing/foaming.

References

1. Serrero G (2003) Autocrine growth factor revisited: PC-cell-derived growth factor (progranulin), a critical player in breast cancer tumorigenesis. Biochem Biophys Res Commun 308:409–413
2. Serrero G, Ioffe OB (2003) Expression of PC-cell-derived growth factor in benign and malignant human breast epithelium. Hum Pathol 34:1148–1154
3. Tkaczuk KR, Yue B, Zhan M et al (2011) Increased circulating level of the survival factor GP88 (Progranulin) in the serum of breast cancer patients when compared to healthy subjects. Breast Cancer 5:155–162
4. Edelman MJ, Feliciano J, Yue B et al (2014) GP88 (Progranulin): a novel tissue and circulating biomarker for non-small cell lung carcinoma. Hum Pathol 45:1893–1899
5. Yamamoto Y, Takemura M, Serrero G et al (2014) Increased serum GP88 (Progranulin) concentrations in rheumatoid arthritis. Inflammation 37:1806–1813

Chapter 8

Immunohistochemical Detection of Progranulin (PGRN/GP88/GEP) in Tumor Tissues as a Cancer Prognostic Biomarker

Ginette Serrero and David Hicks

Abstract

Immunohistochemistry is a well characterized and robust staining technique for the identification and localization of proteins in biological tissues. The following methodology describes the process for identifying progranulin (PGRN/GP88/GEP) in the cellular cytoplasm of target tissues such as tumor biopsies. Such tissue is collected from cancer patients and stabilized using formalin-fixed paraffin embedding after which 4–5 μm sections of the tissue are mounted on positively charged microscope slides. After deparaffinization using ethanol, the tissue is processed through an antigen retrieval process using citric acid at pH 6.4. A monoclonal antibody (Clone AG03) against human PGRN/GP88/GEP is then contacted with the tissue before a horseradish peroxidase (HRP) conjugated antibody is employed to detect AG03 bound to PGRN/GP88/GEP. Subsequent application of the HRP substrate results in color generation. Using a light microscope, a trained user can "score" the percentage of cells stained and intensity of such staining. Scoring ranges from 0 (no staining) to 3+ (strong cytoplasmic staining in >10% tumor cells). When scoring is correlated to patient outcome, we have demonstrated that a patient with tumor PGRN/GP88/GEP staining of "3+" has a fivefold increased risk of disease recurrence within 5 years.

Key words Immunohistochemistry, Immunostaining, Progranulin, GP88, PCDGF, GEP, Acrogranin, Cancer, Tumor, Prognostic

1 Introduction

The 88 KDa glycoprotein progranulin (PGRN/GP88/GEP) also known as GP88, PC cell-derived growth factor (PCDGF), granulin-epithelin precursor (GEP), or acrogranin has been identified by a biological screen as playing a critical role in the biological process of tumorigenesis, cancer cell development, invasiveness, and survival [1]. Included in NIH's Atlas of Genetics and Cytology in Oncology and Hematology, PGRN/GP88/GEP has been implicated in a number of cancers and has been shown to be present in aggressive cancers, while it is not normally expressed in corresponding normal tissue or benign disease [2]. In addition, PGRN/GP88/GEP con-

Andrew Bateman et al. (eds.), *Progranulin: Methods and Protocols*, Methods in Molecular Biology, vol. 1806,
https://doi.org/10.1007/978-1-4939-8559-3_8, © Springer Science+Business Media, LLC, part of Springer Nature 2018

tains a 17-amino acid signal peptide targeting it for secretion. This compendium of evidence has promoted interest in the development of reagents and methods for the detection and quantification of PGRN/GP88/GEP in tumor tissue and body fluids of cancer patients with the specific aim of determining whether measurement of GP88/ PGRN can be used as a disease biomarker for prognosis and monitoring of disease status and/or therapy efficacy.

The use of immunohistochemistry (IHC) and enzyme-linked immunosorbent assays (ELISA) to detect proteins in tissues and body fluids, respectively, are well-established laboratory tests. Specific reagent and methods aimed at detecting PGRN/GP88/ GEP in tumor tissues by IHC and their applications in pathological studies will be presented in the following sections. The method used to measure PGRN/GP88/GEP in bodily fluids by ELISA will be described in a separate chapter.

Immunohistochemistry is a technique that uses a labeled antibody to enable visualization of the presence, distribution, and localization of the target antigen within cells and within tissue sections of a tumor. In general, IHC staining techniques allow the visualization of antigens via the sequential application of a specific antibody to the antigen (primary antibody), a secondary antibody to the primary antibody (link antibody, an antibody-enzyme complex), and a chromogenic substrate with interposed washing steps. The enzymatic activation of the chromogen results in a visible reaction product at the antigen site. The specimen may then be counterstained and sealed using a coverslip. Intensity of staining and/or percentage of stained cells are determined either by visual scoring using a light microscope or by using a digital imaging system linked to an appropriate algorithm.

Tissue samples for use in IHC can be prepared in a number of ways but in general are either formalin-fixed and paraffin-embedded or fresh frozen sections. The tissue is cut into thin sections and can be prepared on individual slides, or several sections from one or more tumors can be arranged on a single slide for comparative analysis. Slides can then be processed and stained manually, or using automated systems. Stained sections can be reviewed and "scored" by trained pathologists or using digital image capture systems which can then be assessed using image recognition algorithms.

Since the development of IHC in the mid-1900s, methodology and reagent improvements have made IHC a reliable and sensitive tool used in most diagnostic and research laboratories. IHC data obtained by using specific tumor biomarkers can enable pathologists to determine if a cancer is benign or malignant while also assisting in determining cell type and origin of a metastasis and to provide prognostic information based on anatomical differentiation that can be useful in the detection, diagnosis, or management of a specific disease. Additionally, IHC examination of specific biomarkers can be used as surrogate to determine drug candidate efficacy during the drug development process. However, no matter the target within cells or tissues being investigated, it is vital that the antibody used to detect a target is specific and suitable for the intended IHC application. In the below methodology for the detection of PGRN/GP88/GEP expression by IHC, a monoclonal antibody (clone AG03 developed by A&G Pharmaceutical) was specifically selected for use as primary antibody, validated for IHC techniques and used in pathological studies. Pathological studies described below correlating GP88 expression in breast and lung cancer to clinical outcomes have established the clinical utility of measuring GP88 tissue expression and have demonstrated that GP88 expression is a prognostic factor for early recurrence thereby emphasizing the importance of measuring GP88 tissue expression in cancer tissues.

2 Materials Required

Five micron tissue sections cut from paraffin-embedded tissue blocks.

Positively charged microscope slides (e.g., Superfrost®/Plus and Snowcoat X-tra™).

Light microscope.

Manual staining equipment such as reagent troughs, staining jars, etc.

Reagent alcohol/ethanol, 100%, 95%, and 70%.

Antigen retrieval solution (10 mM sodium citrate pH 6.4).

Mayer's hematoxylin or equivalent.

Deionized or distilled water.

PBST wash buffer (4.3 mM Na_2HPO_4, 1.4 mM KH_2PO_4, 137 mM NaCl, 2.7 mM KCl, 0.05% Tween 20).

Xylene.

Coverslips.

Coverslip mounting solution.

Steamer or water bath.

Humidity chamber.

Absorbent wipes.

Small tubes.

Blank slides.

Whatman filter paper.

Primary mouse antibody (10 μg/mL)—Mouse monoclonal anti-human PGRN/GP88/GEP antibody (clone AG03), IgG1 κ in PBS buffer, containing stabilizing protein and preservative.

Secondary antibody—Goat anti-mouse IgG conjugated to HRP (horseradish peroxidase) formulated in Tris-HCl, containing stabilizing protein and an antimicrobial agent.

Blocking Solution—Contains 0.3% hydrogen peroxide, sodium azide, and levamisole.

Substrate Reagent A (chromogen)—Contains 3,3′-diaminobenzidine.

Substrate Reagent B (diluent)—Formulated in buffer solution (pH 7.5), containing hydrogen peroxide and a preservative.

NOTE: Substrate Reagents A and B are sold by Dako as Product #K3468.

Positive control slides.

Negative control slides.

NOTE: The control slides are available from A&G Pharmaceutical Inc. The negative and positive control slides contain sections of human breast cancer cell lines expressing no or high levels of GP88, respectively.

3 Methods

In general, the IHC procedure to examine GP88 expression in human tissue sections will use a primary antihuman GP88 (progranulin) mouse monoclonal antibody (Clone 6B3), a horseradish peroxidase (HRP) tagged secondary anti-mouse IgG antibody, a peroxidase blocker to quench the endogenous peroxidase activity, and a chromogenic substrate. The antibody 6B3 is automated platform agnostic (*see* **Note 1**), and the present chapter will provide methodologies applicable to manual staining and automatic staining using any of the typical IHC automated staining systems.

The primary mouse AG03 antibody raised against human progranulin in our laboratory binds to human PGRN expressed in the cytoplasm of carcinoma cells fixed on slides. This step is followed by the addition of a peroxidase-conjugated secondary antibody that binds to the primary antibody. The specific primary antibody-secondary antibody complex is then visualized with an optimally diluted chromogenic substrate. The sample slide is counterstained and sealed with a coverslip. Each of these steps has specific incubation times and is interposed with specific washing steps. The stained slide is examined using a light microscope. Scoring is performed, and results are interpreted by a trained pathologist (*see* **Note 2**).

Specimen Preparation

It is recommended that tissues should be sectioned (4–6 micron thickness) and mounted onto positively charged microscope slides. Consult appropriate handbook for further details on specimen preparation (*see* **Note 3**).

Citrate buffer has been shown to be suitable for use as an antigen retrieval agent.

3.1 Detailed Method for GP88 IHC Staining

All reagents should be equilibrated at room temperature (20°–25 °C) before use. It is recommended that a positive and negative control slides be included in each staining run to verify integrity of the staining (*see* below in Subheading 3.2).

1. Deparaffinize FFPE slides at room temperature:
   (a) Xylene, three times, 2–3 min each
   (b) 100% ethanol, three times, 2–3 min each.
   (c) 95% ethanol, three times, 2–3 min each.
2. Rinse and maintain slides in purified water and move to antigen retrieval steps using either steamer or water bath method (*see* **Note 4**).
   (a) *Steamer antigen retrieval method:*
      - Preheat steamer to boiling.
      - Boil antigen retrieval buffer using microwave. Make sure it does not over boil.
      - Equilibrate retrieval buffer in slide jars by placing in steamer for 15 min.
      - Place slides in a rack and immerse into the retrieval buffer in slide jars, and leave for 35 ± 1 min.
      - Unplug steamer, and remove slides in rack to slow running cold water for 1–2 min to cool the slides.
      - Rinse slides with PBST, four times (1 min for each rinse).
      - Go to **step 3**.
   (b) *Water bath antigen retrieval method:*
      - Preheat water bath to 94 °C.
      - Boil antigen retrieval buffer.
      - Equilibrate retrieval buffer in the water bath for 15 min.
      - Place slides in rack into the preheated retrieval buffer, and leave for 25 min.
      - Remove slides in rack to slow running cold water for 1–2 min to cool the slides.
      - Rinse the slides with PBST, four times (1 min for each wash).
      - Go to **step 3**.
3. Remove slides, carefully wipe around each specimen on slides with an absorbent wipe to avoid unwanted dilution of the reagents, and place horizontally in a humidity chamber. NEVER allow specimens to dry.

4. Apply 1–2 drops of the Blocking Solution to each slide, and incubate for 5–7 min at room temperature.
5. Rinse the slides in PBST, four times (1 min for each wash).
6. Carefully wipe around each specimen with an absorbent wipe. Never allow the specimen to dry.
7. Carefully place two drops (100–200 μL) of the primary mouse antibody onto the specimen surface making sure that the specimen is well covered. (Large specimen sections or tissue arrays may require additional fluid to provide coverage and avoid desiccation.)
8. Check fluid coverage of all slides, and close the humidity chamber. Incubate for 30–35 min at room temperature.
9. At the end of the incubation, remove slides and briefly drain before transferring to a slide rack. Wash in PBST buffer four times, each wash for 3 min, and change buffer for each wash.
10. Carefully wipe around the specimen on the slide with an absorbent wipe. Never let the specimen dry out. Carefully place two drops of the secondary mouse antibody onto the specimen surface making sure that the specimen is well covered. (Large specimen sections or tissue arrays may require additional fluid to provide coverage and avoid desiccation.)
11. Check fluid coverage of all slides, and close the humidity chamber. Incubate for 30–35 min at room temperature.
12. Prepare freshly diluted chromogenic substrate solution. Transfer 1 mL of Substrate Reagent B (diluent) to a small tube, and add 1 drop (20 μL) of Substrate Reagent A (chromogen) and mix. Each 1 mL aliquot is sufficient for ten specimen sections.
13. At the end of the incubation from **step 11**, remove slides and briefly drain before transferring to a slide rack. Wash in PBST buffer four times, each wash for 3 min, and change buffer for each wash.
14. Carefully wipe around specimen with an absorbent wipe. Never allow specimen to dry. Carefully apply to each slide sufficient freshly diluted chromogenic substrate solution (prepared in **step 12**) to cover the specimen and incubate at room temperature for 8–10 min.
15. Remove slides and briefly drain before transferring to a slide rack. Rinse in distilled water for 1 min; repeat two times.
16. Carefully wipe around specimen with an absorbent wipe and counterstain with freshly prepared Mayer's Hematoxylin Solution for 2–4 min; ensure the tissue section is completely covered with counterstain. Rinse with tap water until nuclei are blue. Remove excess water and air dry. Add 1–2 drops of mounting solution and coverslip.

Table 1
Scoring of GP88 staining pattern (*see* Note 5)

Staining pattern	GP88 staining score
No staining, or faint staining, is observed, or cytoplasmic staining is observed in less than 10% of the tumor cells	0
A weak cytoplasmic staining is observed in more than 10% of tumor cells	1+
A weak to moderate cytoplasmic staining is observed in more than 10% of the tumor cells	2+
A strong cytoplasmic staining is observed in more than 10% of the tumor cells	3+

17. Slides are examined under a light microscope and scored according to the guide provided in Table 1. Integrity of the run will be checked by examining the positive and negative slides for appropriate staining and absence of false-positive or false-negative stain.

3.2 Quality Control

Differences in tissue processing and technical procedures in the user's laboratory may produce significant variability in results (*see* **Note 6**); hence regular in-house performance controls are required in addition to the positive and negative control slides provided. Consult the quality control guidelines of the College of American Pathologists (CAP) or the NCCLS Quality Assurance for Immunocytochemistry, approved guideline for additional information.

Positive control slide: A known positive control slide is provided in the A&G's kit and should be used in each staining run, and the intense, specific (3+) cytoplasmic staining is indicative of correctly prepared tissues and proper staining techniques.

Negative control slide: A known negative control slide is provided in the kit and should be used in each staining run, and the no, or faint, (0) cytoplasmic staining is indicative of correctly prepared tissues and proper staining techniques.

3.3 Interpretation of Staining

3.3.1 Positive Tissue Control

The positive control slide stained with the primary mouse antibody should be examined first to ascertain that all reagents are functioning properly. The presence of a strong, intense (3+) reddish-brown reaction product in the cytoplasm is indicative of positive reactivity. If the positive control slide fails to demonstrate positive staining, any results with the test specimens should be considered invalid.

3.3.2 Negative Tissue Control

The negative control slide should be examined after the positive control slide to verify the specificity of the staining of GP88 by the primary mouse antibody. The absence, or faint, (0) staining in the negative control slide confirms the lack of antibody cross-reactivity to cells/cellular components. If specific staining (false-positive staining) occurs in the negative control slides, results with the patient specimens should be considered invalid.

Nonspecific staining, if present, usually has a diffuse appearance. Sporadic staining of connective tissue may also be observed in sections from excessively formalin-fixed tissues. Use intact cells for interpretation of staining results. Necrotic or degenerated cells often stain nonspecifically.

Depending on the incubation length and potency of the hematoxylin used, counterstaining will result in a pale to dark blue coloration of the cell nuclei. Excessive or incomplete counterstaining may compromise proper interpretation of results (*see* **Note 6**).

3.3.3 Patient Tissues

Examine patient specimens stained with primary mouse antibody last. As with any immunohistochemical test, a negative result means that the antigen was not detected, not that the antigen was absent in the cells/tissue assayed (*see* **Note 7**). For the determination of GP88 protein expression, only cytoplasmic staining intensity in invasive ductal carcinoma should be evaluated [2]. Use the scale in Table 1 to score the staining pattern. For the evaluation of the immunohistochemical staining and scoring, using a 10× microscope objective is appropriate. Nuclear staining should be considered as nonspecific staining and is not to be included in the assessment of cytoplasmic staining. The interpretation of patient specimen evaluations should be performed by a pathologist.

In some cases, histiocytes surrounding tumor tissues have been shown to have positive staining for GP88, which appears to be specific. Care should be taken to identify histiocytes, which are infiltrating in and around tumor cells, and not count them as specific staining of the cancer tissue. Usually, such histiocyte staining will be easy to interpret, since histiocytes with a similar staining pattern are also seen in areas of more intense inflammation adjacent to the area of the tumor. In addition, even histiocytes which have apparently wrapped around tumor cells, such that it may be difficult to determine whether it is histiocyte or tumor cell staining, usually occur in <10% of the total tumor cell population and would therefore not be significant in the overall scoring of the tumor. Figure 1 shows typical staining achieved in breast tumors, ranging from 0 to 3+, while Fig. 2 shows a similar pattern of staining achieved in lung carcinoma using a lung cancer tissue microarray.

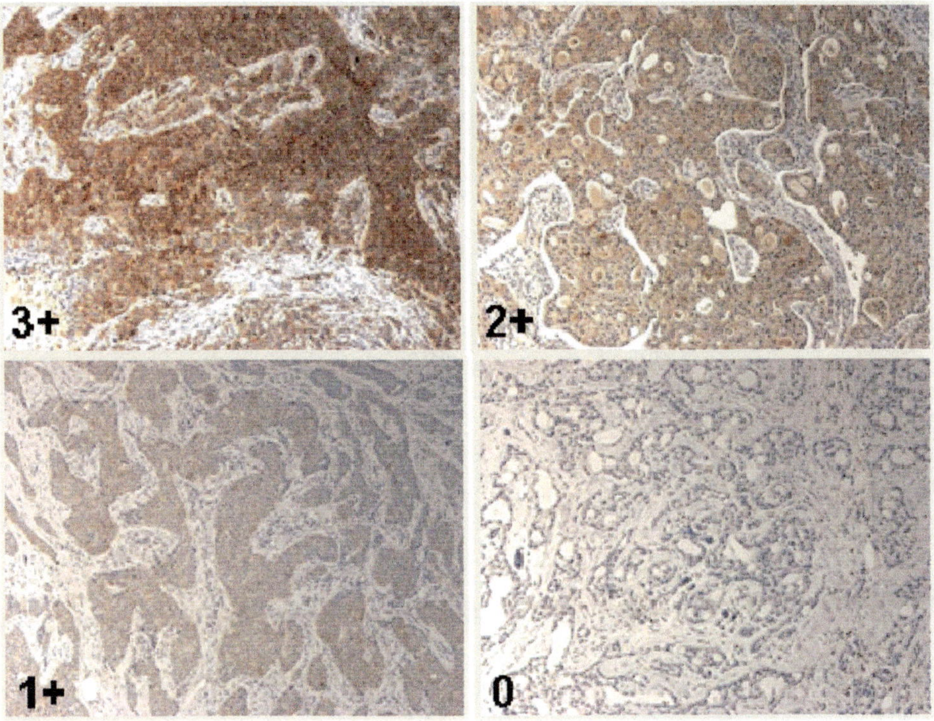

Fig. 1 Photomicrographs of ER+ IDC tissue sections stained for GP88 and scored using the system described in Table 1

4 Pathological Applications of PGRN/GP88/GEP Staining in Breast and Lung Cancer

4.1 PGRN/GP88/GEP Is a Prognostic Factor of Recurrence in Primary Estrogen Receptor-Positive Invasive Ductal Carcinoma

IHC studies of GP88 expression in FFPE tissue sections have shown that GP88 was expressed in invasive ductal carcinoma (IDC) and ductal carcinoma in situ (DCIS); whereas it was not expressed in normal mammary epithelial tissues. Benign disease and lobular carcinoma were mostly negative [2]. Since GP88 appeared as a critical driver of tumor aggressiveness and estrogen independence [3], it was hypothesized that GP88 expression may be a prognostic marker for ER^+ IDC. For this purpose, two retrospective multisite clinical studies examined GP88 expression by immunohistochemistry (IHC) analysis in paraffin-embedded tumor tissues in correlation with patients' survival outcomes using the method described above with a Dako automated stainer. The training trial established a GP88 cutoff value associated with decreased disease-free (DFS) and overall (OS) survivals. The validation study verified the GP88 cutoff value and compared GP88 prognostic information with other prognostic factors in multivariate analysis [4]. The results demonstrated that GP88 expression is associated with a statistically significant increase in recurrence risk for ER^+ IDC patients. The training trial established that GP88

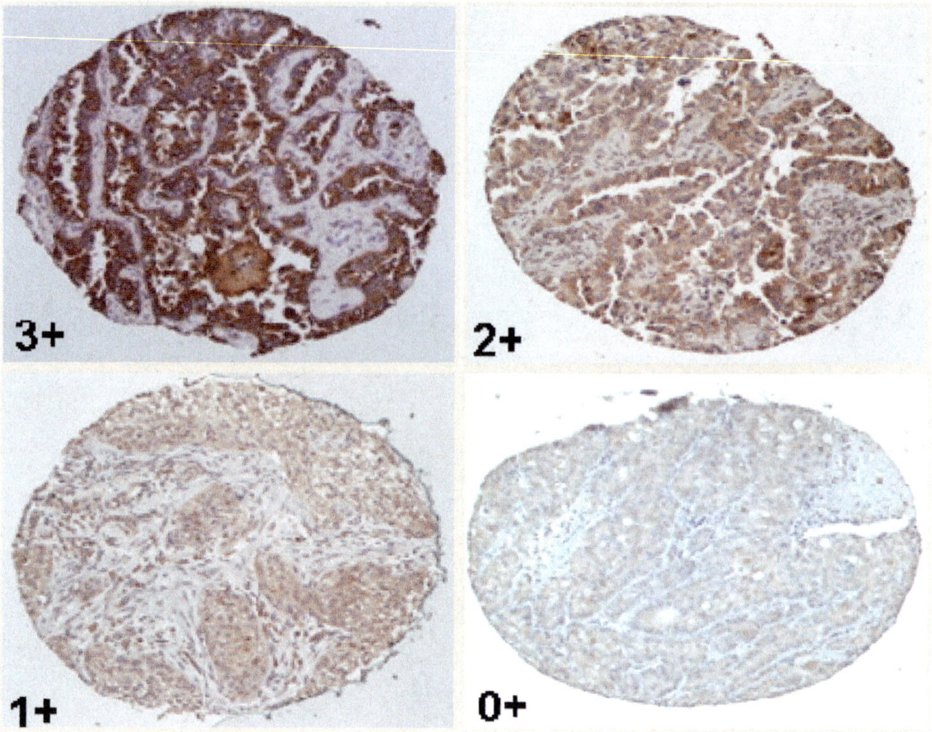

Fig. 2 Photomicrographs of NSCLC tumor tissue microarray stained for GP88 and scored using the system described in Table 1

IHC score of 3+ was associated with decreased DFS ($p = 0.0004$) and OS ($p = 0.0036$). The independent validation study verified that GP88 3+ was associated with a 5.9-fold higher hazard of disease recurrence and a 2.5-fold higher mortality hazard compared to patients with tumor GP88 <3+. Figures 3 and 4 show the Kaplan-Meier survival curves (DFS and OS) for GP88 score (3+) when compared to GP88 score <3+ and illustrates the dramatic drop in survival for high GP88 expressors. Multivariate analysis indicated that GP88 was an independent risk predictor even after considering lymph node status, tumor size, tumor grade, and disease stage. These data demonstrate that the survival factor PGRN/GP88/GEP is a prognostic factor, predictive of recurrence risk and increased mortality for patients with ER^+ IDC, and independent from other prognostic markers.

4.2 PGRN/GP88/GEP Is a Prognostic Factor of Recurrence in Non-small Cell Adenocarcinoma

The prognostic value of GP88 tissue expression was also established by pathological studies in lung cancer. GP88 expression was determined by immunohistochemistry in lung tumor tissues. GP88 was significantly expressed in adenocarcinoma and squamous cell carcinoma in contrast to normal lung or small cell lung cancer. GP88 IHC tissue expression was then examined in non-small cell carcinoma (NSCLC) patients with localized (stage I–II) and locally

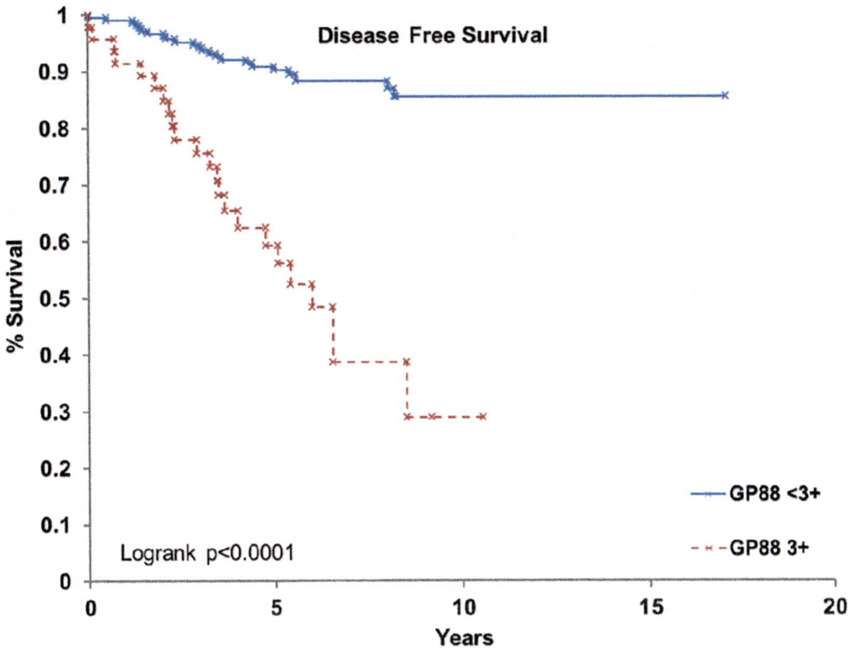

Fig. 3 Comparing disease-free survival for ER+ IDC with GP88 IHC Scores 3+ vs <3+

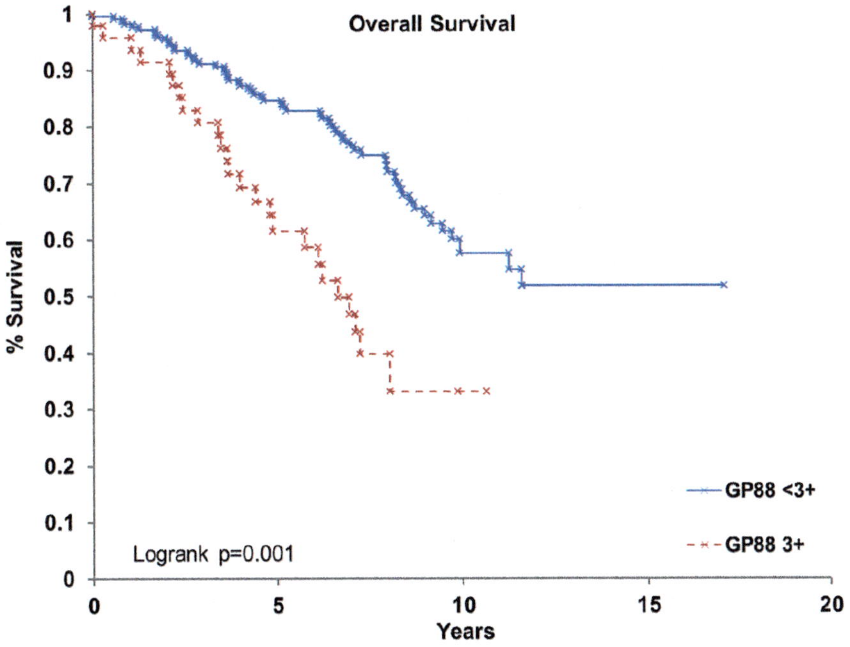

Fig. 4 Comparing overall survival for ER+ IDC with GP88 IHC Scores 3+ vs <3+

advanced disease (IIIa–IIIb) and correlated with clinical outcome [5]. There was a statistically significant association of GP88 expression (GP88 IHC score 3+ vs. <3+) with decreased survival for patients with localized resected NSCLC with HR (hazard ratio) = 2.28 ($p = 0.0076$) for disease-free survival and HR = 2.17 ($p = 0.014$) for overall survival. A statistically significant decrease in progression-free survival (HR = 2.9, $p = 0.022$) for GP88 scores of 3+ vs. <3+ was observed for locally advanced disease after chemoradiotherapy. This study demonstrates GP88 tissue expression as a prognostic biomarker in localized non-small cell lung carcinoma.

5 General Conclusion

This IHC GP88 kit and antibody usage should allow investigating the possible prognostic capability of GP88 tissue expression in other cancers for which managing risk at an early time is important and for other diseases where abnormal GP88 expression is observed.

6 Notes

General Advice

1. The proprietary antibodies and commercial reagents used to perform this methodology are available commercially from the author. The reagents described in this chapter have been validated to provide appropriate staining on automated staining platforms from Dako, Ventana, and Leica.
2. Immunohistochemistry is a multistep diagnostic process that requires specialized training in the selection of the appropriate reagents; tissue selection, fixation, and processing; preparation of the IHC slide; and interpretation of the staining results.
3. Tissue staining is dependent on the handling and processing of the tissue prior to staining.
4. Ensure the antigen retrieval buffer is at the appropriate temperature throughout the entire retrieval process. If the temperature is allowed to fluctuate, the staining of target protein may be compromised.
5. Troubleshooting (*see* Table 2).
6. Improper fixation, freezing, thawing, washing, drying, heating, sectioning, or contamination with other tissues or fluids may produce artifacts, antibody trapping, or false-negative results. Inconsistent results may be due to variations in fixation and embedding methods or to inherent irregularities within the tissue. Excessive or incomplete counterstaining may compromise proper interpretation of results.

Table 2
Troubleshooting

Problem	Possible cause	Suggested change
1. No staining of slides	(a) Reagents not used in proper order (b) Sodium azide in wash buffers	(a) Review order of reagents used (b) Use fresh preparation of wash buffers
2. Weak staining of slides	(a) Insufficient antigen retrieval (b) Inappropriate fixation (c) Reagent incubation times too short	(a) Verify that antigen retrieval was performed appropriately (b) Determine if tissue is not over-fixed or that a suitable fixative was not used (c) See methodology
3. High-background staining	(a) Paraffin not completely removed (b) Insufficient slide rinsing (c) Sections dried during staining procedure (d) Inappropriate fixation (e) Nonspecific binding of reagents to tissue	(a) Use fresh deparaffinization solutions, and follow procedure (b) Use fresh wash buffers (c) Use humidity chamber. Wipe only three to four slides at a time between steps in procedure (d) Determine that appropriate fixative was used (e) Check specimen fixation method and presence of necrosis
4. Excessively strong specific staining	(a) Inappropriate fixation method used (b) Inappropriate antigen retrieval method used (c) Incubation times too long (d) Inappropriate wash buffer used	(a) Determine that appropriate fixatives and fixation methods are used (b) Determine that appropriate antigen retrieval method was used (c) See methodology (d) Use recommended wash buffers

7. Tissues from persons infected with hepatitis B virus and containing hepatitis B surface antigen (HBsAg) may exhibit nonspecific staining with horseradish peroxidase [6]. Reagents may demonstrate unexpected reactions in previously untested tissues. The possibility of unexpected reactions even in tested tissue groups cannot be completely eliminated due to biological variability of antigen expression in neoplasms. False-positive results may be seen due to non-immunological binding of proteins or substrate reaction products. They may also be caused by pseudo-peroxidase activity (erythrocytes) or endogenous peroxidase activity (cytochrome C).

References

1. Serrero G (2003) Autocrine growth factor revisited: PC-cell-derived growth factor (progranulin), a critical player in breast cancer tumorigenesis. Biochem Biophys Res Commun 308:409–413
2. Serrero G, Ioffe OB (2003) Expression of PC-cell-derived growth factor in benign and malignant human breast epithelium. Hum Pathol 34:1148–1154
3. Tangkeangsirisin W, d Serrero G (2004) PC cell-derived growth factor (PCDGF/GP88, progranulin) stimulates migration, invasiveness and VEGF expression in breast cancer cells. Carcinogenesis 25:1587–1592
4. Serrero G, Hawkins DM, Yue B et al (2012) Progranulin (GP88) tumor tissue expression is associated with increased risk of recurrence in breast cancer patients diagnosed with estrogen receptor positive invasive ductal carcinoma. Breast Cancer Res 14:R26
5. Edelman MJ, Feliciano J, Yue B et al (2014) GP88 (progranulin): a novel tissue and circulating biomarker for non-small cell lung carcinoma. Hum Pathol 45:1893–1899
6. Omata M, Liew CT, Ashcavai M et al (1980) Nonimmunologic binding of horseradish peroxidase to hepatitis B surface antigen. A possible source of error in immunohistochemistry. Am J Clin Pathol 73:626–632

Chapter 9

Analysis of Progranulin-Mediated Akt and MAPK Activation

Shi-Qiong Xu, Simone Buraschi, Ryuta Tanimoto, Manuela Stefanello, Antonino Belfiore, Renato V. Iozzo, and Andrea Morrione

Abstract

Progranulin has emerged in recent years as an important regulator of various biological functions including cell proliferation, wound healing, motility, and protection from apoptosis. Progranulin is also critical for transformation as established in several cancer models.

Progranulin biological responses elicit through the activation of the Akt and MAPK pathways, which are critical for progranulin downstream signaling.

In this chapter various experimental approaches aiming at detecting progranulin-mediated Akt and MAPK activation will be discussed.

Key words Progranulin, SDS PAGE, MAP kinase, Akt, Electrophoresis, PathScan, Nitrocellulose membrane, Confocal analysis, Immunofluorescence

1 Introduction

The growth factor progranulin, also known as proepithelin, PCDGF, granulin-epithelin precursor, or acrogranin, is a secreted glycoprotein which plays a critical role in cell proliferation, wound healing, and anchorage-independent growth in several cancer model systems [1–3].

We have recently established that progranulin plays a critical role in genitourinary cancers by promoting urothelial and prostate cancer cell motility, invasion, and anchorage-independent growth [4–8].

Progranulin activates Akt and MAPK pathways, which are critical for progranulin-induced responses in both urothelial and prostate cancer cells [4–8].

Here, we describe the methods utilized in our laboratory to detect progranulin-dependent activation of the Akt and MAPK signaling pathways by Western immunoblot [9–11] and confocal analysis [12].

Andrew Bateman et al. (eds.), *Progranulin: Methods and Protocols*, Methods in Molecular Biology, vol. 1806, https://doi.org/10.1007/978-1-4939-8559-3_9,

2 Materials

Prepare solutions with Millipore deionized or equivalent grade water (dH_2O).

1. *1× Dulbecco's phosphate-buffered saline (DPBS)* (Corning Cellgro, 21-031-CV).
2. *10× Tris-buffered saline (TBS) (BIO-RAD, #170-6435) and 1× TBST*: For 1 L 1× TBS: add 100 mL 10× TBS to 900 mL dH_2O and mix. For 1 L 1× TBST: add 100 mL 10× TBS to 900 mL dH_2O, 1 mL Tween® 20 (SIGMA, #P1379-500ML).
3. *2×/4× SDS sample buffer*: For 2× sample buffer, add 50 μL 2-mercaptoethanol to 950 μL 2× Laemmli sample buffer (BIO-RAD, #161-0737); for 4× sample buffer, add 100 μL 2-mercaptoethanol to 900 μL 4× Laemmli sample buffer (BIO-RAD, #161-0747) and mix.
4. *1× NP-40 lysis buffer*: Tris–HCl (50 mM, pH 7.4, NaCl (150 mM), EDTA (5 mM), and NP-40 (1%) (Boston BioProducts, #BP-119).
5. *Halt protease and phosphatase inhibitor cocktail (100×):* (Thermo Scientific, 1,861,281).
6. *10× Tris-glycine SDS running buffer*: (BIO-RAD, #1610772). For 1 L 1× running buffer: add 100 mL 10× running buffer to 900 mL dH_2O and mix.
7. *10× Tris-glycine transfer buffer*: (BIO-RAD, #161-0771). For 1 L 1× transfer buffer: add 100 mL 10× transfer buffer to 200 mL methanol +700 mL dH_2O and mix.
8. *Nonfat dry milk/blotting-grade blocker*: (BIO-RAD, #170-6404).
9. *Bovine serum albumin (BSA)*: (Fisher Scientific, #BP1600-100).
10. *Western blocking buffer:* The blocking buffer for Western analysis is 1× TBST with 5% w/v nonfat dry milk for total protein blotting or bovine serum albumin for kinase activation blotting; for 300 mL, add 15 g nonfat dry milk or 15 g BSA to 300 mL 1× TBST and mix well.
11. *Wash buffer*: (BIO-RAD, #170-6435) 1× TBST.
12. *Primary antibody dilution buffer*: 1× TBST with 5% BSA; for 20 mL, add 1.0 g BSA to 20 mL 1× TBST and mix well.
13. *BenchMark prestained protein ladder, (Premixed Format)*: (Invitrogen, #10748-010).
14. *Blotting Membrane and Paper*: This protocol has been optimized for nitrocellulose blotting membranes. Pore size 0.45 μm is generally recommended (Amersham Protran,

#10600002). Fisherbrand chromatography paper (thick) has been used for transfer (Fisher Scientific, #05-714-4).

15. *Primary antibodies:* Akt antibody (Cell Signaling, #9272). P44/42 MAP kinase antibody (Cell Signaling, #9102). Anti-paxillin monoclonal antibody and anti-FAK polyclonal antibody (BD Biosciences PharMingen, #610619 & #610087). PathScan® Multiplex Western Cocktail I containing phospho-Akt (Ser473) and phospho-p44/42 MAPK (Erkl/2) (Thr202/Tyr204) (Cell Signaling, #5301S).
16. *Secondary antibody conjugated to HRP*: Goat anti-rabbit IgG, peroxidase conjugate-linked antibody (CalBiochem, #DC0 3 L); goat anti-mouse IgG, peroxidase conjugate-linked antibody (CalBiochem, #DC0 2 L).
17. *Detection reagent*: ECL™ Western blotting detection reagent (Amersham™ #RPN2106).
18. *Stripping buffer: (a) Restore™ Western blot stripping buffer*: (Thermo Scientific, #21059). (b) Alternatively, in our laboratory, we use a stripping buffer composed of Tris base, SDS, and β-mercaptoethanol. For 100 mL, mix 0.76 g Tris base, 2 g SDS, and 700 μL β-mercaptoethanol. Bring to 100 mL with deionized H_2O. Adjust pH to 6.8 with 1 N HCl.
19. *X-ray film:* Premium Autoradiography Film (Denville Scientific Inc., #E3018).
20. *Fixative:* 4% Paraformaldehyde can be made from powder or 16% formaldehyde ampule (Fisher Scientific, #PI-28906; *see* **Note 1**) by 1:4 dilution with 1× DPBS.
21. *Secondary antibodies used in confocal analysis*: Alexa Fluor 647 (blue) goat anti-mouse, Alexa Fluor 488 (green) goat anti-mouse, and Alexa Fluor 594 goat anti-rabbit (red) (Molecular Probes, Invitrogen, #A21235, #A11017, and #A11072).
22. *Purification of human recombinant progranulin*

    Human recombinant progranulin was purified from conditioned media of 293-EBNA cells stably transfected with a cDNA plasmid expressing 6His-tagged human progranulin. Serum-free conditioned medium was concentrated with polyethylene glycol, dialyzed, and purified on Ni-NTA resin eluted with 250 mmol/L imidazole as previously described [13]. The purified recombinant His-tagged progranulin protein was used in all experiments described below.
23. *SFM*: DMEM supplemented with 0.1% bovine serum albumin and 50 μg/mL of transferrin (Sigma-Aldrich, St. Louis, MO).
24. *Protein assay dye reagent concentrate* (Bio-Rad, #500-0006).
25. *Progranulin antibody*: Anti-progranulin polyclonal antibody from US Biological Life Sciences (Swampscott, MA, USA, # P9182-35B).

26. *Confocal blocking buffer*: The blocking buffer used in confocal analysis is 1x DPBS with 1% bovine serum albumin and 0.5% Triton X-100.
27. *Mounting media for fluorescence*: The mounting media can be purchased from Vector Labs, Inc. (Burlingame, CA, #H-1000). You can also prepare a mounting media that will minimize photo bleaching as follows:
    (a) Add 100 mg of DABCO (Sigma, #D2522) to 1 mL of buffer and dissolve.
    (b) Add 9 mL of glycerol (Glycerin) and gently invert until the two mix thoroughly.
28. *Nail polish:* Clear nail polish from Electron Microscopy Sciences (Hatfield, PA, Cat# 72180).

3 Methods

3.1 Detection of Akt and MAPK Activation with Western Immunoblotting [9–11, 14]

3.1.1 Sample Preparation, Loading, Running, and Transfer to Nitrocellulose Membrane

1. Serum-starved the cells for 24 h in serum-free media (SFM).
2. Stimulate the cells with recombinant progranulin (20 ~ 40 nM) for time points ranging from 10 min to 2 h.
3. Aspirate media and wash cells with 1× DPBS.
4. Lyse cells by adding 1× NP-40 lysis buffer supplemented with protease and phosphatase inhibitors (75 μL/well for six-well plates or 250 μL for 10 cm diameter plates). After 10 min, scrape the cells off the plate, and transfer the extract to an Eppendorf Tube. Keep on ice.
5. Vortex for a few times in 5 ~ 10 min.
6. Spin down at 20,100 × *g* for 10 min, and collect the supernatant.
7. Measure the protein concentration by using the Protein Assay Dye Reagent Concentrate (Bio-Rad, #500-0006).
8. Mix protein lysate containing 50 μg protein with equal volume of 2× SDS sample buffer, heat to 95–100 °C for 5 min, and cool on ice.
9. Microcentrifuge for 5 min.
10. Load sample onto 12% SDS-PAGE gel (Bio-Rad, #161-1156) (*see* **Note 2**).
11. Run the gel in Bio-Rad chamber (Mini-PROTEAN 3 Cell) with running buffer at 135 V for 1 h or until desired separation is achieved.
12. Electrotransfer to nitrocellulose membrane (Amersham Protran, #10600002) at 110 V for 1 h.
13. Proceed to next section: Membrane Blocking and Antibody Incubation.

3.1.2 Membrane Blocking and Antibody Incubation

1. After transfer, wash nitrocellulose membrane with 25 mL TBST for 5 min at room temperature. Incubate membrane in 20 mL of Western blocking buffer (*see* **Note 3**) for 1 h at room temperature.
2. Wash three times for 5 min each with 15 mL of TBST.
3. Incubate membrane and primary antibody (PathScan® Multiplex Western Cocktail I 1:200 dilution in 1× TBST with 5% BSA as recommended in the product datasheet) in 10 mL primary antibody dilution buffer with gentle agitation overnight at 4 °C. This reagent allows the simultaneous detection of P-Akt and P-ERKs on a single blot.
4. Wash the membrane three times for 10 min each with 15 mL of TBST.
5. Incubate membrane with secondary goat anti-rabbit IgG, peroxidase conjugate-linked antibody (CalBiochem, #DC0 3 L at 1:3000) in 10 mL of Western blocking buffer (*see* **Note 3**) with gentle agitation for 45 min at room temperature.
6. Wash three times for 10 min each with 15 mL of 1× TBST.
7. Proceed with detection (*see* below).

3.1.3 Detection of Phosphorylated Akt and MAP Kinase Proteins

1. Wash membrane three times for 10 min in TBST.
2. Prepare 1× ECL™ (Amersham™, #RPN2106) by mixing 2× reagent A and 2× reagent B at a 1:1 ratio.
3. Incubate the membrane for 1 min in the ECL solution, remove excess but keeping the membrane wet, wrap in plastic, and expose to X-ray film followed by development with various exposure time.

3.1.4 Membrane Reprobing for Total Akt and MAPK Protocol

Reprobing of an existing membrane is a convenient means to sequentially immunoblot for multiple proteins when only a limited amount of sample is available. However, it should be noted that for the best possible results, a fresh blot is always recommended. With each reprobing of a membrane, there is potential for increased background signal as well as loss of specificity. Additionally, it is recommended that you verify the removal of leftover signals from previous incubations. This can be done by re-exposing the membrane with ECL after the stripping procedure.

1. Wash membrane four times for 5 min each in TBST. Best results are obtained if the membrane is not allowed to dry.
2. Incubate membrane for 30 min at room temperature in stripping buffer A or at 50 °C in stripping buffer B (*see* Subheading 2) with gentle agitation.
3. Wash membrane thoroughly six times for 5 min each in TBST.

4. To assure that the original signal is removed, wash membrane twice for 5 min each with 10 mL of TBST. Incubate membrane with ECL with gentle agitation for 1 min at room temperature. Drain membrane of excess developing solution. Do not let dry. Wrap in plastic wrap and expose to X-ray film.
5. Wash membrane again four times for 5 min each in TBST.
6. The membrane is now ready for reprobing.
7. Incubate membrane in 10 mL 5% milk Western blocking buffer with a mixture of rabbit anti-Akt or anti-p42/44 antibodies at 1:1000 dilution for total Akt or MAP kinase proteins as recommended in the product datasheet, with gentle agitation overnight at 4 °C.
8. Wash three times for 10 min each with 15 mL of TBST.
9. Incubate membrane with secondary goat anti-rabbit IgG, peroxidase conjugate-linked antibody (CalBiochem, #DC03Lat 1:3000) in 10 mL of Western blocking buffer (*see* **Note 3**) with gentle agitation for 45 min at room temperature.
10. Wash three times for 10 min each with 15 mL of TBST.
11. Proceed with detection (*see* Subheading 3.1.3). A representative result is shown in Fig. 1.

3.2 Double Staining Confocal Microscopy Protocol [12, 15, 16]

1. *Cultured cells.* Cells are grown on glass coverslips (Corning, 22 mm sq.).
2. *Fixation and permeabilization.*

   Different fixatives can have various effects on cellular structure and antigen preservation. Formaldehyde is a highly reactive, cell-permeable agent that is used as a reversible cross-linking agent for proteins and nucleic acids by coupling neighboring primary amines within the cell or as a general cell-fixing agent for imaging-based applications. Cell permeability allows the reagent to readily enter living cells, enabling intracellular applications, such as protein interaction discovery or bioimaging.

   (a) Incubate the cells with fixative for 30 min at room temperature followed by permeabilization with 0.1% Triton X-100 in 1× DPBS for 10 min.
3. *Colocalization analysis.*

   This is a generalized antibody labeling procedure [12]. You will need to optimize your primary and secondary antibody concentrations along with incubation times as needed.

   (a) Incubate sample for 15 min in 1× DPBS with 0.5% BSA for blocking.

   (b) Rinse several times in 1× DPBS with 0.5% BSA.

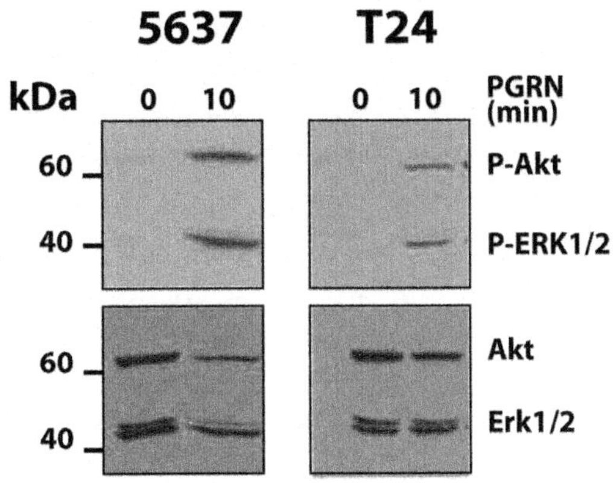

Fig. 1 Progranulin-mediated activation of the Akt/MAPK pathway in urothelial cancer cells. 5637 and T24 cells were serum-starved for 24 h and then stimulated with recombinant progranulin (20 nM) for 10 min. The activation of Akt and ERK1/2 was analyzed by immunoblotting using phospho-specific antibodies anti-phospho-Akt (Ser473) and anti-phospho-p44/42 MAPK (Erk1/2) (Thr202/Tyr204) using the PathScan® Multiplex Western Cocktail I. Blots were stripped and reprobed for total Akt and ERK1/2 level using anti-Akt and anti-ERK1/2 polyclonal antibodies (Cell Signaling)

(c) Add primary antibody solutions (see below) on top of the coverslip.

(d) Incubate with primary antibodies in 1× DPBS with 0.5% BSA for 1 h at room temperature. If you are doing multiple antibodies, it is possible to mix them together and incubate. In our analysis, we used anti-progranulin polyclonal antibodies (US Biologicals, Swampscott, MA, USA) and anti-paxillin monoclonal antibodies, both at 1:100 dilution.

(e) After incubation with the primary antibody, wash again in 1× DPBS with 0.5% BSA. In our study, secondary antibodies were goat anti-mouse IgG Alexa Fluor® 488 and goat anti-rabbit IgG Alexa Fluor® 594 antibodies (Invitrogen) [8]. Incubate the sample in fluorescent secondary antibodies for 1 h at room temperature. It is advisable to incubate secondary antibodies sequentially to minimize cross-reactivity if multiple secondary antibodies are used. Direct labeling agents (i.e., DAPI, Hoescht, phalloidin, etc.) can be added at any point during the secondary antibody labeling.

3.3 Triple Staining Confocal Analysis

1. Cells are plated onto coverslips, allowed attached overnight in an incubator at 37 °C with 10% CO_2.
2. Wash three times with PBS, and then serum starved for 24 h in an incubator at 37 °C with 10% CO_2.

3. Starved cells are stimulated for 10 and 30 min with 2 μg/mL of purified progranulin.
4. After two washes in1× DPBS, cells are fixed in 4% paraformaldehyde for 10 min.
5. Incubate in confocal blocking buffer (1× DPBS/1% bovine serum albumin, 0.5% Triton X-100 for 30 min at 37 °C.
6. Coverslips are incubated with any antibody of choice in the confocal blocking buffer for 30 min 37 °C. In previous work from our laboratory [4], we used 1:100 diluted anti-paxillin monoclonal antibody (BD PharMingen) and anti-active MAPK p42/44 polyclonal antibody (Cell Signaling).
7. After washing five times with confocal blocking buffer, incubate again in confocal blocking buffer for 30 min at 37 °C with 1:100 diluted (in confocal blocking buffer) secondary antibodies Alexa Fluor 647 (blue) and Alexa Fluor 488 (green) in confocal blocking buffer.
8. The third additional staining involves a repetition of the protocol after the double staining. FAK was stained in our laboratory [4] using anti-FAK polyclonal antibodies (BD PharMingen) using the same incubation conditions and duration as in **step 6**.
9. Secondary antibodies are Alexa Fluor 594 (red; Molecular Probes).
10. The coverslips are washed twice with PBS and mounted using SlowFade mounting medium (Molecular Probes).
11. Coverslips were analyzed at the Kimmel Cancer Center Confocal Microscopy Core Facility with a Zeiss Axiovert 200 M cell live microscope using the Metamorph Image Acquisition and Analysis software (Universal Imaging) using a 63×, 1.3 oil immersion objective of a Zeiss LSM-78 confocal laser-scanning microscope.
12. The images are merged using the software "Adobe Photoshop."

3.4 Mounting Coverslips on Slides

1. Clean a slide with alcohol followed by rinsing with dH_2O and drying with a Kimwipes (Kimberly-Clark Kimtech Science Brand).
2. Place ~10 μL of mounting media on the center of a slide.
3. Take your coverslip, eliminate any excess buffer, and carefully place the coverslip with cells facing down on the small spot of mounting media. Seal using clear fingernail polish.

3.5 Confocal Analysis

1. Confocal analysis was carried out using a 63×, 1.3 oil immersion objective of a Zeiss LSM-78 confocal laser-scanning microscope. To quantify colocalization of the two proteins,

Z-stack series were acquired maintaining the same slice interval (*see* **Note 4**).

2. All images were analyzed using ImageJ and Adobe Photoshop CS6 (Adobe Systems).
3. Line-scanning plots were performed utilizing SigmaPlot software 8.0 (SPSS Science).

4 Notes

1. Pierce™ Formaldehyde Ampules, Methanol-free is available at Fisher Scientific™ (PI-28906). See more at: https://www.fishersci.com/shop/products/thermo-scientific-pierce-formaldehyde-ampules-methanol-free/p-3599306#sthash.u0vck0nz.dpuf).
2. Loading of prestained molecular weight markers (Fisher Scientific/Invitrogen, 15 μL/lane) to verify electrotransfer and determine molecular weights is recommended.
3. Western blocking buffer volumes are for 10 cm × 10 cm (100 cm^2) membranes; for different sized membranes, adjust volumes accordingly. For kinase activation blot 5% BSA is used, while 5% nonfat dry milk is used for total protein blot instead.
4. After immunostaining, the glass slides can also be analyzed using a phase-contrast inverted fluorescence microscope (Nikon Eclipse TE 2000-U) connected to a digital camera and a video imaging system. Data can be determined by Journal software version 7.0rl (Metamorph), performed in Kimmel Cancer Center Bioimaging Resource. The values were evaluated statistically with SigmaPlot 8.0 (SPSS Science).

Acknowledgments

Our laboratories are supported by the Benjamin Perkins Bladder Cancer Fund, the Schorsch Family Fund for Innovative Medical Research and Care, National Institutes of Health Grants RO1 CA164462 (A.M., R.V.I.), Associazione Italiana per la Ricerca sul Cancro (AIRC) (grant n. 10625/12), AIRC project Calabria 2015, and Fondazione Cassa di Risparmio di Calabria e Lucania PON01_01078 (A.B.).

References

1. He Z, Ismail A, Kriazhev L et al (2002) Progranulin (PC-cell-derived growth factor/ acrogranin) regulates invasion and cell survival. Cancer Res 62:5590–5596
2. He Z, Bateman A (2003) Progranulin (granulin-epithelin precursor, PC-cell-derived growth factor, acrogranin) mediates tissue repair and tumorigenesis. J Mol Med 81:600–612
3. He Z, Ong CH, Halper J et al (2003) Progranulin is a mediator of the wound response. Nat Med 9:225–229
4. Monami G, Gonzalez EM, Hellman M et al (2006) Proepithelin promotes migration and invasion of 5637 bladder cancer cells through the activation of ERK1/2 and the formation of a paxillin/FAK/ERK complex. Cancer Res 66:7103–7110
5. Monami G, Emiliozzi V, Bitto A et al (2009) Proepithelin regulates prostate cancer cell biology by promoting cell growth, migration, and anchorage-independent growth. Am J Pathol 174:1037–1047
6. Lovat F, Bitto A, Xu SQ et al (2009) Proepithelin is an autocrine growth factor for bladder cancer. Carcinogenesis 30:861–868
7. Tanimoto R, Morcavallo A, Terracciano M et al (2015) Sortilin regulates progranulin action in castration-resistant prostate cancer cells. Endocrinology 156:58–70
8. Xu SQ, Buraschi S, Morcavallo A et al (2015) A novel role for drebrin in regulating progranulin bioactivity in bladder cancer. Oncotarget 6:10825–10839
9. Shaul Y, Seger R (2006) The detection of MAPK signaling. Curr Protoc Mol Biol (Wiley Online Library) Chapter 18:Unit 18-12
10. Gallagher S, Winston SE, Fuller SA et al (2008) Immunoblotting and immunodetection. Curr Protoc Immunol Chapter 8:Unit 8–10
11. Pankov R (2004) Determination of Akt/PKB signaling. Curr Protoc Cell Biol (Wiley Online Library) Chapter 14:Unit 14–16
12. Zinchuk V, Grossenbacher-Zinchuk O (2011) Quantitative colocalization analysis of fluorescence microscopy images. Curr Protoc Cell Biol 62:Unit 4-19 11-14
13. Gonzalez EM, Mongiat M, Slater SJ et al (2003) A novel interaction between perlecan protein core and progranulin: potential effects on tumor growth. J Biol Chem 278:38113–38116
14. Yao Z, Seger R (2001) Immunological detection of phosphorylation. Curr Protoc Cell Biol (Wiley Online Library) Chapter 14:Unit 14–12
15. Mansfield JR, Hoyt C, Levenson RM (2008) Visualization of microscopy-based spectral imaging data from multi-label tissue sections. Curr Protoc Mol Biol (Wiley Online Library) Chapter 14:Unit 14–19
16. Rappaz B, Wiseman PW (2013) Image correlation spectroscopy for measurements of particle densities and colocalization. Curr Protoc Cell Biol (Wiley Online Library) Chapter 4:Unit 427, 21–15

Chapter 10

Mouse Monoclonal Antibodies Against Progranulin (PGRN/GEP) as Therapeutics in Preclinical Cancer Models

Chi Wai Yip, Phyllis F. Y. Cheung, Nicholas C. L. Wong, Sze Wai Fung, and Siu Tim Cheung

Abstract

The use of monoclonal antibody (mAb) has become a unique means of targeted therapy for human cancers. mAb-based therapies have shown survival benefits by applying alone or in combination with chemotherapeutics. Being a humanized biomolecule with exquisite target specificity, mAb demonstrated effects in a relatively lower dose range with limited off-target harm to the patients. Nowadays, novel targets involved in tumorigenic mechanisms and biomarkers expressed exclusively on cancer cell surface are being constantly discovered. The potential effects of their specific mAb could be investigated in the preclinical cancer model. In this chapter, we outlined our experimental procedures in determining the feasibility of novel mAb in the preclinical cancer model, with an example of progranulin (PGRN/GEP) mAb against hepatocellular carcinoma (HCC) tumor in mouse model. This chapter included the establishment of subcutaneous and orthotopic HCC tumor in mouse model, the injection of the mouse monoclonal antibody in combination with cytotoxic chemotherapeutics, the assessment of tumor development, and the analyses of the molecular changes of the tumor cells.

Key words Monoclonal antibody, HCC, Targeted therapy, Animal model, Granulin-epithelin precursor

1 Introduction

Targeted therapies are mostly accomplished by monoclonal antibodies or small molecule inhibitors. Monoclonal antibodies inhibit the targeted molecules from interacting with their binding partners and processing their functional roles in carcinogenesis. Additionally, the binding of cancer cells by the monoclonal antibodies can lead to apoptosis, antibody-dependent cellular cytotoxicity (ADCC), and complement-dependent cytotoxicity (CDC) [1, 2]. Due to these promising effects, targeted therapy has become an essential part of treatments in many cancers, leading to a reduced reliance on drugs that non-discriminately kill tumors and host cells. Nevertheless, clinical studies also showed further survival benefits when combining chemotherapeutics with mAb [3–5].

Andrew Bateman et al. (eds.), *Progranulin: Methods and Protocols*, Methods in Molecular Biology, vol. 1806,
https://doi.org/10.1007/978-1-4939-8559-3_10,

We have established a subcutaneous xenograft mouse model using liver cancer cell line without luciferase [6]. Subsequently, for the establishment of a preclinical cancer model with orthotopic xenograft, we used luciferase-expressing liver cancer cell line for the inoculation [7]. The expression of luciferase is useful for monitoring orthotopic xenograft noninvasively throughout the experimental period since the tumor size is often hard to be assessed without sacrificing the mouse. A high correlation between tumor volume and the luciferase bioluminescence has been observed in many cell lines [8, 9], where the bioluminescent signal is recoded as radiance: p/s/cm^2/sr (photons per second per square centimeter per steradian). On the other hand, tumor volume of subcutaneous xenograft could be estimated from the measurement of caliper without terminating the mouse, using the formula $(L \times S^2) \times \pi/6$, where L represents the long dimension and S represents the short dimension in the measurement. This estimation assumes the height of the tumor is equal to the short dimension and therefore may not reflect the tumor volume with high accuracy [10, 11]. Therefore, luciferase-expressing cell line could also be used for subcutaneous xenograft to enhance the accuracy of the measurement of tumor volume.

The combination of immunotherapy and chemotherapy has once been considered incompatible as chemotherapies at high dose could induce immunosuppression [12]. However, accumulating evidence suggests that combinations of these treatments give improved clinical outcomes than the single treatment [3–5]. The cellular and molecular mechanisms of these synergic effects have been delineated. Chemotherapy supports the effect of immunotherapy by inducing tumor cell death or enhancing tumor cell sensitivity to cytotoxic T lymphocyte lysis [13, 14]. On the other hand, through inducing apoptosis, immunotherapy modulates the tumor's sensitivity to chemotherapy [7, 15, 16]. Therefore, incorporating immunotherapy-chemotherapy combinations into studies using preclinical cancer models would not only lead to a better understanding on the value or compatibility of the combination but could also reveal the underlying molecular mechanisms of the synergic effects by tumor cell analyses, hence provide the rationales for future direction of combination strategies in clinical trials.

In particular, a kinase inhibitor-based targeted therapy, sorafenib, has been approved by FDA to treat patients with advanced hepatocellular carcinoma (HCC). Clinical studies demonstrated survival benefits of about 3 months in advanced HCC patients [17, 18]. With the currently moderate effect of sorafenib, chemotherapy is still widely employed to treat advanced HCC, while chemoresistance is frequently observed from the patients [19, 20]. Therefore, there is a demand of novel therapeutic targets and treatment approaches for this aggressive disease. We have previously reported the expression profiles of liver cancer showing overexpression of progranulin in more than 70% of HCCs [21, 22], where it controls proliferation, invasion, and tumorigenicity

[23, 24]. Progranulin was considered primarily as a growth factor [25]; however, evidence is emerging that progranulin has crucial role on regulating cancer stem cell properties and chemoresistance [26, 27]. We have shown that progranulin antibody inhibited the growth of HCC tumor in subcutaneous [6] and orthotopic [7] model systems. In this chapter, we outlined our experimental procedures in testing the progranulin mAb in preclinical cancer model of HCC. These included the establishment of subcutaneous and orthotopic HCC tumor in mouse model, the injection of the mouse mAb with combination of cytotoxic chemotherapeutics, the assessment of tumor development, and the analyses of the molecular changes in the tumor cells.

2 Materials

2.1 Cell Culture

1. Hep3B human liver cancer cell line (American Type Culture Collection, ATCC, Manassas, VA).
2. Advanced minimum essential medium (AMEM, Life Technologies).
3. Heat-inactivated fetal bovine serum (FBS, Life Technologies).
4. GlutaMAX™ Supplement (Life Technologies).
5. Trypsin-EDTA (0.25%, Life Technologies).
6. Lipofectamine 2000 (Life Technologies).
7. Firefly luciferase expression vector (pcDNA3.1, Life Technologies).
8. Geneticin (G418, Life Technologies).
9. D-Luciferin (Gold Biotechnology, St Louis, MO, USA).
10. Trypan blue stain (Life Technologies).
11. Hemocytometer.

2.2 Tumor Inoculation

1. Syringe (1 mL) and needle (23G).
2. Male BALB/c nu/nu (nude) mice.
3. Isoflurane.
4. Digital caliper.
5. Pentobarbitone sodium.
6. Dissection instruments.
7. Saline.
8. Surgical blade.
9. 6–0 silk suture (Perma-hand™, Ethicon Inc., USA).
10. Cotton swabs.
11. Gauze and 70% ethanol.
12. Surgical sealant (Coseal, Baxter, USA).

2.3 Randomization and Injection of mAb and Chemo-therapeutics

1. Saline.
2. Cisplatin.
3. Progranulin mAb (A23 [6]).
4. Insulin syringe (29G needle and 0.5 cc syringe, BD biosciences).

2.4 Assessment of Tumor Development

1. Pentobarbitone sodium.
2. D-Luciferin (Gold Biotechnology, St Louis, MO, USA).
3. Insulin syringe (BD biosciences).

2.5 Harvest of the Tumor Cells

1. Isoflurane.
2. Dissecting sets.
3. Saline.
4. Surgical blade.
5. Gauze and 70% ethanol.
6. 40 μm cell strainer (BD Biosciences).
7. Type IV collagenase (Sigma-Aldrich).
8. ACK lysis buffer: 0.15 M NH_4Cl, 10 mM $KHCO_3$, 0.1 mM Na_2EDTA in H_2O, pH 7.2–7.4.
9. AMEM, 10%FBS, 1× GlutaMAX™.

2.6 mRNA Expression

1. Lysis buffer, DNaseI, stop buffer, reverse transcriptase, RT buffer (TaqMan® Gene Expression Cells-to-C_T™ Kit from Life Technologies).
2. TaqMan® Universal Master Mix II (Life Technologies).
3. TaqMan® Gene Expression Assays (18S rRNA, progranulin and ABCB5, Life Technologies).

2.7 Signaling Analysis by Western Blot

1. Lysis buffer (Cell Signaling Technology, Boston, MA, USA, freshly add 1 mM PMSF).
2. DC™ Protein Assay (Bio-Rad Life Science).
3. Blocking Buffer (5% skim milk in Tris-buffered saline with 0.05% Tween 20).
4. Antibodies against signaling molecules (ERK1/2, AKT, pERK1/2, and pAKT, Cell Signaling Technology).
5. Enhanced chemiluminescence (GE Healthcare Life Sciences).

2.8 Cell Apoptosis Assay

1. FITC-conjugated annexin V (BD Biosciences). Store at 4 °C.
2. 10× Annexin V Binding Buffer (BD Biosciences). 0.1 M Hepes/NaOH (pH 7.4), 1.4 M NaCl, 25 mM $CaCl_2$. Store at 4 °C. For 1× working solution, dilute 1 part of the 10× Annexin V Binding Buffer to nine parts of distilled water.
3. Propidium iodide (BD Biosciences).

2.9 Flow Cytometric Analysis

1. 0.1% Saponin in PBS (freshly add the powder to PBS prior to use and keep at 4 °C).
2. 4% formaldehyde in PBS (dissolve paraformaldehyde powder in PBS at 60 °C, add 1 N NaOH dropwise until the powder completely dissolve. Cool down the solution and adjust the pH to 7 by adding 1 N HCl. Filter the solution and keep at 4 °C).

2.10 Colony Formation

1. Fixative (15 mL glacial acetic acid, 40 mL ethanol, 445 mL H_2O).
2. Crystal violet solution (Sigma-Aldrich, St. Louis, MO).

2.11 Spheroid Formation

1. Dulbecco's Modified Eagle's Medium/Nutrient Mixture F-12 Ham (DMEM/F12) medium (Life Technologies, Carlsbad, CA), supplemented with 20 ng/mL human recombinant EGF (Life Technologies), 10 ng/mL human recombinant bFGF (Life Technologies), 4 μg/mL insulin (Sigma-Aldrich) and B27 (1:50) (Life Technologies). Freshly prepare.
2. Ultra-low attachment plate (Corning, *Tewksbury,MA*).

3 Methods

3.1 Preparation of Tumor Cells

1. Seed Hep3B cells at 1×10^5 in a 6-well plate with complete AMEM (10% FBS and supplemented with GlutaMAX™).
2. Incubate cells for 2 days in 37 °C until confluence achieves 80%. Change the medium with 2.5 mL fresh complete AMEM.
3. Mix 2 μg firefly luciferase gene containing plasmid DNA with 5 μg Lipofectamine in 0.25 mL plain AMEM. Stand the mixture in room temperature for 20 min.
4. Add the transfection mix to a well evenly and incubate the cells for 6 h. Change the medium with fresh complete AMEM, and further incubate the cells for 2 days.
5. Pass the cells at 1:6 with trypsin-EDTA, and complete AMEM into a new 6-well plate. Incubate the cells for 1 day, and change the medium with complete AMEM containing 0.4 mg/mL G418. Refresh the selective medium every 2 days until colonies formed.
6. Transfer 4–8 colonies into different wells in a 24-well plate with trypsin-EDTA. Incubate the cells with selective medium. Pass the cells when confluent.
7. Reduce the amount of G418 to 0.2 mg/mL when selection completes. Split the cells to 24-well with 1250, 2500, and 5000 cells, and incubate the cells for 2 days.
8. Aspirate the medium, and add plain AMEM containing 150 μg/mL luciferin. Incubate the cells at 37 °C for 10 min.

9. Transfer the plate to the 37 °C chamber of the IVIS100 imaging system (Caliper Life Sciences, Hopkinton, MA, USA), and expose for 5, 20, and 60 s.
10. Detach the cells by trypsin-EDTA, and count the cell numbers in each well. Plot a correlation curve between cell numbers and bioluminescent signal.
11. Select the colony that show strong bioluminescent signal in every cells and good correlation for the inoculation. Stock the cells in liquid nitrogen.

3.2 Inoculation of Tumor Cells

1. Prepare male BALB/c nu/nu (nude) mice when they are 4–5 weeks old at the time for inoculation (*see* **Note 1**).
2. Maintain the firefly luciferase gene-transfected Hep3B in complete AMEM with 0.2 mg/mL G418.
3. Seed 2×10^6 cells with complete AMEM without G418 in a T175 flask. Change medium 2 days after passage. Detach the cells by trypsin-EDTA on day 3. Collect the cells with complete AMEM, and wash the cells thrice with plain AMEM.
4. Resuspend the cells with plain AMEM to 1×10^7/mL.
5. Keep the mice under anesthesia with inhalation of isoflurane. Inject 1×10^6 cells subcutaneously into the right flank of the nude mice (*see* **Note 2**).
6. For determining the initial establishment of the tumor, please follow Subheading 3.4.
7. Measure the subcutaneous tumor by digital caliper every week until the tumor reached around 1 cm in diameter.
8. Sacrifice the mouse by isoflurane inhalation.
9. Remove the subcutaneous tumor, and briefly wash it in saline on a sterilized petri dish.
10. Mince the tumor with sterile surgical blades into 1 mm^3 cubes, and soak them in saline.
11. Prepare another group of 4–5 weeks old nude mice. Keep the mice under anesthesia with intraperitoneal (i.p.) injection of 54 mg/kg pentobarbitone sodium.
12. Use adhesive tape or rubber band to maintain a fix posture for the mouse on a working platform without hurting it.
13. Sterilize the skin at the operation area with iodophor (e.g., Betadine) scrub and 70% ethanol with gauze. Cut open the skin with surgical scissors to make an incision of about 1.5 cm wide just above the liver.
14. Use sterile cotton swabs to gently pull out the left lobe of the liver from the open cut (Fig. 1).

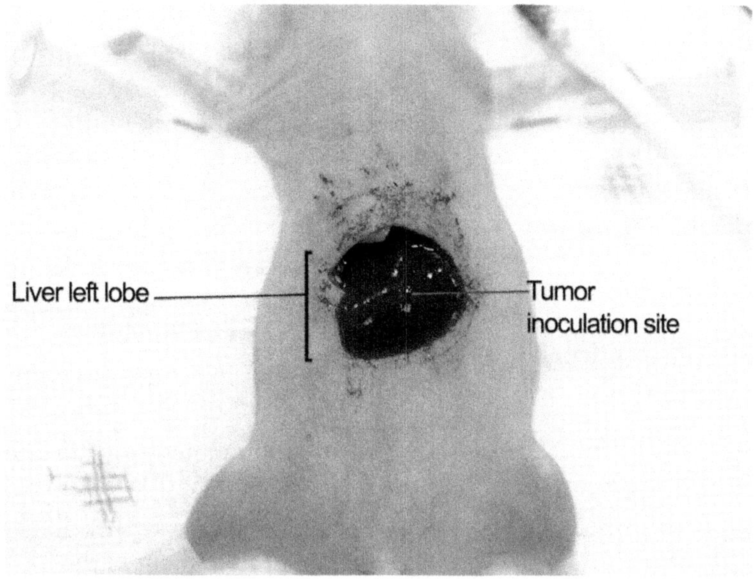

Fig. 1 Orthotopic HCC xenografting. The position of xenograft implantation at the left lobe of a nude mouse

15. Incise a small hole (1 mm diameter) at the left liver lobe by the tip of forceps. Introduce one piece of the subcutaneous tumor promptly to the incision.
16. Close the incision with surgical sealant. Wait 30–60 s for the glue to form gel. Pull the skin by forceps, and push the liver back gently by cotton swab.
17. Sew the skin by a 6–0 silk suture. Clean the skin by saline and gauze. Place the mice on a heating carpet until fully recover.

3.3 Randomization and Injection of mAb and Chemotherapeutics

1. At 1 week and 2 weeks after inoculation of the orthotopic tumor, monitor the establishment of the tumor by in vivo imaging of the bioluminescent signal according to Subheading 3.4.
2. After the second imagining, randomize the mice with Hep3B intrahepatic xenografts showing the region of interest (ROI) greater than 1×10^6 p/s/cm^2/sr into four groups (saline control, mAb, cisplatin, and combination), with at least five mice per group (*see* **Note 3**).
3. Compare the means and standard deviations of the ROI of the four groups for being not significantly different ($p > 0.05$) before the experiment starts.
4. Apply i.p. injection of progranulin mAb and cisplatin on separate days at 10 mg/kg/week and 3.75 mg/kg/week, respec-

tively, with the insulin syringe. Inject the mice in control group with the same volume of saline (*see* **Note 4**).

5. Apply the injections weekly for 2–6 weeks. Monitor the tumor development every week until harvesting the tumors.

3.4 Assessment of Tumor Development

1. During the experiment period, monitor the tumor size weekly by the in vivo IVIS100 imaging.
2. Apply i.p. injection of 54 mg/kg pentobarbitone sodium and i.p. injection of 150 mg/kg of D-luciferin at the same time with separate sets of insulin syringe (*see* **Notes 5** and **6**).
3. At 15 min after injection, place the mice of a group into the 37 °C chamber.
4. Detect the bioluminescent signal intensity at different exposure times, such as 5, 20, and 60 s.
5. Gather two or three representative mice from each group together, and place them in the 37 °C chamber to allow comparison among groups (Fig. 2).
6. Detect the bioluminescent signal at different exposure times.
7. Place the mice back to the cages, and monitor them until they are conscious.

3.5 Tissues and Organs Collection

1. Terminate the mice by isoflurane inhalation.
2. Dissect the mouse, and collect the blood from the main bloodstream into polypropylene tubes (*see* **Note 7**).
3. Remove the liver with surgical scissors and forceps. Wash the liver briefly in saline.
4. Take a photo of the livers and tumors. Separate the xenograft tumor from the liver.
5. Fix a portion of the tumor in a cassette by formalin, and freeze another portion in a vial by liquid nitrogen. Store the fixed cassette and frozen vial of liver and other organs of interest (*see* **Note 8**).
6. Mince each tumor with sterile surgical blades into approximately 1 mm^3 cubes. Collect the tumor pieces into a 50 mL plastic tube.
7. Digest the pieces of each tumor with type IV collagenase for 5 min at 37 °C.
8. Centrifuge the mixture at 50 × *g* for 5 min at 4 °C.
9. Collect the supernatants as cells released from the tumors.
10. Repeat **steps 7–9** twice.
11. Filter the supernatants through a 40 μm cell strainer.

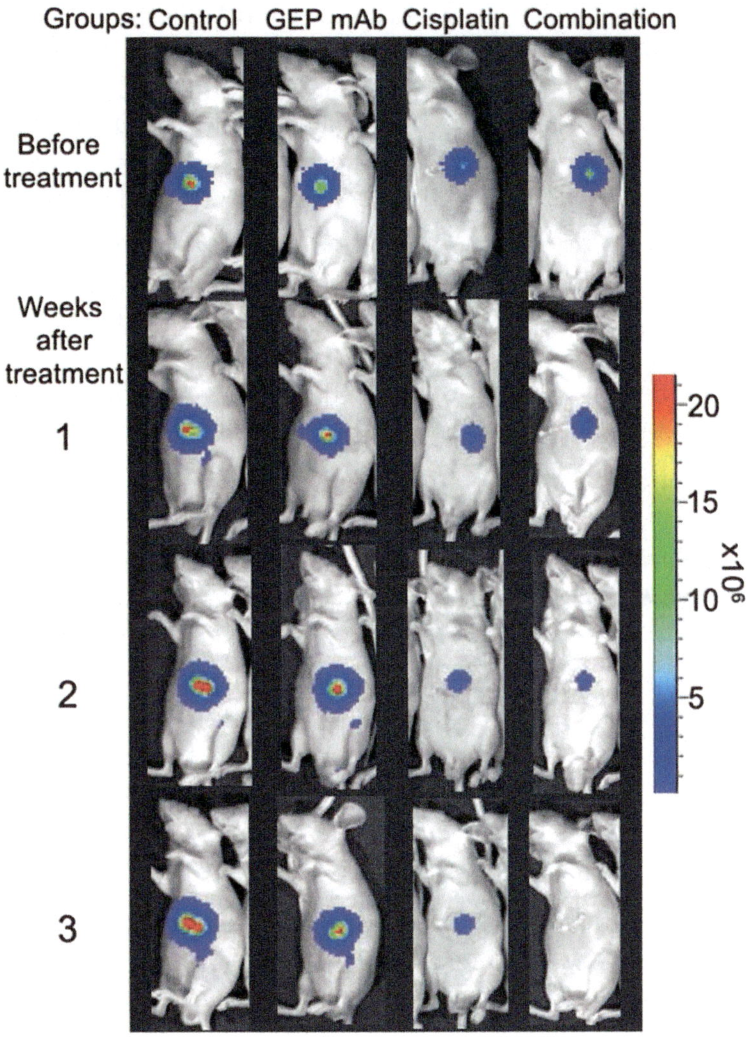

Fig. 2 Orthotopic HCC model using nude mice. Hep3B xenografts were treated with progranulin mAb in combination with cisplatin. Representative mice with bioluminescent signal at various time points were shown. Reproduced from Mol Cancer Ther 2014 [7]

12. Centrifuge the cells at 800 rpm for 5 min at 4 °C.
13. Remove the supernatant.
14. Lyse the red blood cells with 10 mL ACK lysis buffer for 1 min at room temperature.
15. Add 20 mL complete AMEM to the mixture, and centrifuge the cells at 800 rpm for 5 min at 4 °C.
16. Assess the cell viability by trypan blue exclusion staining and count cells.

3.6 mRNA Expression

1. Take 0.5–1 × 10^4 cells from the tumor.
2. Wash the cells with 500 μL PBS. Pellet the cells, and resuspend the cells with 50 μL lysis buffer (100 μL lysis buffer freshly mixed with 1 μL DNaseI; cell-to-C_T kit).
3. Stand the cells at room temperature for 5 min. Add 5 μL stop buffer (cell-to-C_T kit) and mix with pipetting up and down.
4. Stand the lysate at room temperature for 2 min.
5. Perform reverse transcription at 37 °C for 60 min and 95 °C for 5 min, where 5–10 μL lysate can be included in a 50 μL reaction.
6. Dilute the cDNA sixfold. In a 20 μL reaction using master mix II, add 4 μL diluted cDNA and probe sets. Perform real-time QPCR on the target molecule and other genes of interest such as drug transporters, and use 18S rRNA as internal control.

3.7 Signaling Analysis by Western Blot

1. Wash 2–5 × 10^5 cells with 1 mL PBS.
2. Lyse the cells by adding 50–100 μL lysis buffer with pipetting up and down. Incubate the lysate at 4 °C for 15 min.
3. Centrifuge the lysate at 20,000 × *g* for 10 min at 4 °C, and collect the supernatant.
4. Quantify the concentration of the lysate by *DC*™ protein assay, and load 10 μg of protein in SDS sample buffer to a well of 10% SDS-PAGE.
5. Run the electrophoresis at 100 V voltage for 1.5 h. Electrotransfer the proteins onto polyvinylidene fluoride membranes.
6. Block the membrane with 5% skim milk, and incubate the membrane with primary antibodies overnight at 4 °C.
7. Wash the membrane, and apply secondary antibody conjugated with horseradish peroxidase. Incubate the membrane at room temperature for 1 h.
8. Wash the membrane and add enhanced chemiluminescence reagent (GE Healthcare Life Sciences). Incubate the membrane in room temperature for 5 min.
9. Develop the signal to a film.

3.8 Cell Apoptosis Assay

1. Transfer 1 × 10^5 cells to a flow tube.
2. Wash cells once with PBS and then once with 1× binding buffer at 800 rpm for 5 min at 4 °C.
3. Incubate 1 × 10^5 cells with 5 μL of FITC annexin V and 2.5 μL of propidium iodide for 15 min at room temperature in dark.

4. Add 200 μL of 1× binding buffer to the cells and subject to flow cytometric analysis in FACSCalibur (BD Biosciences) within 1 h.

3.9 Flow Cytometric Analysis

1. Wash the cells with PBS. Every reaction needs ~0.5–1.5 × 10^5 cells, which varies for the numbers of markers included.
2. For intracellular markers, permeabilize the cells with 2 mL ice-cold 0.1% saponin in PBS. Incubate the cells on ice for 20 min and vortex every 10 min. For surface markers, skip this step and proceed to **step 3**. For co-staining of both intracellular and surface markers, stain the surface markers prior to permeabilization.
3. Incubate the cells with 200 μL fluorescein-conjugated antibodies or equal amount of corresponding isotypic control antibodies with concentrations recommended by the manufacturers. For detection of intracellular markers, use 0.1% saponin in PBS to dilute the antibodies and to wash the cells.
4. Subject to flow cytometric analysis in FACSCalibur (BD Biosciences).

3.10 Colony Formation

1. Take 1000 viable cells released from the tumor and seed in a 6-well, and allow them to grow for 10 days.
2. Remove the medium and rinse the cells with 3 mL PBS for three times. Add 3 mL fixative and stand at room temperature for 10 min.
3. Remove the fixative and rinse the cells with 3 mL PBS for three times. Add 3 mL crystal violet solution and incubate for 10 min at room temperature.
4. Remove the crystal violet solution, and rinse the cells with 3 mL PBS for three times. Count the colonies under the microscope in five fields per well.

3.11 Spheroid Formation

1. Take 1000 viable cells released from the tumor, and seed in a 6-well of an ultra-low attachment plate with serum-free supplemented DMEM/F12 medium.
2. Replenish with 100 μL of supplemented medium every other day.
3. Count the spheroids (≥50 μm) under the microscope after 1 month.

4 Notes

1. In addition to nude mouse, severe combined immunodeficiency (SCID) mouse may also be used. Note that SCID mouse should be shaved at the region of surgical operation before tumor inoculation. Their hair generally does not affect the bioluminescence.
2. For subcutaneous tumor inoculation, Matrigel™ (BD Biosciences, high concentration, standard growth factors) could be used to enhance the rate of forming xenograft tumor, especially for those cell lines which are not highly tumorigenic. When using Matrigel™, mix 1:1 with the cells in plain medium and keep on ice. Injection of the cell number can be up to 1×10^7 in a maximum final volume of 300 μL.
3. For the orthotopic implantation, the rate of establishing tumors with a suitable size (ROI > 1×10^6 p/s/cm^2/sr) in 2 weeks is usually above 70%. To ensure sufficient animals per group, inoculate 50% extra mice for the target number of mice for the whole study. For example, if one needs 10 mice for a study, one would need to inoculate 15 mice, with 70% take rate; thus it would result in 10–12 mice available in 2 weeks with comparable tumor size for the study.
4. The intraperitoneal injections of mAb, chemotherapeutics, and pentobarbitone/luciferin should be spread throughout a week to prevent interference from each other. For example, injection of mAb could be scheduled on Monday, cisplatin on Wednesday, and pentobarbitone/luciferin on Friday. And repeat this cycle from next Monday. This would avoid the mice from overstressing and provide enough time for the absorption or clearance of the reagents before injection of another one. Injections could be scheduled in the morning that the mice are monitored afterwards for unexpected effects of the injections.
5. Injection of pentobarbitone and luciferin should be performed at the same time in two shots. The whole procedure should be performed quickly to avoid difference among mice and across groups. Both pentobarbitone and luciferin take effect about 10–15 min after injections. At this time, the mice should be constantly under anesthesia and will show relatively constant bioluminescent signals for 15–20 min.
6. In our experience, intraperitoneal injections are suitable for both subcutaneous xenograft and orthotopic xenograft on liver. The mAb from peritoneal cavity could reach both tumor sites, demonstrated by our previous experiment where biotinylated mAb injected through i.p. was detected from both

subcutaneous and orthotopic xenografts by immunohistochemistry (unpublished data).

7. From the blood taken from Subheading 3.5, serum or plasma can be extracted for the detection of the change of the target molecule, if necessary.
8. Formalin-fixed xenograft tumor tissue collected from Subheading 3.5 can be analyzed by immunohistochemistry for apoptosis and proliferative index with in situ cell death detection kit POD (Roche Diagnostics, Indianapolis, IN, USA) and Ki-67 staining (BD Bioscience), respectively.

Acknowledgments

This study was supported in part by Hong Kong Research Grants Council (GRF 764111 and 764112) and Health and Medical Research Fund (01121566).

References

1. Cheung PF, Yip CW, Wong NC et al (2014) Granulin-epithelin precursor renders hepatocellular carcinoma cells resistant to natural killer cytotoxicity. Cancer Immunol Res 2:1209–1219
2. Scott AM, Wolchok JD, Old LJ (2012) Antibody therapy of cancer. Nat Rev Cancer 12:278–287
3. Pujade-Lauraine E, Hilpert F, Weber B et al (2014) Bevacizumab combined with chemotherapy for platinum-resistant recurrent ovarian cancer: the AURELIA open-label randomized phase III trial. J Clin Oncol 32:1302–1308
4. Tan AR, Swain SM (2003) Ongoing adjuvant trials with trastuzumab in breast cancer. Semin Oncol 30:54–64
5. Van Cutsem E, Kohne CH, Hitre E et al (2009) Cetuximab and chemotherapy as initial treatment for metastatic colorectal cancer. N Engl J Med 360:1408–1417
6. Ho JC, Ip YC, Cheung ST et al (2008) Granulin-epithelin precursor as a therapeutic target for hepatocellular carcinoma. Hepatology 47:1524–1532
7. Wong NC, Cheung PF, Yip CW et al (2014) Antibody against granulin-epithelin precursor sensitizes hepatocellular carcinoma to chemotherapeutic agents. Mol Cancer Ther 13:3001–3012
8. Jenkins DE, Hornig YS, Oei Y et al (2005) Bioluminescent human breast cancer cell lines that permit rapid and sensitive in vivo detection of mammary tumors and multiple metastases in immune deficient mice. Breast Cancer Res 7:R444–R454
9. Zabala M, Alzuguren P, Benavides C et al (2009) Evaluation of bioluminescent imaging for noninvasive monitoring of colorectal cancer progression in the liver and its response to immunogene therapy. Mol Cancer. https://doi.org/10.1186/1476-4598-8-2
10. Tomayko MM, Reynolds CP (1989) Determination of subcutaneous tumor size in athymic (nude) mice. Cancer Chemother Pharmacol 24:148–154
11. Euhus DM, Hudd C, LaRegina MC et al (1986) Tumor measurement in the nude mouse. J Surg Oncol 31:229–234
12. Weir GM, Liwski RS, Mansour M (2011) Immune modulation by chemotherapy or immunotherapy to enhance cancer vaccines. Cancers 3:3114–3142
13. Correale P, Botta C, Cusi MG et al (2012) Cetuximab +/− chemotherapy enhances dendritic cell-mediated phagocytosis of colon cancer cells and ignites a highly efficient colon cancer antigen-specific cytotoxic T-cell response in vitro. Int J Cancer 130:1577–1589
14. Ramakrishnan R, Huang C, Cho HI et al (2012) Autophagy induced by conventional chemotherapy mediates tumor cell sensitivity to immunotherapy. Cancer Res 72:5483–5493

15. Radfar S, Wang Y, Khong HT (2009) Activated CD4+ T cells dramatically enhance chemotherapeutic tumor responses in vitro and in vivo. J Immunol 183:6800–6807
16. Shabbits JA, Hu Y, Mayer LD (2003) Tumor chemosensitization strategies based on apoptosis manipulations. Mol Cancer Ther 2:805–813
17. Llovet JM, Ricci S, Mazzaferro V et al (2008) Sorafenib in advanced hepatocellular carcinoma. N Engl J Med 359:378–390
18. Cheng AL, Kang YK, Chen Z et al (2009) Efficacy and safety of sorafenib in patients in the Asia-Pacific region with advanced hepatocellular carcinoma: a phase III randomised, double-blind, placebo-controlled trial. Lancet Oncol 10:25–34
19. Thomas MB, O'Beirne JP, Furuse J et al (2008) Systemic therapy for hepatocellular carcinoma: cytotoxic chemotherapy, targeted therapy and immunotherapy. Ann Surg Oncol 15:1008–1014
20. Yeo W, Mok TS, Zee B et al (2005) A randomized phase III study of doxorubicin versus cisplatin/interferon alpha-2b/doxorubicin/fluorouracil (PIAF) combination chemotherapy for unresectable hepatocellular carcinoma. J Natl Cancer Inst 97:1532–1538
21. Chen X, Cheung ST, So S et al (2002) Gene expression patterns in human liver cancers. Mol Biol Cell 13:1929–1939
22. Cheung ST, Chen X, Guan XY et al (2002) Identify metastasis-associated genes in hepatocellular carcinoma through clonality delineation for multinodular tumor. Cancer Res 62:4711–4721
23. Cheung ST, Wong SY, Lee YT et al (2006) GEP associates with wild-type p53 in hepatocellular carcinoma. Oncol Rep 15:1507–1511
24. Cheung ST, Wong SY, Leung KL et al (2004) Granulin-epithelin precursor overexpression promotes growth and invasion of hepatocellular carcinoma. Clin Cancer Res 10:7629–7636
25. Yip CW, Cheung PF, Leung IC et al (2014) Granulin-epithelin precursor interacts with heparan sulfate on liver cancer cells. Carcinogenesis 35:2485–2494
26. Cheung PF, Cheng CK, Wong NC et al (2011) Granulin-epithelin precursor is an oncofetal protein defining hepatic cancer stem cells. PLoS One 6:e28246
27. Cheung ST, Cheung PF, Cheng CK et al (2011) Granulin-epithelin precursor and ATP-dependent binding cassette (ABC)B5 regulate liver cancer cell chemoresistance. Gastroenterology 140:344–355

Chapter 11

Methods to Analyze the Role of Progranulin (PGRN/GEP) on Cancer Stem Cell Features

Phyllis F. Y. Cheung and Siu Tim Cheung

Abstract

Emerging evidence suggests that tumors are hierarchically organized with a distinct subpopulation called cancer stem cells (CSCs) lying at the apex of the hierarchy. These cells are not only responsible for tumor initiation and progression but also endowed with stem cell properties, including self-renewal, chemoresistance, and tumor initiation. Although existing therapies can initially eliminate the bulk population of tumor, the stem cell properties of CSCs enable them to survive and repopulate the tumor, resulting in disease relapse. Recently, our group has shown that progranulin (PGRN/GEP) defined a hepatic cancer stem cell subpopulation in hepatocellular carcinoma. This subpopulation demonstrated enhanced ability for colony and spheroid formation, chemoresistance, and tumor initiation. In this chapter, we describe the methods used to isolate the progranulin+ subpopulation and analyze their CSC properties.

Key words Cancer stem cell, Tumor-initiating cell, Chemoresistance, Granulin-epithelin precursor, Self-renewal

1 Introduction

Hepatocellular carcinoma (HCC) is the fifth most common cancer worldwide with high mortality rate and very poor prognosis [1]. The majority of HCC patients are diagnosed at advanced stage, which precludes curative treatment such as surgical resection and liver transplantation. Chemotherapy is the second-line treatment for HCC, but overall response rate is unsatisfactory due to the highly chemoresistant nature of the tumor [1]. HCC is considered a heterogeneous disease due to its multiple molecular profiles and varied clinical outcomes [2]. The heterogeneous nature of HCC and the lack of appropriate biomarkers have impeded patient prognosis and treatment.

For years, tumor cell heterogeneity has been explained by the clonal evolution model [3]. However, recent evidence suggests that tumors are hierarchically organized with a distinct subpopulation called cancer stem cells (CSCs) lying at the apex of the

Andrew Bateman et al. (eds.), *Progranulin: Methods and Protocols*, Methods in Molecular Biology, vol. 1806,
https://doi.org/10.1007/978-1-4939-8559-3_11,

hierarchy. CSCs are considered as the "root" of cancers, which are not only responsible for tumor progression but also endowed with stem cell properties, including self-renewal, chemoresistance, and tumor-initiating ability [4]. Although existing therapies usually could eliminate the bulk population of tumor initially, some of the cells could survive, repopulating the tumor and resulting in disease relapse [5]. The properties of CSCs partly explained the phenomena. The concept of CSCs has been documented in diverse malignancies including breast [6], colon [7], and HCC [8–11]. The identification, isolation, and characterization of hepatic CSC are therefore critical for improving the clinical management and treatment of HCC.

Progranulin (PGRN/GEP) is a pluripotent growth factor regulating fetal development, tissue repair, and tumorigenesis in various cancers [12]. Our group demonstrated that progranulin was overexpressed in over 70% of human HCC, and its expression was associated with aggressive HCC features in patients. Functional studies revealed its role in regulating HCC cell proliferation, invasion, tumorigenicity, and chemoresistance [13–15]. Moreover, neutralization of progranulin could inhibit the growth of established HCC in mouse model [16]. Recently, we have demonstrated that progranulin was a hepatic oncofetal protein defining a CSC population in HCC [10]. Progranulin-expressing cells possessed greater ability in colony and spheroid formation, chemoresistance, and enhanced tumor-initiating ability in immunodeficient mice.

Defining properties of cancer stem cells include self-renewal ability, chemoresistance, and tumor-initiating ability. Self-renewal is a hallmark of stem cell. There are two main in vitro assays used to assess self-renewal, the colony formation and spheroid formation assays. Colony formation assay involves seeding for a small number of cells onto culture plates and monitoring colony formation after a defined time period. This assay is based on the assumptions that only cells that are able to self-renew can form large colonies from a small titers of cells [17]. The number of colonies present is correlated with the self-renewal capacity of the cells. Spheroid formation assay involves the culture of cells in a condition in which stem cells are preferentially maintained. Clonal multicellular spheroid aggregates form in low adherence culture plates with defined serum-free medium. This method is now widely used for the enrichment of CSCs from many types of cancer, including HCC. Spheroids originate from a small fraction of the tumor cell population and can be serially passaged [18, 19], allowing investigation for self-renewal capacity and further enrichment of CSCs.

Resistance to standard chemotherapeutic agents is one of the most clinically relevant properties of CSCs. Chemoresistance in CSCs can result from several distinct mechanisms, including alterations of tumor cell cycle checkpoints, impairment of tumor apoptotic pathways, repair of damaged cellular targets, and reduced drug accumulation in tumor cells [5, 20]. Among those, decreased intracellular drug accumulation is a well-studied mechanism of chemoresistance and has been shown to result at least partially from tumor cell expression of the ATP-dependent drug efflux transporters [20]. To assess chemoresistance, drug accumulation and apoptosis of cells upon exposure to chemotherapeutic agents are measured. Cells showing low level of intracellular drug accumulation and apoptosis are considered to be chemoresistant and thereby possess CSC property. Doxorubicin is a chemotherapeutic agent commonly used in HCC. Due to its intrinsic fluorescence (emission maximum at 560–590 nm), intracellular doxorubicin accumulation can be detected by flow cytometry at FL2 spectrum and can therefore be used for drug accumulation assay.

Serial transplantation in animal models is the gold standard assay to evaluate both self-renewal and tumor-initiating ability of CSCs [21]. In transplantation assays, low titers of cells are injected into immunodeficient mice. Cells that are able to form primary tumors at low titers are proposed to have CSC phenotypes [21, 22]. To demonstrate self-renewal ability, cells are isolated from the primary tumors and then transplanted into a second recipient animal, i.e., serial transplantation [21, 23]. Matrigel has been commonly used in xenotransplantation assays. As a reconstituted matrix, it was primarily used to study invasion of tumor cells [24]. It was later found to dramatically facilitate tumorigenesis of various cancer cells [25, 26]. However, it is criticized that Matrigel may provide tumor cells with additional proliferative or survival signals that they may not encounter in their usual microenvironment. Despite the potentially more permissive environment provided by Matrigel, it is still agreed that only CSCs, but not the negative counterpart from primary tumors, are able to form tumors in recipient mice.

In this chapter, the methods used to isolate and analyze the above cancer stem cell properties of progranulin+ subpopulation are described.

2 Materials

2.1 Cell Sorting Components

1. Anti-progranulin antibody for cell sorting: FITC-conjugated mouse antihuman progranulin antibody [16]. Store at 4 °C.
2. Anti-FITC microbeads (Miltenyi Biotec Inc., Bergisch Gladbach, Germany). Store at 4 °C.
3. LS columns (Miltenyi Biotec Inc.).
4. Cell sorting buffer: PBS supplemented with 2% BSA.

2.2 Post-sorting Analysis Components

1. Trypan blue (Life Technologies, Carlsbad, CA) for assessing cell viability.
2. Anti-progranulin antibody for post-sorting analysis: mouse antihuman progranulin antibody (R&D systems, Minneapolis, MN). Store at 4 °C.
3. PE-conjugated anti-mouse secondary antibody (Dako, Carpinteria, CA).
4. 0.1% saponin. Dissolve 0.5 g saponin (Merck Millipore, Billerica, MA) in PBS. Use within 1 month. Store at 4 °C.

2.3 Colony Formation Components

1. Advanced Minimum Essential medium (AMEM) supplemented with 1% FBS (Life Technologies). Store at 4 °C.
2. Fixative: 15 mL glacial acetic acid, 40 mL ethanol, 445 mL H_2O.
3. Crystal violet solution (Sigma-Aldrich, St. Louis, MO).

2.4 Spheroid Formation Components

1. Medium

   Dulbecco's Modified Eagle's medium/nutrient mixture F-12 Ham (DMEM/F12) medium (Life Technologies), supplemented with 20 ng/mL human recombinant EGF (Life Technologies), 10 ng/mL human recombinant bFGF (Life Technologies), 4 μg/mL insulin (Sigma-Aldrich), B27 (1:50) (Life Technologies). Freshly prepare.
2. Ultra-low attachment plate (Corning, Tewksbury, MA).

2.5 Chemoresistance Components

1. AMEM supplemented with 1% FBS (Life Technologies). Store at 4 °C.
2. Doxorubicin.
3. FITC-conjugated annexin V (BD Biosciences, San Jose, CA). Store at 4 °C.
4. 10× Annexin V Binding Buffer (BD Biosciences). 0.1 M Hepes/NaOH (pH 7.4), 1.4 M NaCl, 25 mM $CaCl_2$. Store at 4 °C. For 1× working solution, dilute one part of the 10× Annexin V Binding Buffer with nine parts of distilled water.

2.6 In Vivo Tumorigenicity Components

1. BALB/c athymic nude mice.
2. Roswell Park Memorial Institute (RPMI) 1640 medium (Life Technologies). Store at 4 °C.
3. Matrigel™ (BD Biosciences). Store in single-use aliquots at −20 °C; thaw on ice before use.
4. 1 mL syringe with 23-gauge needle.
5. Isoflurane.

2.7 Serial Transplantation of Xenograft Tumors

1. Type IV collagenase (Sigma-Aldrich).
2. AMEM (Life Technologies). Store at 4 °C.
3. 40 μm cell strainer (BD Biosciences).
4. ACK lysing buffer: 0.15 M NH_4Cl, 10 mM $KHCO_3$, 0.1 mM Na_2EDTA, H_2O, pH 7.2–7.4.

3 Methods

3.1 Cell Sorting

1. Harvest cells and centrifuge them at 800 rpm for 5 min at 4 °C.
2. Resuspend the cells in cell sorting buffer and count the cells.
3. Resuspend 1×10^7 cells in 100 μL cell sorting buffer, and incubate the cells with 10 μL FITC-conjugated mouse antihuman progranulin antibody per 1×10^6 cells for 45 min at 4 °C in dark (*see* **Note 1**).
4. Wash the cells once with cell sorting buffer at 800 rpm for 5 min at 4 °C.
5. Resuspend 1×10^7 cells in 100 μL cell sorting buffer and incubate with 10 μL anti-FITC magnetic microbeads per 1×10^7 cells for 30 min at 4 °C in dark.
6. Wash the cells once with cell sorting buffer at 800 rpm for 5 min at 4 °C.
7. Resuspend 1×10^7 cells in 1 mL cell sorting buffer.
8. Place column in the magnetic field of a suitable MACS Separator.
9. Prepare column by rinsing with 2 mL cell sorting buffer.
10. Apply cell suspension onto the column.
11. Collect the flow-through as the negative control fraction.
12. Wash the column with 5 mL cell sorting buffer.
13. Remove column from the separator, and place it on a suitable collection tube.
14. Add 5 mL of buffer onto the column. Immediately flush out fraction with the magnetically labeled cells by firmly applying the plunger, supplied with the column.
15. After cell isolation, assess the cell viability by trypan blue staining and count the cells (*see* **Note 2**).
16. Collect 1×10^6 cells from each sorted population to assess purity by flow cytometry.
17. Centrifuge the cells at 800 rpm for 5 min at 4 °C.
18. Permeabilize the cells with 1 mL 0.1% saponin for 20 min, on ice.
19. Centrifuge the cells at 800 rpm for 5 min at 4 °C.

20. Incubate the cells with 10 μL mouse antihuman progranulin antibody per 1×10^6 cells for 45 min at 4 °C.
21. Wash the cells with 4 mL 0.1% saponin at 800 rpm for 5 min at 4 °C.
22. Incubate the cells with 1 μL PE-conjugated rabbit anti-mouse antibody for 30 min at 4 °C in dark.
23. Wash the cells with 4 mL 0.1% saponin at 800 rpm for 5 min at 4 °C.
24. Subject the cells to flow cytometric analysis.

3.2 Colony Formation Assay

1. Seed the freshly isolated cells at a density of 1000 cells/6-well in 2 mL of AMEM supplemented with 1% FBS, and allow them to grow for 10 days.
2. Remove the medium and rinse the cells with 3 mL PBS for three times.
3. Add 3 mL fixative to fix the cells for 10 min at room temperature.
4. Remove the fixative and rinse the cells with 3 mL PBS for three times.
5. Add 3 mL crystal violet solution for 10 min at room temperature.
6. Remove the crystal violet and rinse the cells with 3 mL PBS for five times.
7. Count the colonies (≥10 cells) under the microscope in five fields per well.

3.3 Spheroid Formation Assay

1. Seed the freshly isolated cells at a density of 1000 cells/6-well in 2 mL serum-free supplemented DMEM/F12 medium and seeded into ultra-low attachment plate.
2. Replenish with 100 μL of supplemented medium every other day.
3. Count the spheroids (≥50 μm) under the microscope after 1 month. An example of spheroids derived from progranulin-expressing cells is shown in Fig. 1.

3.4 Chemoresistance Assays

3.4.1 Doxorubicin Accumulation

1. Treat the 5×10^4 cells with or without doxorubicin at 0.5 μg/mL in 0.5 mL of AMEM supplemented with 1% FBS for 24 h.
2. Subject the cells to flow cytometry and detect the intracellular doxorubicin accumulation at FL2 spectrum.

3.4.2 Apoptosis

1. Treat 5×10^4 cells with or without doxorubicin at 0.5 μg/mL in 0.5 mL of AMEM supplemented with 1% FBS for 24 h.
2. Wash cells once with 4 mL PBS and then once with 4 mL 1× Annexin V Binding Buffer at 800 rpm for 5 min at 4 °C.

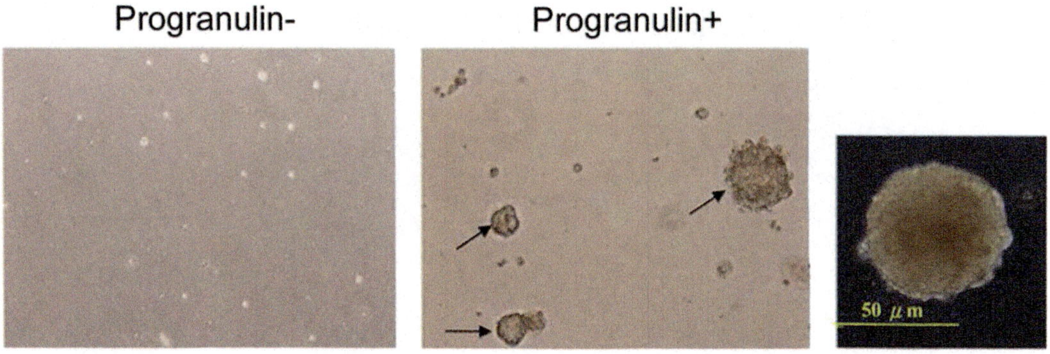

Fig. 1 Spheroid culture. Spheroids derived from progranulin- and progranulin+ cells after 1 month culture. The right panel shows the size of a spheroid derived from progranulin+ cells

3. Incubate 5×10^4 cells with 5 μL of FITC Annexin V for 15 min at room temperature in dark.
4. Add 400 μL of 1× Annexin V Binding Buffer to the cells and subject to flow cytometric analysis within 1 h.

3.5 In Vivo Tumorigenicity Assays

3.5.1 Transplantation of Sorted Cells into Immunodeficient Mice (See Note 3)

1. Wash the freshly isolated cells with 20 mL RPMI medium twice at 800 rpm for 5 min at 4 °C.
2. Resuspend the cells in 1 mL RPMI medium and count the cells.
3. Thaw Matrigel on ice and mix with cell suspension at 1:1 ratio.
4. Inject the cell suspension subcutaneously into the right and left flanks using a 23-gauge needle. The injection volume should be 100 μL for each side.
5. Monitor tumor growth. An example of tumor formation from isolated progranulin+ cells is shown in Fig. 2.

3.5.2 Serial Transplantation of Xenograft Tumors

1. Dissect the tumor-bearing mice and collect the xenograft tumor.
2. Cut the tumors with sterile surgical blades into approximately 1 mm^3 cubes.
3. Digest the tumors with type IV collagenase for 5 min at 37 °C.
4. Centrifuge the mixture at $50 \times g$ for 5 min at 4 °C.
5. Collect the supernatants as cells released from the tumors.
6. Repeat **steps 3** and **4** twice.
7. Filter the supernatants through a 40 μm cell strainer.
8. Centrifuge the cells at 800 rpm for 5 min at 4 °C.
9. Remove the supernatant.
10. Lyse the red blood cells with 10 mL ACK lysing buffer for 1 min at room temperature.
11. Add 20 mL AMEM to the mixture and centrifuge the cells at 800 rpm for 5 min at 4 °C.

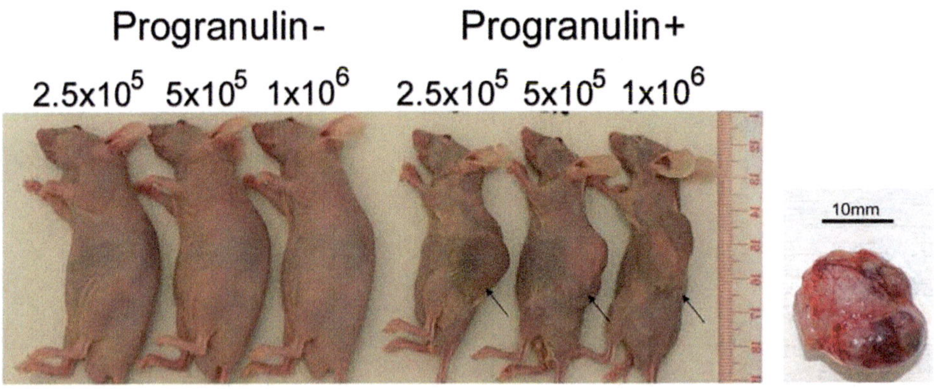

Fig. 2 Tumorigenicity assay. Nude mice injected subcutaneously with progranulin– and progranulin+ cells after 8 weeks. The right panel shows the subcutaneous tumors derived from 2.5×10^5 progranulin+ cells. Reproduced from PLoS One [10]

12. Assess the cell viability by trypan blue exclusion staining and count cells.
13. Proceed to cell sorting (Subheading 3.1) and injected the sorted cells into secondary recipient mice according to Subheading 3.5.1.

4 Notes

1. For cell sorting, cells are sorted based on the surface expression of progranulin, but not on their intracellular expression, because permeabilization is not feasible to keep the cells viable for subsequent functional assays.
2. Up to 10–20% of cells would be nonviable after cell sorting; therefore, viability should be checked before determining cell number.
3. For in vivo tumorigenicity assay, since CSCs represent only a rare population within tumor, the number of cells injected into mice should be further reduced (order of magnitude 10^4 cells or less) to delineate the tumor-initiating efficiency of the cells. In that case, the most sensitive host with highest engraftment rate should be considered, including severe combined immunodeficient (SCID) mice, non-obese diabetic SCID (NOD/SCID) or NOD SCID GAMMA (NSG) mice.

Acknowledgments

This study was supported in part by Hong Kong Research Grants Council (GRF 764111, 764112, and T12-401/13-R), Health and Medical Research Fund (01121566).

References

1. de Lope CR, Tremosini S, Forner A et al (2012) Management of HCC. J Hepatol 56(Suppl 1):S75–S87
2. Thorgeirsson SS, Grisham JW (2002) Molecular pathogenesis of human hepatocellular carcinoma. Nat Genet 31:339–346
3. Nowell PC (1976) The clonal evolution of tumor cell populations. Science 194:23–28
4. Reya T, Morrison SJ, Clarke MF et al (2001) Stem cells, cancer, and cancer stem cells. Nature 414:105–111
5. Dean M, Fojo T, Bates S (2005) Tumour stem cells and drug resistance. Nat Rev Cancer 5:275–284
6. Al-Hajj M, Wicha MS, Benito-Hernandez A et al (2003) Prospective identification of tumorigenic breast cancer cells. Proc Natl Acad Sci U S A 100:3983–3988
7. Ricci-Vitiani L, Lombardi DG, Pilozzi E et al (2007) Identification and expansion of human colon-cancer-initiating cells. Nature 445:111–115
8. Ma S, Chan KW, Hu L et al (2007) Identification and characterization of tumorigenic liver cancer stem/progenitor cells. Gastroenterology 132:2542–2556
9. Yamashita T, Ji J, Budhu A et al (2009) EpCAM-positive hepatocellular carcinoma cells are tumor-initiating cells with stem/progenitor cell features. Gastroenterology 136:1012–1024
10. Cheung PF, Cheng CK, Wong NC et al (2011) Granulin-epithelin precursor is an oncofetal protein defining hepatic cancer stem cells. PLoS One 6:e28246
11. Yang ZF, Ho DW, Ng MN et al (2008) Significance of CD90+ cancer stem cells in human liver cancer. Cancer Cell 13:153–166
12. Bateman A, Bennett HP (2009) The granulin gene family: from cancer to dementia. BioEssays 31:1245–1254
13. Cheung ST, Cheung PF, Cheng KC et al (2011) Granulin-epithelin precursor and ATP-dependent binding cassette (ABC)B5 regulate liver cancer cell chemoresistance. Gastroenterology 140:344–355
14. Cheung ST, Wong SY, Lee YT et al (2006) GEP associates with wild-type p53 in hepatocellular carcinoma. Oncol Rep 15:1507–1511
15. Cheung ST, Wong SY, Leung KL et al (2004) Granulin-epithelin precursor overexpression promotes growth and invasion of hepatocellular carcinoma. Clin Cancer Res 10: 7629–7636
16. Ho JC, Ip YC, Cheung ST et al (2008) Granulin-epithelin precursor as a therapeutic target for hepatocellular carcinoma. Hepatology 47:1524–1532
17. Hellsten R, Johansson M, Dahlman A et al (2011) Galiellalactone inhibits stem cell-like ALDH-positive prostate cancer cells. PLoS One 6:e22118
18. Reynolds BA, Weiss S (1992) Generation of neurons and astrocytes from isolated cells of the adult mammalian central nervous system. Science 255:1707–1710
19. Celia-Terrassa T, Meca-Cortes O, Mateo F et al (2012) Epithelial-mesenchymal transition can suppress major attributes of human epithelial tumor-initiating cells. J Clin Invest 122:1849–1868
20. Gottesman MM, Fojo T, Bates SE (2002) Multidrug resistance in cancer: role of ATP-dependent transporters. Nat Rev Cancer 2:48–58
21. Visvader JE, Lindeman GJ (2008) Cancer stem cells in solid tumours: accumulating evidence and unresolved questions. Nat Rev Cancer 8:755–768
22. Quintana E, Shackleton M, Sabel MS et al (2008) Efficient tumour formation by single human melanoma cells. Nature 456:593–598
23. Clarke MF, Fuller M (2006) Stem cells and cancer: two faces of eve. Cell 124:1111–1115
24. Bissell MJ, Labarge MA (2005) Context, tissue plasticity, and cancer: are tumor stem cells also regulated by the microenvironment? Cancer Cell 7:17–23
25. Mehta RR, Graves JM, Hart GD et al (1993) Growth and metastasis of human breast carcinomas with Matrigel in athymic mice. Breast Cancer Res Treat 25:65–71
26. Henson B, Li F, Coatney DD et al (2007) An orthotopic floor-of-mouth model for locoregional growth and spread of human squamous cell carcinoma. J Oral Pathol Med 36:363–370

Chapter 12

Methods to Study the Role of Progranulin in the Tumor Microenvironment

Moshe Elkabets and Samuel Brook

Abstract

Accurate measurement of progranulin (PGRN) in the circulation and in the tumor microenvironment is essential for understanding its role in cancer progression and metastasis. This chapter describes a number of approaches to measure the transcription level of the *GRN* gene and to detect and analyze PGRN expression in cancer cells and in the local environment of the tumor, in mouse and human samples. These validated protocols are utilized to investigate the functional role of PGRN in cancer. Finally, we discuss strategies to investigate the functions of PGRN in tumors using genetically modified mouse models and gene silencing techniques.

Key words Progranulin, *GRN*, Tumor microenvironment, Immunohistochemistry, Immunofluorescence, Flow cytometry, qPCR, ELISA

1 Introduction

Cytokines, chemokines, and growth factors are proteins that regulate fundamental biological processes such as cell proliferation, differentiation, and migration of normal and tumor cells. Progranulin (PGRN, also named PC cell-derived growth factor) is a growth factor that plays an important role in developmental biology and in physiological and pathological conditions such as cancer and Alzheimer's disease. Increased expression of PGRN is observed across multiple types of cancer and is associated with poor prognosis in breast [1], liver [2], and ovarian [3] cancer patients. Furthermore, inhibition of this protein has recently been shown to enhance the antitumor efficacy of chemotherapeutic agents in hepatocellular cancer [4]. A growing body of evidence highlights the importance of PGRN in cancer progression and its potential as a therapeutic target [5].

The ability to detect and quantify PGRN and *GRN* in clinical tissue samples or pre-clinical models is required for studying its biological role and exploring its potential as a biomarker and

Andrew Bateman et al. (eds.), *Progranulin: Methods and Protocols*, Methods in Molecular Biology, vol. 1806,
https://doi.org/10.1007/978-1-4939-8559-3_12,

therapeutic target. This chapter provides a number of methods to measure levels of PGRN in cells using common laboratory equipment and reagents. We describe immunohistochemical (IHC) and immunofluorescent (IF) staining techniques to detect and visualize PGRN expression in tissue sections and an intracellular staining (ICS) technique to detect PGRN within cells using flow cytometry. We also provide a protocol for an enzyme-linked immunosorbent assay (ELISA) to measure levels of PGRN secreted into the tumor microenvironment. In addition, we present a protocol that uses Quantitative PCR (qPCR) to detect transcript levels of the *GRN* gene. Finally, we describe several strategies to study the role of PGRN in the tumor microenvironment using genetically engineered mouse models and gene silencing/knockdown.

IHC and IF are immunostaining methods used to detect the presence of specific antigens, e.g., proteins or other molecular markers in a tissue sample (slide-mounted tissue section) [6]. Detection of a target protein is accomplished by exposing tissue samples to a primary antibody tagged with a fluorophore or an enzyme such as peroxidase, which catalyzes a color-producing reaction, allowing the interaction between the antibody and the target antigen to be visualized. Expression patterns and intracellular localization can then be assessed with light or fluorescent microscopy. These assays are useful in studying functional processes in the context of intact cellular architecture as the morphology and tissue structure of the embedded or fixed sample is preserved. These assays are widely used in the clinic as diagnostic and prognostic tools and provide therapeutic information to guide the course of treatment. For example, IHC is used in breast cancer to determine the presence of membrane receptors such as estrogen and progesterone receptor, which serve as biomarkers to determine the suitability of certain molecular targeted therapies. Staining for PGRN in murine or human tissue by IHC or IF enables visualization of localized expression in the cell and in the tissue. Together, these methods of detecting and measuring PGRN are used to gain insight into its functional role in the tumor microenvironment.

ICS is an immunofluorescence-based technique that uses tagged/labeled antibodies to detect intracellular proteins by flow cytometry [7]. It is possible to measure multiple cellular antigens simultaneously with flow cytometry by employing multicolor immunofluorescence staining, a detection technique that allows the investigator to identify the type of cells expressing a protein by staining for specific surface markers as well as intracellular proteins of interest. Using flow cytometry to detect intracellular PGRN in specific cell types is a useful method for studying the heterogeneous population of the tumor microenvironment and presents an opportunity to test the induction or reduction of PGRN expression following stimulus/treatment on a single cell basis. Furthermore, this technique enables analysis of a large number of cells and a large number of samples in a relatively short amount of time. For optimal measurement of secreted factors such as PGRN, blocking the protein transport machinery with transport inhibitors like Brefeldin A is highly recommended.

ELISA is an immuno-enzymatic technique that uses antibodies to capture and quantify proteins of interest by using the antibody sandwich principle [8]. The bases of the wells of a 96-well microplate are coated with a primary antibody that recognizes an epitope/antigen on the protein of interest, thereby immobilizing the protein on the plate for detection and quantification. After all non-specifically bound molecules are washed away, a labeled secondary antibody is added to detect different epitopes on the same protein. In the majority of ELISA assays, detection is mediated by a third antibody bioconjugated to the enzyme horseradish peroxidase (HRP). Upon incubation of the substrate, the enzymatic activity of the peroxidase produces a color-changing reaction, and the rate of color development is proportional to the amount of protein present in each sample and is quantifiable using a spectrophotometer. The specificity and sensitivity of detection is the major advantage of this assay. This method is used to measure levels of circulating PGRN in the plasma of cancer patients and tumor-bearing mice and reflects the amount of PGRN expressed at the tumor site [1, 9].

qPCR is a technique based on the polymerase chain reaction (PCR) that detects and quantifies amplified DNA molecules (amplicons) in real time [10]. The amplicons generated by the PCR reaction are detected and measured by fluorescent dyes such as SYBR green. This technique is generally used to measure the levels of mRNA of the gene of interest by comparing to transcript levels of control reference genes (e.g., housekeeping genes *GAPDH* and *ACTB*). The calculation for relative expression is based on threshold cycle (Ct), which is the number of cycles at which the fluorescence exceeds the logarithmic curve. This methodology was used to investigate the role *GRN* in biological processes such as wound response and inflammation [11, 12].

2 Materials

2.1 Detection of PRGN in Human and Mouse Tissue by Immunohistochemistry (IHC) and Immunofluorescence (IF)

2.1.1 Tissue Fixation and Sample Preparation

Paraffin- or OCT (optimal cutting temperature; frozen)-embedded tissue sections need to be mounted on gelatin-coated histological slides.

1. Manual slide staining system (Thermo Fisher Scientific catalog number A78500001).
2. High-density polyethylene staining jar.
3. Microwave or pressure cooker.
4. De-waxing agent (Xylene).
5. Antigen unmasking/antigen retrieval solution.

6. 10 mM sodium citrate buffer: 2.94 g/L of dH_2O and titrate pH to 6 and add 0.5 mL of 100% Tween 20.
7. Formaldehyde fixative solutions.
   (a) IHC: 85 mM Na_2HPO_4, 75 mM KH_2PO_4, 4% paraformaldehyde.
   (b) IF: 85 mM Na_2HPO_4, 75 mM KH_2PO_4, 4% paraformaldehyde and 14% (v/v) saturated picric acid, pH 6.9.
8. Automation buffer.
   (a) Genetex Immunostain Wash Buffer (catalog number GTX73344).
   (b) PBS (0.137 M NaCl, 0.05 M NaH_2PO; pH 7.4).
9. Matched antibodies pairs: primary antibody (a) for capturing PGRN and secondary antibody (b) for detection.
   (a) Mouse anti-PGRN (Rat IgG) (R&D catalog number MAB25571).
   (b) Human anti-PGRN (Rabbit IgG) (Atlas Antibodies catalog number HPA028747).
10. Incubation buffer or immunostain blocker.
    (a) 1% bovine serum albumin, 1% normal donkey serum, 0.3% Triton® X-100, and 0.01% sodium azide in PBS.
    (b) Genetex Immunostain blocker (catalog number GTX73323).
11. Hydrogen peroxide (1%,v/v) in methanol.
12. Redusol or blocking buffer (1% horse serum in PBS).
13. Slide staining tray.
14. Secondary antibody kit.
    (a) VECTASTAIN Elite ABC Kit—Rat IgG (Vector Laboratories catalog number PK-6104) (for IHC).
    (b) Fluorophore-conjugated anti-Rat IgG antibody (Invitrogen) (for IF).
15. Antifade Mounting Medium and DAPI (for IF): ProLong® Gold Antifade Mountant with DAPI (Invitrogen catalog number P36931).
16. Enhancer reagent—High-Sensitivity Streptavidin-HRP conjugate (Pierce™ catalog number 21130).
17. Chromogen kit—HRP (horseradish peroxidase) substrate for IHC: aminoethylcarbazole kit (AEC) (Invitrogen).
18. Hematoxylin (Vector Laboratories catalog number H3404) for IHC.

19. Mounting media (for IHC) Dako Faramount (Dako, catalog number s3025).
20. PAP water-resistant adhesive pen.
21. Glass coverslips.

2.2 Detection of Intracellular PGRN by Flow Cytometry

1. 96-well V-shaped bottom plates or FACS tubes (or round bottom tube, high clarity polypropylene).
2. FACS buffer—1% FBS/FCS (fetal bovine or calf sera) in PBS.
3. Anti-FcgII/III receptors—antibody blocker.
4. Antibodies.
   (a) Surface markers (e.g., Grl for neutrophils, CD3 for T cells), do not use rat antibodies for surface markers in this assay because of the primary antibody of PGRN.
   (b) Anti-mouse PGRN (Rat IgG) (R&D Systems catalog number MAB25571) (1:200) and anti-Rat Alexa 488 (Invitrogen) (1:500).
5. Intracellular Fixation and Permeabilization Buffer Kit set (eBioscience catalog number 88-8824-00).
6. Dead cell exclusion—Live/Dead Fixable Aqua cell stain (Life Technologies catalog number L34957).

2.3 Enzyme-Linked Immunosorbent Assay (ELISA) to Measure Levels of PGRN in Murine and Human Plasma

1. DuoSet Ancillary Reagent Kits (R&D Systems).
2. 1× PBS- 137 mM NaCl, 2.7 mM KCl, 8.1 mM Na_2HPO_4, 1.5 mM KH_2PO_4, pH 7.2–7.4 (sterile filtered through 0.2 μm pore membrane).
3. Wash buffer—0.05% Tween20 in PBS, pH 7.2–7.4.
4. Blocking buffer—1% BSA in PBS, pH 7.2–7.4.
5. Multichannel pipette.
6. Microplate sealing tape.

To detect mouse PGRN using a kit (such as R&D Systems catalog number DY2557) that includes:

1. Recombinant mouse PGRN.
2. Capture antibody—anti-mouse PGRN primary.
3. Detection antibody—biotinylated—anti-mouse PGRN antibody.
4. Streptavidin-HRP.

To detect human PGRN using a kit (such as R&D Systems catalog number DY2420) that includes:

1. Recombinant human PGRN.
2. Capture antibody—Antihuman PGRN.

3. Detection antibody—Biotin-conjugated antihuman PGRN.
4. Streptavidin-HRP.

2.4 Quantitative Real-Time Polymerase Chain Reaction (qPCR) to Measure Murine GRN Expression in Cells

1. RNeasy Mini Kit (Qiagen).
2. High-Capacity cDNA Reverse Transcription Kits (Applied Biosystems, Thermo Fisher Scientific, catalog number 4368814).
3. UltraPure™ DNase/RNase-free distilled water.
4. MicroAmp® 96-/384-well plates or PCR Strip Tubes, and adhesive films.
5. Probes.
   (a) TaqMan gene expression X20 (Applied Biosystems, Thermo Fisher Scientific).
      - Human *GRN*, catalog number Hs00963707_gl; Human GAPDH, Hs02758991_gl.
      - Mouse *GRN*, catalog number Mm01245914_gl; Mouse GAPDH, Mm03302249_gl.
      - TaqMan Master Mixes (Applied Biosystems, catalog number 4369016).

   (b) SYBR-green.
      - Fast SYBR Green Master Mix (Applied Biosystems, catalog number 4385617).
6. Primers (OriGene).
   (a) Human *GRN* HP225819, Human GAPDH HP205798.
   (b) Mouse *GRN* MP205735, Mouse GAPDH MP205604.

3 Methods

3.1 Detection of Expression and Localization of PRGN in Human and Murine Tissues by Immuno-histochemistry (IHC) and Immuno-fluorescence (IF)

This protocol describes step-by-step instructions on how to stain for PGRN in paraffin-embedded or frozen tissues. See Fig. 1 *for a summary of the IHC/IF protocol.*

Day 1: Tissue Preparation and Primary Antibody Staining

Specimen preparation for paraffin-embedded tissue sections:

1. Confirm proper tissue fixation (*see* **Note 1**).
2. Label slides with a pencil or other marker that is not soluble in ethanol.
3. Prepare the slide staining system with two jars containing fresh xylene and two jars with 100% and 95% ethanol, respectively.
4. Insert slides gently into the slide staining rack (*see* **Note 2**). Dip the rack of slides into the first jar of xylene for 1 min, and

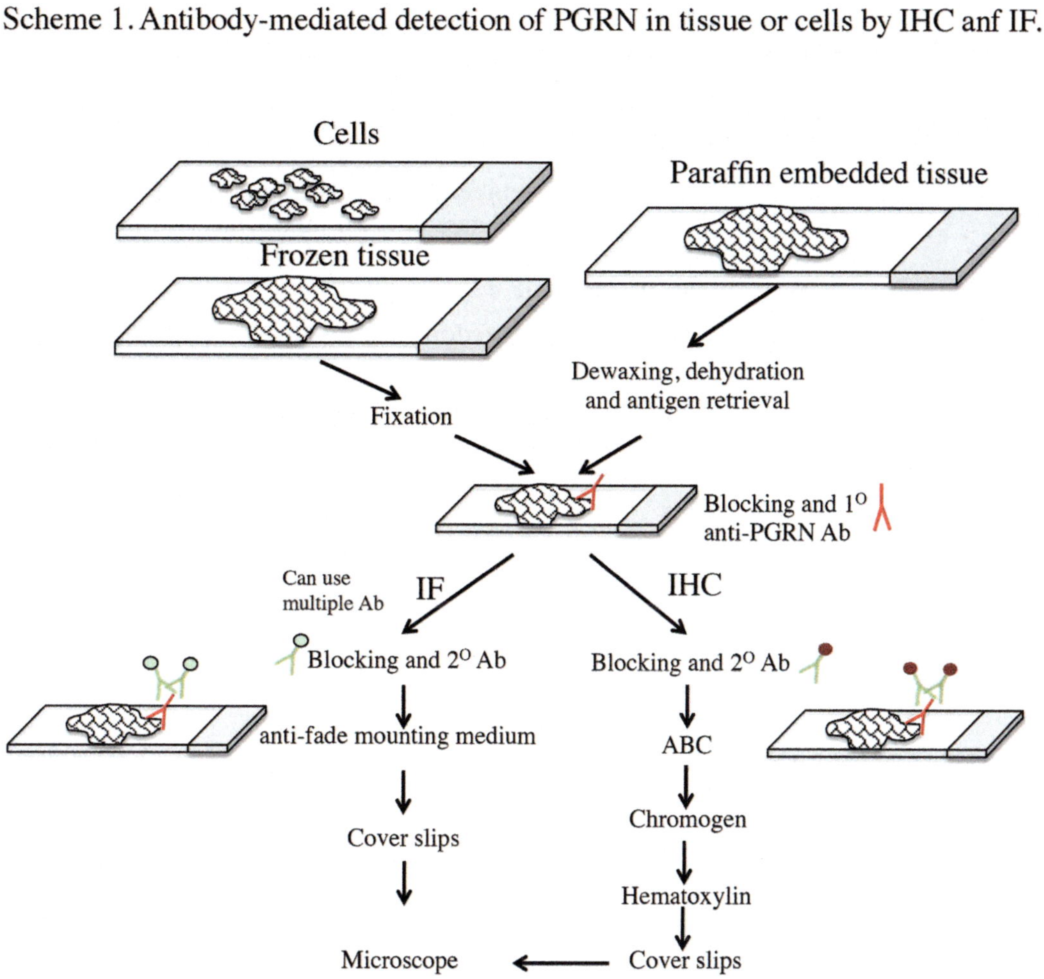

Fig. 1 Antibody-mediated detection of PGRN in tissue or cells by IHC and IF

then transfer the rack into the second jar of xylene for 2 min (*see* **Note 3**). When handling xylene, be sure to work in a fume hood with proper PPE (personal protective equipment).

5. Dip the rack of slides in 100% alcohol for 5 min.
6. Dip the rack of slides in 95% alcohol for 5 min.
7. Dip the rack of slides in deionized H_2O for 5 min.
8. Fill the high-density polyethylene staining jar with 10 mM sodium citrate buffer, and then transfer the slides into the jar and close the lid so the steam can release.
9. Warm the staining jar in a microwave or pressure cooker for 10 min. For microwaves, place the staining jar inside a Pyrex dish filled with 0.5 L of H_20, and heat at maximum power for 20 min or until the water boils. For pressure cooker, follow the

pressure cooker manufacturer's instructions and cook the staining jar for 3 min.

10. Continue to Subheading 3.1.1

Specimen preparation for frozen sections:

1. Tissue was cryo-embedded in OCT media and sectioned with a cryostat at a thickness of 5–10 μm. Serial sections were collected onto gelatin-coated histological slides and then stored at −80 °C (*see* **Note 4**).
2. Remove cryostat sections from −80 °C freezer and allow to return to room temperature.
3. Add 100 μL of cold fixation buffer (or enough to cover all tissue) onto each slide, and incubate slides in fixation buffer and incubate at −20 °C for 20 min or at 4 °C for 10 min (*see* **Note 5**).
4. Once tissue sections are properly prepared and fixed, proceed to Subheading 3.1.1 for IHC or Subheading 3.1.2 for IF.

3.1.1 PGRN Detection by Immunohistochemistry (IHC)

IHC Protocol—Day 1 (cont'd): Staining with anti-PGRN primary antibody

1. Use the PAP-pen to draw a border on the gelatin-coated slide outlining sections. The marking from the PAP-pen creates a hydrophobic barrier around the sections and minimizes loss/waste of antibodies and staining reagents.
2. Add 50 μL Redusol (or other blocking buffer) and incubate at 40 °C for 2 min or at room temperature for 30 min. Incubating in blocking buffer is a critical step in the staining protocol because it prevents non-specific staining/binding between the primary antibodies and the tissue.
3. Immerse slides into automation buffer five times.
4. Add 100 μL of hydrogen peroxide and incubate slides at 40 °C for 2 min or at room temperature for 10 min. This process is needed to quench the endogenous peroxidase.
5. Immerse slides into automation buffer five times.
6. Add 50 μL blocking buffer onto each slide and incubate at 40 °C for 2 min or at room temperature for 10 min.
7. To prepare the primary PGRN antibody, add 1 μL of antibody for every 50 μL of blocking buffer and add 50–100 μL of diluted primary antibody onto the tissue sections (*see* **Note 6**).
8. Place the slides into a staining tray and cover with a wet towel paper or 10 mL of water and incubate overnight at 4 °C (*see* **Notes 7** and **8**).

IHC Protocol—Day 2: Secondary Antibody Incubation and PGRN Visualization

1. Immerse slides in automation buffer five times.
2. Prepare the secondary antibody in blocking buffer by adding three drops of blocking serum to 10 mL of automation buffer in the mixing bottle (yellow label). Shake well (*see* **Note 9**).
3. Add one drop (50 μL) of the biotinylated antibody to the 10 mL of premade blocking buffer and mix well. Adjust the amount of biotinylated antibody based on the volume of blocking buffer.
4. Add 50–100 μL of the diluted secondary antibody onto each tissue, and incubate the slides for 1 h at room temperature or at 4 °C for 2 h.
5. When there are 30 min remaining in the secondary antibody incubation, prepare the VECTASTIN ABC solution: add 5 mL of Genetex Immunostain Wash Buffer into the ABC reagent mixing bottle, and then add two drops of REAGENT A and two drops of REAGENT B into the ABC reagent mixing bottle and mix by inverting several times. Allow mixing bottle to stand at room temperature for 30 min.
6. Wash slides by immersing in automation buffer 5× for 5 min per wash (*see* **Note 10**).
7. Add two drops of ABC solution onto each tissue section and incubate for 30 min at room temperature.
8. Wash slides by immersing in automation buffer five times for 2 min each.
9. Add one drop of enhancer solution on the tissues for 2 min. (The enhancer solution amplifies the detection signal of PGRN using the streptavidin-HRP conjugate peroxidase enhancer.) Then, immerse slides in automation buffer 5× for 2 min per wash.
10. Prepare the chromogen solution: To a 1.5 mL Eppendorf tube, add 1 mL of dH_2O, and then add one drop of Reagent A and invert the tube several times to mix. Then, add one drop of Reagent B and Reagent C and mix again (*see* **Note 11**).
11. Add 100 μL of the chromogen solution onto the tissue sections, and incubate at room temperature protected from light. Monitor the staining by checking the sections every 3–5 min until brown color appears (*see* **Note 12**).
12. Immerse the slides in automation buffer5× for 2 min. After the final wash, rinse the slides with dH_2O to remove any remaining buffer or staining reagents.

13. Remove any excess water from slides, and then incubate sections with hematoxylin for 5–45 s. Then dip the slides in water for 10 s, and repeat this step until the water is colorless. (This nuclear counterstaining step improves the visualization of the tissue morphology by making each cell's nucleus more easily identifiable.)
14. Tap gently on the edge of the slide to remove all drops of water, and place slides (face up) on a paper towel for at least 10 min until slides/sections have dried completely.
15. Prepare the supplies for mounting and open the coverslip box.
16. Add 2–3 drops of mounting media onto each slide, and carefully place the coverslip over the tissue section(s). Allow the slides to remain flat overnight before stacking and storing away the slides (*see* **Note 13**).
17. Proceed to PGRN visualization by microscope.

3.1.2 PGRN Detection by Immunofluorescence (IF)

IF Protocol—Day 1 (cont'd): Staining with Primary Antibodies

1. Use the PAP-pen to draw a border on the gelatin-coated slide outlining sections. The marking from the PAP-pen creates a hydrophobic barrier around the sections and minimizes loss/waste of antibodies and staining reagents.
2. Add Redusol or (other blocking buffer) and incubate at 40 °C for 2 min or at room temperature for 30 min. This is a critical step in the staining protocol because it prevents non-specific staining/binding of the primary antibodies to the tissue.
3. Immerse the slides into automation buffer five times.
4. Add 100 μL of hydrogen peroxide, and incubate slides at 40 °C for 2 min or at room temperature for 10 min. This process is needed to quench endogenous peroxidase.
5. Immerse slides into automation buffer five times.
6. Add 50 μL blocking buffer onto slides, and incubate at 40 °C for 2 min or at room temperature for 10 min.
7. To prepare the primary PGRN antibody, add 1 μL of antibody for every 50 μL of blocking buffer, and add 50–100 μL of diluted primary antibody onto the tissue sections (*see* **Note 14**).
8. To prepare the antibodies that recognize the cell surface, add 1–2 μL (follow the manufacturer's recommendation) of antibody for every 50 μL of blocking buffer, and add 50–100 μL of diluted primary antibody onto the tissue sections (*see* **Note 15**).
9. Place the slides into a staining tray and cover with a wet paper towel or 10 mL of water and incubate overnight at 4 °C (*see* **Notes 16** and **17**).

IF Protocol—Day 2: Secondary Antibody Incubation and Visualization of PGRN Expression

1. Prepare a 1:200 dilution of fluorophore-conjugated anti-rat secondary antibody (to detect PGRN) in dilution buffer.
2. Add 100 μL of diluted secondary antibody onto the tissue, and incubate for 30 min at room temperature (*see* **Note 18**).
3. Immerse slides in automation buffer five times.
4. Add 2–3 drops of antifade mounting medium onto the sections ensuring the entire tissue is covered by the mounting medium, and carefully place a coverslip onto the slide.
5. Leave the slides with mounting medium in −20 °C for several hours.
6. Proceed to PGRN visualization by fluorescent microscope.

3.2 Intracellular Staining (ICS) of PGRN for Flow Cytometric Analysis

This protocol describes an intracellular staining (ICS) technique that uses flow cytometry to detect proteins inside individual cells. Although the following steps are specific to PGRN detection in immune cells, this protocol can be modified to measure intracellular protein levels in a number of cell types. This assay is capable of detecting both surface markers and intracellular proteins and makes it possible to identify the type(s) of cells producing/secreting PGRN. This protocol will not provide information regarding the surface markers (*see* **Notes 19** and **20**). *See* Fig. 2 *for a summary of this protocol.*

1. Isolate lymphocytes from whole blood, tissue (spleen/ lymph node), or bone marrow; full protocol can be found in reference (13) (*see* **Note 21**).
2. Count and seed cells in a V-shape 96-well plate at a concentration of 1 × 10^6 cells in 190 μL of cold PBS (serum-free) per well.
3. To label dead cells, add 10 μL of the diluted Fixable Aqua cell stain to each well of the 96-well plate, and then cover the plate to protect from light and incubate for 15–30 min (*see* **Note 22**).
4. Centrifuge the plates for 5 min at 1500 rpm (~250 × *g*). Then carefully remove the supernatant using a multichannel aspirator/vacuum, and add 200 μL of cold PBS to each well. Centrifuge again and repeat the PBS wash.
5. Prepare blocking antibody (anti-FcgII/III receptors) in FACS buffer as recommended on the antibody spreadsheet, and add 50 μL of the diluted blocking antibody per well and incubate on ice for 30 min.
6. Dilute the surface antibodies (2×) in 50 μL per sample, and add this volume to the samples (containing the blocking antibody), and then incubate on ice for additional 30 min.

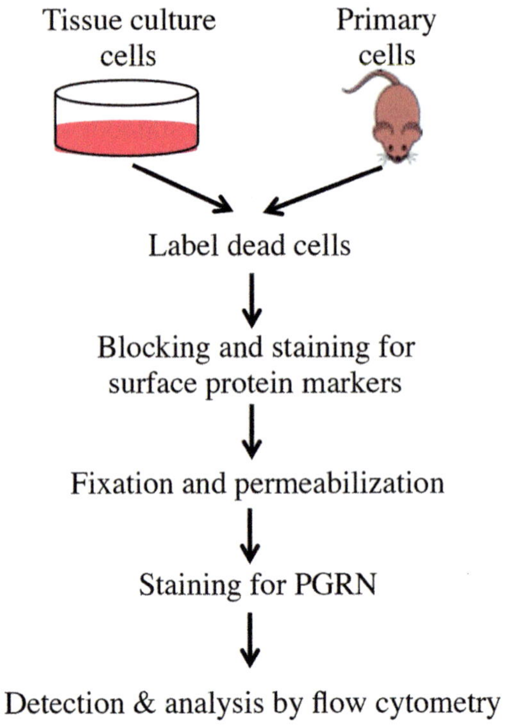

Fig. 2 Detection of intracellular PGRN using flow cytometry

7. Centrifuge the plate for 5 min at 1500 rpm (~250 × *g*), remove supernatant, and add 200 μL of cold PBS (*see* **Note 23**).
8. Add 100 μL of Cytofix/Cytoperm solution to each well and resuspend the cells by pipetting the samples. Incubate the plate for 10–20 min at 4 °C for optimal fixation and permeabilization of the cells (*see* **Note 24**).
9. Add 200 μL of 1× Perm/Wash solution (includes in the Intracellular Fixation and Permeabilization Buffer Kit set), centrifuge plate at 1500 rpm (~250 × *g*) for 5 min, and remove supernatant (*see* **Note 25**).
10. Prepare anti-mouse PGRN (primary antibody) at a dilution of 1:200 per sample in 1× Perm/Wash solution. Then add 50 μL of diluted PGRN antibody to each well, and incubate on ice for 30 min (*see* **Note 26**).
11. Add 100 μL of 1× Perm/Wash solution, centrifuge plate at 1500 rpm (~250 × *g*) for 5 min, and remove supernatant (*see* **Note 27**).

12. Prepare a 1:500 dilution of anti-rat Alexa Fluor-conjugated secondary antibody in 1× Perm/Wash solution, and then add 50 μL to each well and incubate on ice for 15 min.
13. Add 100 μL of 1× Perm/Wash solution, centrifuge plate at 1500 rpm (~250 × *g*) for 5 min, and remove supernatant.
14. Resuspend cells in 200 μL of Perm/Wash buffer, and proceed to measure PGRN using flow cytometry.

3.3 Detection of Secreted PGRN by Enzyme-Linked Immunosorbent Assay (ELISA)

This protocol explains how to perform an ELISA to measure levels of secreted murine or human PGRN. The ELISA is a high sensitivity assay that uses antibodies to measure the amount of protein or secreted factors present in blood plasma or cell supernatant at low concentrations (detection can occur at the picogram range). This assay is a multistep procedure that includes plate preparation, antibody capturing, and antigen detection. See Fig. 3 *for a summary of the ELISA protocol.*

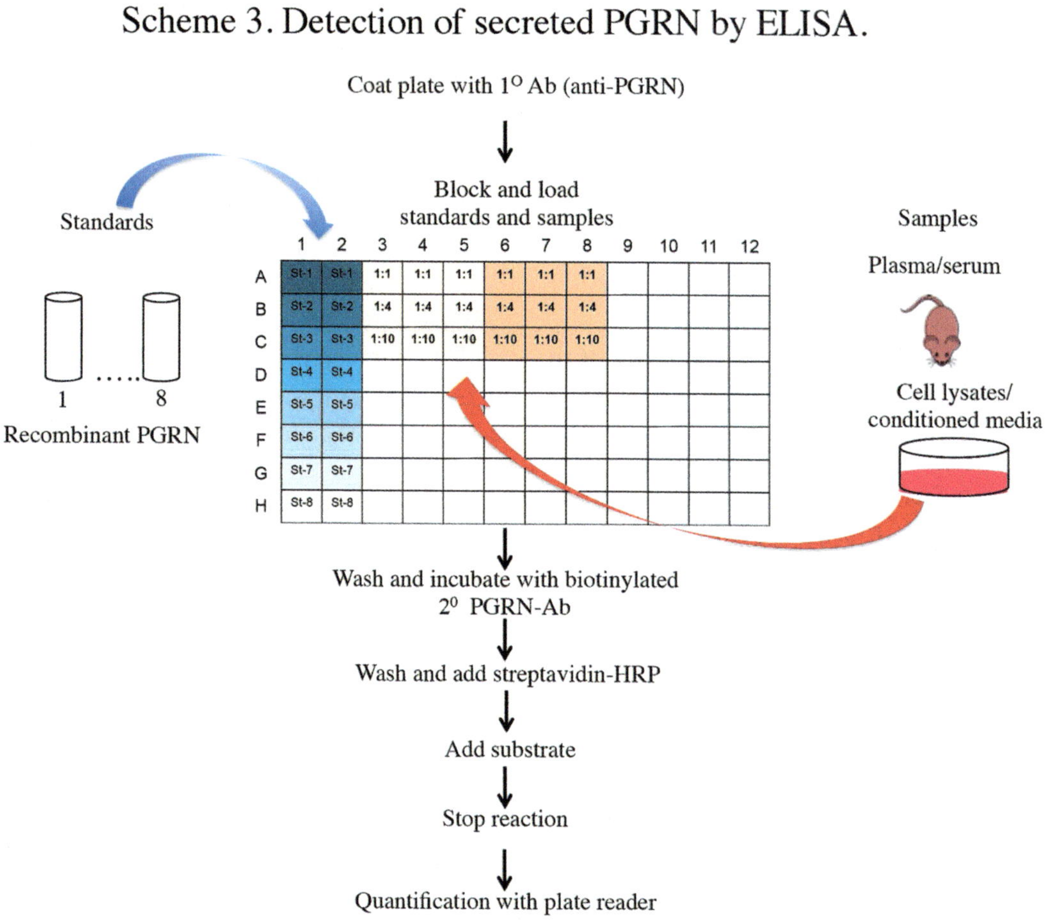

Fig. 3 Detection of secreted PGRN by ELISA

IMPORTANT: To avoid degradation of PGRN, the samples should be stored at −80 °C.

ELISA Protocol—Day 1: Plate coating

1. Dilute anti-PGRN capture antibody in PBS to a final concentration of 1–4 μg/mL as the manufacturer's protocol recommend, and then add 100 μL to each well of the 96-well microplates (*see* **Note 28**).
2. Seal the plate with an adhesive strip (microplate sealing tape), and incubate overnight at room temperature.

ELISA Protocol—Day 2: Plate Preparation and Sample Loading

1. Wash microplate by filling each well with 200 μL wash buffer (0.05% Tween20 in PBS, pH 7.2–7.4), and then aspirate or tap it against clean paper towels or absorbent lab paper. Repeat this process 3–4 times for optimal performance.
2. Use a multichannel pipette to add 200 μL of blocking buffer per well, and then incubate plate for 2 h at room temperature.
3. For the standards, prepare eight Eppendorf tubes (labeled #1–8), and add 500 μL blocking buffer to tubes #2–8. In tube #1, add recombinant PGRN in 800 μL of blocking buffer to make a final concentration of 5 ηg/mL. Perform serial dilutions (1:1 ratio) by taking 500 μL of PGRN standard from tube #1 into tube #2, pipet well, and then transfer 500 μL from tube #2 into tube #3, etc., *until reaching tube #7* (*see* **Note 29**).
4. For optimal measurement of PGRN in plasma, prepare several dilutions of the sample with blocking buffer (*see* **Note 30**).
5. Load plate with the standards (tubes #1–8, loaded in order of increasing concentrations) and the samples in triplicate (100 μL per well).

   *For an example of plate loading: standard samples (St), sample #1 and sample #2 labeled in orange and dilution is listed (*Fig. 3*).*
6. After loading the samples, cover and seal the plate with an adhesive strip to prevent evaporation, and incubate overnight at 4 °C.

ELISA Protocol—Day 3: Detection and Measurement of PGRN

1. Wash the microplate with loaded standards and samples by filling each well with 200 μL wash buffer and aspirating and then patting the plate dry on a clean paper towel or absorbent pad/surface. Repeat this wash 3–4 times.
2. Prepare the biotinylated anti-PGRN detection (secondary) antibody solution in blocking buffer at the concentration rec-

ommended by the manufacturer of the antibody, mix well, and then add 100 μL to each well of the plate. Use a new adhesive strip to seal the plate, and incubate for 2 h at room temperature.

3. Wash the microplate 3–4 times as described in **step 1** above.
4. Dilute streptavidin-HRP in blocking buffer to a concentration of 0.5 μg/mL, and add 100 μL to each well of the plate. Cover the plate to protect from light and incubate for 20 min at room temperature (*see* **Note 31**).
5. Wash microplate 3–4 times as described in **step 1** above.
6. Prepare substrate solution by mixing (1:1) Reagent A and Reagent B of DuoSet Ancillary Reagent Kit, and then add 100 μL to each well of the plate. Incubate at room temperature in the dark for 10–30 min. Check plate periodically and monitor the color-changing reaction; stop the reaction once a strong blue coloration appears in the standard curve samples with high concentrations. To stop the reaction, add 50 μL of stop solution (in the DuoSet Ancillary Reagent Kit) to each well and mix well (*see* **Note 32**).
7. After stopping the reaction, measure optical density at 450 nm and 540/570 nm.

3.4 Measuring GRN mRNA Expression in Human and Mouse Cells with Quantitative Real-Time Polymerase Chain Reaction (qPCR)

*Quantitative real-time PCR (RT-PCR/qPCR) is a technique in molecular biology used for quantitative measurement of mRNA levels and can provide both relative and absolute values of expression. Two common detection reagent systems used to quantify gene expression using qPCR are fluorescent dyes (*e.g.*, SYBR green) and fluorescent-tagged oligonucleotides (*e.g.*, TaqMan). Below, we describe how to measure and quantify GRN mRNA in cells obtained from tissue culture, freshly sorted cells, and in paraffin-embedded or frozen tissue using both fluorescent dye and fluorescent-tagged oligonucleotide detection systems. In this protocol, qPCR is used to measure gene expression in cDNA samples derived by reverse transcription of extracted RNA isolated from cell samples (see* **Notes 33–35**).

1. For SYBR green: optimize the primer concentration as described [14] and prepare a 20× stock of primers.
2. Thaw cDNA samples and primers or TaqMan gene expression reagents on ice.
3. Calculate the number of reactions required for each sample cDNA and the selected total reaction volume.
4. Based on the selected total reaction volume, dilute the cDNA sample with DNase-free Eppendorf tubes at an appropriate concentration as shown in Table 1 (*See* **Note 36**).

Table 1
Sample preparation for qPCR reaction

Total volume per sample (μL)	Dilution of cDNA	2× SYBR Master Mix or 2× TaqMan Master Mix	20× Primers or 20× TaqMan gene expression
10	10 ηg/4.5 μL	5 μL	0.5 μL
20	10 ηg/9 μL	10 μL	1 μL
40	10 ηg/18 μL	20 μL	2 μL

5. To each reaction well, add the appropriate amount of 2× Master mix (SYBR or TaqMan) and 20× primers to the each gene/probe (*GRN*, A*CTB*), as indicated in Table 1 (*see* **Note 37**).
6. For example in a total reaction volume of 20 μL, add 9 μL of diluted cDNA (10 ηg) per well, and after loading all samples, use an adhesive strip to seal the plate and centrifuge for 30 s at 500 rpm.
7. Perform qPCR using the appropriate thermocycle settings which should be adjusted based on the volume of the master mix and the primers/probes in the reactions.

3.5 Approaches to Investigate the Role of PGRN in Tumor Microenvironment

PGRN is expressed in and secreted from multiple cell types in cancer including tumor cells, immune cells (lymphocyte and myeloid lineage), and blood vessels. The expression of PGRN has been observed in a number of cancer types including ovarian and breast and was associated with poor prognosis in breast cancer patients. This section summarizes several strategies for studying PGRN derived from tumor cells or from surrounding cells in the tumor microenvironment.

3.5.1 GRN Knockout and Transgenic Mice

Genetically engineered knockout (KO) mice and Cre-recombinase transgenic mice are powerful tools widely used to model human disease and investigate the function of genes in vivo. Genetic KO mice are designed to prevent translation of a gene of interest in all organs, whereas transgenic *loxP* mice models are created to breed with Cre-expressing mice so the expression of a gene will be eliminated only in a specific tissue. Cre-lox recombination is a powerful tool in conditional gene targeting, making it possible to generate tissue-specific knockout of target genes. For example, when *GRN*-Flox transgenic mice are bred with mice expressing Cre- (such as Tie2 mice), recombination of the mutated *GRN* results in deletion of PGRN expression only in endothelial-specific tyrosine receptors (e.g., Tie2-expressing cells). The *GRN* KO and Flox transgenic mice are healthy and fertile and do not display developmental/

Table 2
Genetically engineered mouse models to study *GRN* function in cancer

Source	Strain	Tg/KO	*GRN*
Jackson	B6.129	Flox	Deletion of whole coding sequence
Jackson	B6.129	Flox	Deletion of exon 2–13
Jackson	C57BL/6	Flox	Deletion of exon 1–4
Taconic	C57BL/6	Flox	Deletion promoter and 1–13
Aihao Ding laboratory	C57BL/6	KO	Homozygous PGRN-deficient mice

physiological abnormalities [15], and therefore these models can be used to investigate the role of PGRN in different contexts of cancer. There are a number of strains of *GRN* transgenic mice available for purchase from multiple vendors, and *GRN* KO mice are available from Aihao Ding laboratory (Table 2).

3.5.2 Experimental Models to Investigate the Role of Tumor- and Host-Derived PGRN

Implantation of human tumor cell lines into mice is a robust technique widely used in cancer research to model and study disease progression and metastasis. With this approach, tumor cell features such as the tumorigenicity and invasive capability can be studied in vivo. The role of tumor-derived *GRN* can be assessed by implanting tumor cell lines in which *GRN* expression has been manipulated by knockdown/knockout or overexpression into mice with WT *GRN*.

To investigate the role of host-derived *GRN* in vivo, tumor cells can be injected into *GRN* KO and *GRN* WT mice. Differences in tumor growth between *GRN* KO and *GRN* WT mice can be attributed to *GRN* derived from the host animal/tissue. There are a number of tumorigenic murine cell lines suitable for testing in C57BL/6 *GRN* KO mice: B16, melanoma; EO771, breast; AT3, breast; RM1, prostate; and 3LL, lung cancer cell lines. Tumor cells can be injected intravenously to model and study metastasis or implanted subcutaneously to measure tumor volume and test spontaneous metastasis in C57BL/6 WT and *GRN* KO mice.

Functional consequences of *GRN* gene silencing can be studied in vitro by transfecting tumor cells with RNAi or siRNA targeting *GRN* to generate transient knockdowns. For constant or inducible knockdown of *GRN*, use shRNA. For deletion of *GRN* in tumor cells, CRISPR genomic editing can also be performed. All vectors and reagents are available from a number of companies such as OriGene and Sigma-Aldrich. Upregulation of *GRN* expression in tumor cells can be accomplished by infection or transfection of cells with overexpression vectors. The role of tumor-derived *GRN* in transformation and proliferation of cervical cancer cells was investigated by Lu et al. using overexpression and knockdown vectors to regulate PGRN levels [16].

3.5.3 Generation of Tumor Cell Lines from GRN KO or WT Mice

Chemically induced carcinogenesis is a well-established mutagenic technique used to study the effect of specific genes on tumor initiation and progression. This approach provides an opportunity to monitor the expression of a gene/protein (such as PGRN) at early and late stages of cancer development and progression. Treatment of *GRN* KO mice with either a single injection of 3-methylcholantrane (3MCA) or with two-stage DMBA-TPA carcinogenesis (including a single exposure of the chemical initiator mutagen DMBA (7,12-dimethylbenz[a]anthracene) followed by repeated treatment with the TPA (12-O-tetradecanoylphorbol13-acetate) will result in tumors that do not express *GRN*. These approaches have been wildly used to study the role of host- and tumor-derived secreted factors such as IFN-γ, GM-CSF, and IL-1β on tumor initiation, progression, and metastasis. Cultured tumor cells derived from these KO mice can then be used to investigate the role of various tumor-derived factors (reviewed in [17]).

3.5.4 Adoptive Transfer and Bone Marrow Transplantation of GRN KO and GRN WT Hematopoietic Cells into Mice

To study the role of PGRN derived from immune cells on tumor progression and metastasis, bone marrow transplantation or adoptive transfer of immune cells can be applied. Lethal irradiation of mice and subsequent reconstitution of the bone marrow from a different donor is a well-established methodology to investigate the function of specific genes expressed by hematopoietic stem cells. Moreover, mature immune cells can be activated by adoptive transfer (transplantation) of splenocytes or enriched subpopulations of immune cells into sublethally irradiated mice. Briefly, bone marrow or splenocytes extracted from *GRN* WT, KO, loxP-CRE *GRN* mice are immediately injected into the circulation of irradiated mice. Full protocols on adoptive transfer and bone marrow transplantation have been described [18, 19].

3.5.5 Neutralization and Activation of PGRN In Vivo

There are several ways to control the concentration of PGRN in the tumor microenvironment. Local injection of recombinant PGRN into the tumor site has been shown to increase tumor progression by inducing activation of the stroma [1] and by suppressing inflammatory response by inhibiting neutrophilic inflammation [20]. Another strategy for maintaining a constant concentration of PGRN in mice is with slow release pumps (for specific instructions, visit ALZET.com). These pumps are designed to establish and sustain a constant concentration of circulating compound by releasing at a steady rate over 28 days. As an alternate approach, depletion or neutralization of PGRN can be achieved by injection of anti-PGRN antibodies into the circulation as described [21, 22].

4 Notes

1. Proper tissue preparation/fixation will ensure optimal readout in this assay. For formalin section fixation, incubate in 4% formaldehyde for 3–4 h at room temperature, and then wash 5× with ddH_2O and add 300 μL 70% ethanol to the slides.
2. To avoid tissue loss when loading the slides into the staining chamber, place only one slide per slot, or, if you need two slides per compartment, insert adjacent slides back-to-back to prevent tissue on one slide from coming into contact with the other.
3. When handling xylene, be sure to work in a fume hood with proper PPE.
4. Use a pencil or an ethanol insoluble lab marker to label slides so that markings on the slide do not fade during the washing and staining steps.
5. Adjust the amount of the fixative as necessary based on the tissue size so that all sections are covered in the fixation buffer. Do not allow tissue sections to become dry at any point during the staining procedure.
6. Negative control should include an isotype-matched monoclonal primary antibody: Isotype Rat IgG for detection of mouse PGRN and Isotype Rabbit IgG for detection of human PGRN.
7. It is important to include water/wet paper towel around the slide to provide moisture during the overnight incubation in primary antibody. This minimizes evaporation of the antibody and also prevents sections from drying during the overnight incubation.
8. Staining can also be performed by incubating sections in primary antibody for 2 h at room temperature. However, overnight incubation at 4 °C is recommended for optimal binding to prevent non-specific binding and to minimize background staining.
9. The serum used to make blocking buffer should come from the same species as the biotinylated secondary antibody. Adjust the amount of buffer based on the number of slides.
10. Proper washing at this step removes excess antibody and minimizes background staining. If you encounter high levels of background staining, increase the number of washes.
11. Protect chromogen solution from light by using amber color tubes or by wrapping tubes in aluminum foil.
12. Chromogen reagents are hazardous materials and must be handled inside a fume hood with proper PPE including latex or nitrile gloves and protective eyewear.
13. Bubbles that form during this step can be removed by gently pressing down on the coverslip.

14. Negative control should include an isotype-matched monoclonal primary antibody: Isotype Rat IgG for detection of mouse PGRN and Isotype Rabbit IgG for detection of human PGRN.
15. For simultaneous detection of multiple target antigens ("co-labeling"), chose primary antibodies raised in different host species to avoid cross reactivity.
16. It is important to include water/wet paper towel around the slide to provide moisture during the overnight incubation in primary antibody. This minimizes evaporation of the antibody and also prevents sections from drying during the overnight incubation.
17. Staining can also be performed by incubating sections in primary antibody for 2 h at room temperature. However, overnight incubation at 4 °C is recommended for optimal binding as this method prevents non-specific bindings and minimizes background staining.
18. The slides should be protected from light throughout entire process.
19. Keep cells on ice throughout all steps in this protocol to maintain stable levels of PGRN expression. Cells must also be protected from light during antibody incubation and for all subsequent steps in the protocol.
20. As a positive staining control, activate lymphocytes in vitro with LPS/PMA (25 μg/mL) for 4 h in conjunction with a protein transport inhibitor such as GolgiStop to maximize detectability.
21. To enrich the concentration of the lymphocytes isolated from whole blood, use either red blood cell lysis buffer or the Ficoll-Paque isolation technique. Refer to reference (13).
22. Dead cell labeling must be performed before fixation or permeabilization. Prepare a 20× dilution of Fixable Aqua cell labeling dye in PBS (without FBS). Add 1 μL of the aqua cell stain (20 μL/per mL) for each 19 μL of PBS, and then mix well and protect from light.
23. For optimal results, wash cells again with 200 μL of PBS.
24. For optimal intracellular staining, use Perm/Wash solution throughout the staining protocol to maintain cells in a permeabilized state.
25. For optimal results, add 200 μL of Perm/Wash solution and repeat wash again.
26. To detect the background staining resulting from using anti-mouse PGRN, stain another set of cells with an Isotype Rat

IgG antibody, and use it to normalize for levels of background staining.

27. For optimal results, wash cell again with 100 μL of Perm/Wash solution.
28. When calculating the amount/volume of diluted antibody to prepare, be sure to include the standards in duplicate in your calculations.
29. Serial dilutions (of a protein antigen at known concentrations) are used to create an accurate standard curve to serve as a comparison/reference point when analyzing samples.
30. The optimal dilution should be determined by the investigator. We recommend starting by diluting each sample in blocking buffer at dilution ratios of 1:1, 1:4 and 1:10.
31. Samples/plates must be protected from light for the remainder of this protocol.
32. The intensity of the blue coloration reflects the total concentration of PGRN present in the sample. For accurate quantification, the concentration of the samples should fall within the range of the standard curve.
33. To measure transcript levels of GRN from a small/limited sample of cells (e.g., sorted cells), use an RNA extraction and purification kit such as PicoPure RNA Isolation Kit or Qiagen RNeasy Micro Kit. To obtain RNA from paraffin-embedded tissue or in a small sample of cells from laser capture microdissection, use FFPE RNA extraction kits or RNAqueous®-Micro Total RNA Isolation Kit, respectively.
34. The performance of this assay relies on the quality of the RNA isolated/purified from cell samples (the optical density 260/280 ratio should be greater than 1.8). RNA samples must to be stored at −80 °C.
35. The total volume of sample is dependent on the type of plate. For 96-well plates, the total volume is 20–50 μL, whereas in 384-well plate, a volume of 10–20 μL is used.
36. We recommend performing 3–4 replicates of each reaction. Each plate should include the gene of interest such as GRN and an endogenous control such as GAPDH or ACTIN-B. To screen for contamination, use water (instead of cDNA) as a negative control.
37. To ensure there is enough master mix and primers/TaqMan, include volume needed for all samples plus two extra reactions.

References

1. Elkabets M, Gifford AM, Scheel C et al (2011) Human tumors instigate granulin-expressing hematopoietic cells that promote malignancy by activating stromal fibroblasts in mice. J Clin Invest 121:784–799
2. Cheung ST, Wong SY, Leung KL et al (2004) Granulin-epithelin precursor overexpression promotes growth and invasion of hepatocellular carcinoma. Clin Cancer Res 10:7629–7636
3. Davidson B, Alejandro E, Flørenes VA et al (2004) Granulin-epithelin precursor is a novel prognostic marker in epithelial ovarian carcinoma. Cancer 100:2139–2147
4. Cheung ST, Cheung PF, Cheng CK et al (2011) Granulin-epithelin precursor and ATP-dependent binding cassette (ABC)B5 regulate liver cancer cell chemoresistance. Gastroenterology 140:344–355
5. Bateman A, Bennett HP (2009) The granulin gene family: from cancer to dementia. BioEssays 31:1245–1254
6. Buchwalow IB, Bocker W (2010) Immunohistochemistry: basics and methods. Springer, Berlin, Heidelberg, pp 1–153
7. Loza MJ, Faust JS, Perussia B (2003) Multiple color immunofluorescence for cytokine detection at the single-cell level. Mol Biotechnol 23:245–258
8. Crowther JR (1995) ELISA: theory and practice. Methods Mol Biol 42:1–218
9. Tkaczuk KR, Yue B, Zhan M et al (2011) Increased circulating level of the survival factor GP88 (Progranulin) in the serum of breast cancer patients when compared to healthy subjects. Breast Cancer (Auckl) 5:155–162
10. Schefe JH, Lehmann KE, Buschmann IR et al (2006) Quantitative real-time RT-PCR data analysis: current concepts and the novel "gene expression's CT difference" formula. J Mol Med 84:901–910
11. Ong CH, He Z, Kriazhev L et al (2006) Regulation of progranulin expression in myeloid cells. Am J Physiol Regul Integr Comp Physiol 291:R1602–R1612
12. He Z, Ong CH, Halper J et al (2003) Progranulin is a mediator of the wound response. Nat Med 9:225–229
13. Hay FC, Westwood OM (2002) Practical immunology, 4th edn. Blackwell Science Inc., Malden, MA, pp 1–400
14. Vet JA, Marras SA (2005) Design and optimization of molecular beacon real-time polymerase chain reaction assays. Methods Mol Biol 288:273–290
15. Ahme Z, Sheng H, Xu YF et al (2010) Accelerated lipofuscinosis and ubiquitination in granulin knockout mice suggest a role for progranulin in successful aging. Am J Pathol 177:311–324
16. Lu Y, Zheng L, Zhang W et al (2014) Growth factor progranulin contributes to cervical cancer cell proliferation and transformation in vivo and in vitro. Gynecol Oncol 134:364–371
17. Smyth MJ, Dunn GP, Schreiber RD (2006) Cancer immunosurveillance and immunoediting: the roles of immunity in suppressing tumor development and shaping tumor immunogenicity. Adv Immunol 90:1–50
18. Gray BM, Eaton KA (2012) Adoptive transfer of splenocytes to immunocompromised mice. Methods Mol Biol 921:109–112
19. Duran-Struuck R, Dysko RC (2009) Principles of bone marrow transplantation (BMT): providing optimal veterinary and husbandry care to irradiated mice in BMT studies. J Am Assoc Lab Anim Sci 48:11–22
20. Kessenbrock K, Fröhlich L, Sixt M et al (2008) Proteinase 3 and neutrophil elastase enhance inflammation in mice by inactivating antiinflammatory progranulin. J Clin Invest 118:2438–2447
21. Wong NC, Cheung PF, Yip CW et al (2014) Antibody against granulin-epithelin precursor sensitizes hepatocellular carcinoma to chemotherapeutic agents. Mol Cancer Ther 13:3001–3012
22. Ho JC, Ip YC, Cheung ST et al (2008) Granulin-epithelin precursor as a therapeutic target for hepatocellular carcinoma. Hepatology 47:1524–1532

Part IV

Animal Models to Study the Role of Progranulin in Health and Disease

Chapter 13

Methods to Investigate the Molecular Basis of Progranulin Action on Neurons In Vivo Using *Caenorhabditis elegans*

James Julian Doyle and J. Alex Parker

Abstract

The *C. elegans* nematode is a powerful genetic tool for the study of aging, developmental biology, and neurodegenerative diseases. They are a small and simple model, but its well-known genome and presence of many human orthologs has made it an ideal model to study the effects of PGRN in vivo. Here, we will describe the protocols used by our laboratory to study how PGRN affects *C. elegans* neuronal morphology and locomotion behavior through the use of a loss-of-function model of the nematode otholog of the *GRN* gene, *pgrn-1*. Although these protocols have been used by us to specifically study the *pgrn-1* gene, the experiments are applicable to any of the animal's genes.

Key words *C. elegans*, Neurodegeneration, Motor neurons, Paralysis, Locomotion, Fluorescence

1 Introduction

The nematode *Caenorhabditis elegans* is a model for the study of neurodegenerative diseases and is a tool for the in vivo discovery of novel drug therapeutics and targets [1]. The worm's small size, rapid development, self-mating reproductive process, and well-known genome have made the worm a powerful tool for molecular, genetic, and cellular analyses. Opposed to many of the animal models used today, the simplicity of the worm's 302-cell nervous system (making up 30% of its cellome) has made it an ideal model for the study of neuronal degeneration [1]. Furthermore, the worms are transparent for the entirety of their life cycle which allows for the clear, real-time visualization of neurons expressing fluorescent proteins, permitting the assessment of neuronal health. In our laboratory, we have established the *C. elegans* animal model for the study of age-dependent neurodegenerative disorders.

In order to investigate the in vivo action of PGRN-1 on neuronal function, our laboratory used deletion mutants, RNA interference (RNAi) technology [2–4], fluorescent imaging, as well as a number of motility assays. The RNAi knockdown protocol

Andrew Bateman et al. (eds.), *Progranulin: Methods and Protocols*, Methods in Molecular Biology, vol. 1806,
https://doi.org/10.1007/978-1-4939-8559-3_13,

describes feeding the worms a bacterial strain containing a vector expressing dsRNA against the gene of interest and then monitoring the worms to detect any apparent phenotype. First, the worm populations are synchronized to ensure any phenotypes observed are not due to differences in age and raised on nematode growth media (NGM) and fed *E. coli* OP50 until they reach the fourth larval state (L4). From there, the larvae are transferred onto new plates and fed bacteria containing dsRNA against the *C. elegans pgrn-1* sequence. From there, the worms' motility is assessed for the first 12 days of adulthood.

For fluorescent imaging of neurons in living animals, we can then visualize if the knockdown or knockout of *pgrn-1* results in degeneration of the worm's motor neurons. For this technique, the worms are immobilized in a glycerol solution on an agarose pad atop a microscopy slide. The fluorescence is then analyzed under a fluorescent microscope containing a filter for the fluorophore. Since the worms are naturally transparent, no additional treatment is necessary to visualize the fluorescent signal.

The *C. elegans* nervous system is naturally resistant to RNAi due to the absence of expression of certain proteins, such as SID-1 (systematic RNA interference defective), which regulate the uptake of dsRNA; mutations in this protein lead to RNAi resistance [5]. Given the recent advances in genetics, it has been possible to generate mutant worm strains for *sid-1* but which express the functional protein in specific cell types, rendering them sensitive to RNAi-mediated knockdown only in those cells [6]. Thus, the functional role of a gene product, or interactions between them, in a specific cell type can be determined using this method.

2 Materials

2.1 Preparation of NGM Agar Plates for Worm Maintenance and RNAi Experiments

1. Bacteriological agar (Wisent Bioproducts).
2. Sodium chloride.
3. Casein Digest Peptone (Wisent Bioproducts).
4. Nystatin, 10,000 units/mL in DPBS (Sigma-Aldrich).
5. 50 mg/mL streptomycin stock: dissolve 2.5 g streptomycin sulfate (Fisher Scientific) in 50 mL ddH_2O. Filter through a 0.2 or 0.22 micron filter, and aliquot 1.5 mL in Eppendorf tubes and store at −20 °C.
6. 100 mg/mL ampicillin stock: dissolve 2.5 g ampicillin (Wisent Bioproducts) in 25 mL ddH_2O. Filter through a 0.2 or 0.22 micron filter, and aliquot 1.5 mL in Eppendorf tubes and store at −20 °C.
7. 5 mg/mL cholesterol stock: dissolve 250 mg cholesterol in 50 mL ethanol in a 50 mL Falcon tube. Place in a 37 °C water

bath for a few days until fully dissolved. Aliquot 1 mL into Eppendorf tubes and store at −20 °C.

8. 1 M isopropyl β-D-1-thiogalactopyranoside (IPTG) stock: dissolve 2.381 g of IPTG in 10 mL of water. Filter through a 0.2 or 0.22 micron filter, and aliquot 1 mL into Eppendorf tubes. Store at −20 °C.
9. 1 M $MgSO_4$.
10. 1 M KPO_4.
11. 1 M $CaCl_2$.

2.2 Worm Maintenance

1. Worm pick: carefully break the tip off a glass Pasteur pipette, and cut a small piece of platinum wire (1 in. will suffice). Using a coin, flatten a small section of the end of the platinum wire into a small spatula. Using a pair of tweezers, grasp the flat end of the pick, and insert the opposite end into the broken tip of the Pasteur pipette. Hold this over a flame until the tip of the pipette melts, securing the platinum wire in place.
2. Dissecting microscope (Leica S6E).
3. Incubators (set to 15 °C and 20 °C).
4. NGM agar plates.
5. *E. coli* OP-50 cultures.
6. Ethanol burner.

2.3 PCR Verification of C. elegans Deletion Mutants

1. Worm lysis buffer: for 10 mL, combine 0.5 mL 1 M KCl, 0.1 mL 1 M Tris (pH = 8.3), 0.1 mL 0.25 M $MgCl_2$, 45 μL NP-40 or Triton X-100, 45 μL Tween 20, 1 μL gelatin, and fill up to 10 mL with ddH_2O. Aliquot 200 μL into Eppendorf tubes and store at −20 °C.
2. Proteinase K (Invitrogen): for 20 mg/mL stock, weigh out 0.2 g proteinase K and dissolve in 10 mL ddH_2O. Aliquot 50 μL into Eppendorf tubes and store at −20 °C.
3. 2.5× cresol red PCR dye: dissolve 0.5 g of cresol red into 50 mL ddH_2O to prepare a 1% solution. Dissolve 17 g sucrose into 49 mL ddH_2O, and add 1 mL of the 1% cresol red solution. Aliquot 1 mL into Eppendorf tubes and store at −20 °C.
4. 100 mM dNTP set (Life Technologies): prepared according to manufacturer's instructions.
5. Taq DNA Polymerase with Standard Taq Buffer (New England Biolabs Inc.).
6. Forward and reverse primers (IDT): sequences for *pgrn-1(tm985)* deletion mutants were 5′-GGACCAAGTTGGCGTCAGAAATAG-3′ (Fwd) and 5′-GGTGAGCAGTCTGGAAATTTCA-3′ (Rev). Dissolve primers in an appropriate volume to get a 100 μM stock solution, and use this to make 10 μM working solutions.

7. Professional basic Thermocycler gradient (Biometra).
8. Quick-Load® 1 kb DNA ladder (New England Biolabs Inc.).
9. 1× TAE buffer: for a 50× stock solution, dissolve 242 g Tris base in ddH_2O. Add 57.1 ml glacial acetic acid and 100 mL 0.5 M EDTA (pH = 8.0), and fill up to 1 L with ddH_2O. For a 1× solution, dilute 80 mL in 3920 mL ddH_2O.
10. Agarose.
11. Gel electrophoresis apparatus.

2.4 Worm Synchronization

1. Pipetboy.
2. 10 mL serological pipettes.
3. M9 buffer: weigh out 5 g NaCl, 6 g KH_2PO_4, and 6 g Na_2HPO_4, and add 1 mL of 1 M $MgSO_4$. Fill up to 1 L with ddH_2O (*see* **Note 1**).
4. 5 N sodium hydroxide: in a 50 mL Falcon tube, dissolve 10 g of sodium hydroxide pellets into 50 mL dH_2O.
5. Bleach.
6. Eppendorf centrifuge 5810 R.
7. Vacuum trap.
8. Pasteur pipettes.

2.5 Motility Assay in Liquid Media

1. 96-well plate (Thermofisher).
2. WMicrotracker ONE (Phylumtech).

2.6 Fluorescence Microscopy of Worm Motor Neurons

1. 2% agarose solution.
2. 60% glycerol solution.
3. Glass microscopy slides (Fisher Scientific).
4. Cover slips (VWR International).
5. Paper tape.
6. Inverted microscope (Zeiss Axio Observer Z1).

3 Methods

3.1 Preparation of NGM Agar Plates for Worm Maintenance

Many of the following methods were adapted from the Wormbook (http://www.wormbook.org/)

In a Pyrex glass bottle:

1. For 1 L of NGM agar: weigh out 17 g bacteriological agar, 3 g sodium chloride, and 2.5 g bacteriological peptone.
2. Fill to 1 L with distilled water.
3. Sterilize solution by autoclave.

4. Upon sterilization, cool agar solution to 53 °C, or place in an incubator set to 53 °C with agitation (140–150 rpm) until ready to proceed. Failure to do so will result in solidification of agar.
5. When ready to proceed, add the following supplements to agar solution (volumes are per 1 L, can be adjusted to desired volume):
   (a) 10 mL nystatin suspension.
   (b) 1 mL of 1 M $MgSO_4$.
   (c) 1 mL of 1 M $CaCl_2$.
   (d) 25 mL of 1 M KPO_4.
   (e) 1 mL 5 mg/mL cholesterol.
   (f) 1.5 mL 50 mg/mL streptomycin.
6. Agitate solution to assure homogeneity of added supplements.
7. Pour about 8 mL of agar solution into each plate.
8. Place lids on plates and let solidify overnight.
9. Store at 4 °C until ready to use.

3.2 Preparation of RNAi NGM Plates

In a Pyrex glass bottle:

Steps 1–4 identical as in Subheading 3.1.

5. When ready to proceed, add the following to agar solution (volumes are per 1 L, can be adjusted to desired volume):
   (a) 10 mL nystatin.
   (b) 1 mL of 1 M $MgSO_4$.
   (c) 1 mL of 1 M $CaCl_2$.
   (d) 25 mL of 1 M KPO_4.
   (e) 1 mL cholesterol.
   (f) 1 mL ampicillin.
   (g) 1.5 mL IPTG.
6. Agitate solution to assure homogeneity of added supplements.
7. Pour about 8 mL of agar solution into each plate.
8. Place lids on plate and let agar solidify overnight.
9. Store at 4 °C until ready to use (*see* **Note 2**).

3.3 Worm Maintenance [7]

3.3.1 Preparation of E. coli Cultures (Worm Food)

1. Place 10 mL LB media into cell culture tubes, and add 15 μL streptomycin (*see* **Note 3**) to each culture.
2. Dip a 20 μL pipette tip into a frozen *E. coli* OP-50 (*see* **Note 3**) frozen stock, and place into the culture tube. Use a new tip for each culture tube.

3. Incubate bacterial cultures at 37 °C overnight on a shaker, but for no more than 18 h.
4. Keep bacterial cultures at 4 °C until ready to use.

3.3.2 Inoculation of NGM Agar Plates with E. coli

1. Aspirate *E. coli* bacterial cultures with a 10 mL serological pipette, and dispense a few drops onto an NGM agar/RNAi plate.
2. Use the tip of the pipette to spread the bacteria around the plate.
3. Cover the plate and let sit right-side up overnight to let the LB dry and the bacteria grow.
4. When dry, keep plates at 4 °C until ready for further use.

3.3.3 Maintenance of C. elegans Strains

1. Pick worms, using the worm pick, onto seeded NGM plates, and keep plates in plastic boxes in isothermal incubators (*see* **Note 4**).
2. When very little bacteria remains on the NGM plates, transfer a few worms to a new seeded plate to avoid starvation of the worms (*see* **Note 5**).

3.4 PCR Verification of C. elegans Deletion Mutants

3.4.1 Worm Lysis

1. Add 7 μL 20 mg/mL proteinase K to 200 μL aliquot of worm lysis buffer (make a fresh solution for each lysis).
2. Place 3 μL of this solution into each PCR tube.
3. Using the worm pick, place one worm into each PCR tube.
4. Flash freeze PCR tubes at −80 °C for at least 5 min.
5. Remove tubes from −80 °C, place in thermal cycler, and begin lysis run:

Temperature (°C)	Time (min:s)
60	60:00
95	15:00
15	Pause

3.4.2 PCR Amplification

1. Prepare a master mix for the PCR reaction: for 30 reactions, combine 300 μL 2.5× cresol red, 296.25 μL ddH_2O, 15 μL 100 mM dNTPs, 15 μL of both 10 μM forward and reverse primer, 75 μL Standard Taq Buffer, 3.75 μL Taq DNA polymerase.
2. Dispense 20 μL of this mix into new PCR tubes.
3. Add 7 μL ddH_2O to each PCR tube containing lysed worms, and dispense 1.5 μL of the diluted lyses into each tube containing the PCR mix.
4. Place tubes into thermal cycler with the following run cycle:

	Temperature (°C)	Time (min:s)
35×	94	1:00
	94	0:40
	54	0:40
	72	2:00
	72	10:00
	15	Pause

3.4.3 Agarose Gel Analysis of PCR

1. Prepare 200 mL of a 1% agarose solution and microwave to boil.
2. Add 5 μL ethidium bromide per 100 mL of agarose solution.
3. Cast gel with an appropriate number of combs to make the desire number of wells and let solidify.
4. Load 10 μL of PCR product into each well, and add 5 μL of the ladder.
5. Run electrophoresis apparatus at constant voltage (0.5 V per milliliter of gel volume) for 30 min.
6. Reveal gel under ultraviolet light. Mutant alleles for a deletion are smaller than their wild-type counterparts and will run lower on the gel (*see* Fig. 1).

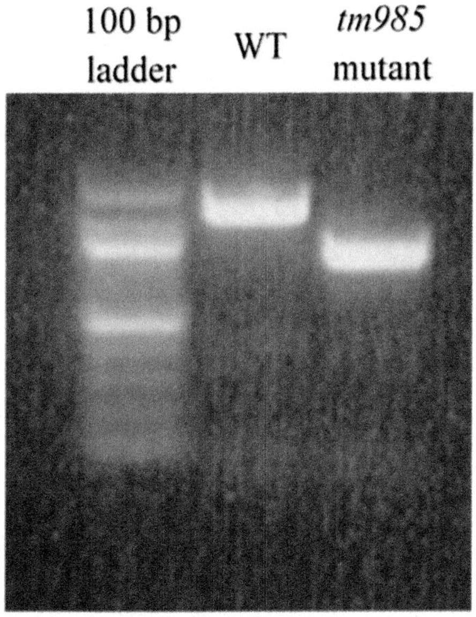

Fig. 1 *pgrn-1(tm985)* deletion allele runs lower than the WT (wild-type) allele on an agarose gel

3.5 Worm Synchronization:

Day 1:

1. Prepare 10–15 standard NGM agar plates and streak with OP50 bacteria.
2. Place 5–8 worms on each plate and let grow.

Day 5:

3. When the plates are covered in gravid adults, wash the plates with M9 buffer to collect the worms with a ratio of 15 mL of buffer per five plates.
4. Place collected worms in 15 mL Falcon tubes.
5. Centrifuge Falcon tubes at 4000 rpm (2900 × *g*) for 4 min to pellet the worms.
6. Aspirate supernatant until 7 mL are left in the Falcon tubes.
7. To each tube, add 3 mL of a 2:1 ratio of bleach and 5 N NaOH.
8. Shake or vortex tubes for 10 s every 2 min for 10 min (*see* **Note 6**).
9. Centrifuge tubes for 4 min at 4000 rpm (2900 × *g*).
10. Aspirate supernatant down to 1 mL, and add 9 mL of M9 buffer to wash the pellet.
11. Repeat **steps 9** and **10** two more times, or until the smell of bleach disappears.
12. After the final wash, aspirate the supernatant down to 0.5 mL.
13. Using a P200 pipettor, resuspend the pellet and transfer it to a standard NGM agar plate without OP50.

Day 6:

14. Once the eggs have hatched, collect larvae in 1 mL of M9 buffer, and transfer to an NGM plate with OP50 using a glass Pasteur pipette (*see* **Note 7**). The worm population will now be synchronized (age-matched).

3.6 Paralysis Test on Agar Plates

Day 0:

1. When the synchronized worm population is in the fourth larval state (L4), prepare three plates with 30 worms per plate for each strain to assay. Store plates in an incubator set to 20 °C.

Days 1–12:

2. The worms should now be in their first day of adulthood. Score worms for paralysis by gently prodding their head with a pick. The worms are considered paralyzed when they are unable to respond to the stimuli, or are only able to move their head (*see* **Note 8**). Worms are to be scored for paralysis over a period of 12 days (*see* **Note 9**).

3.7 Liquid Motility Assay

Day 0:

1. Similarly to Subheading 3.6, transfer over 100 worms in the L4 state from the synchronized population to 1–2 new plates (depending on the thickness of the bacterial lawn). Repeat for all desired strains.

Day 1:

2. In a 96-well plate, pipette 100 μL of M9 buffer into the wells. Prepare three wells per strain.
3. Transfer exactly 30 worms of each strain into the appropriate well, for a total of 90 worms for each strain (*see* **Note 10**).
4. Place the 96-well plate into the MicroTracker and calibrate in such that the acquisition time is set to 600 min (10 h). This step can usually be done overnight.

3.8 Neuro-degeneration Assay

In our laboratory, neurodegeneration assays are done at days 1, 5, and 9 [8]. For this technique, it is essential to use a worm strain which expresses a fluorescent protein in its neurons. In our lab, we commonly use the *unc-47p::GFP* transgene [9, 10] which expresses GFP in its GABAergic motor neurons, and we cross these worms with mutant strains to assess neuronal health over time.

3.8.1 Preparation of Worms for Assay

1. Synchronize desired worm population as per protocol in Subheading 3.4.
2. When worms reach the fourth larval state, transfer 40–50 worms onto three NGM plates (for a total of 120–150 worms) for each strain desired to assay. Perform this step three times to have worms ready to assay at days 1, 5, and 9.
3. For assays at days 5 and 9, transfer the worms onto fresh plates every 2 days to avoid worm starvation and to ensure only age-matched worms remain on plates.

3.8.2 Preparation of Glass Slides for Microscopy

4. Prepare 50 mL of a 2% agarose solution and heat for 45 s in a microwave, or until the agarose is dissolved and the solution has reached a gentle boil.
5. While the solution is heating, place a piece of paper tape on to two glass slides. Place these two slides on either side of a third glass slide (without tape) which will be used for the sample (*see* Fig. 2). The thickness of the tape will aid in balancing a fourth glass slide needed in **step 4**.
6. Cut the end off of a 1000 μL pipette tip, and use this tip to aspirate and dispense a drop of the molten 2% agarose solution onto the center of a glass slide.
7. Immediately place a fourth glass slide on top of the agarose drop, perpendicular to the one with the drop, to flatten it and create a "pad." If this is not done immediately after dispensing the drop, it will solidify in the shape of a drop (*see* **Note 11**).

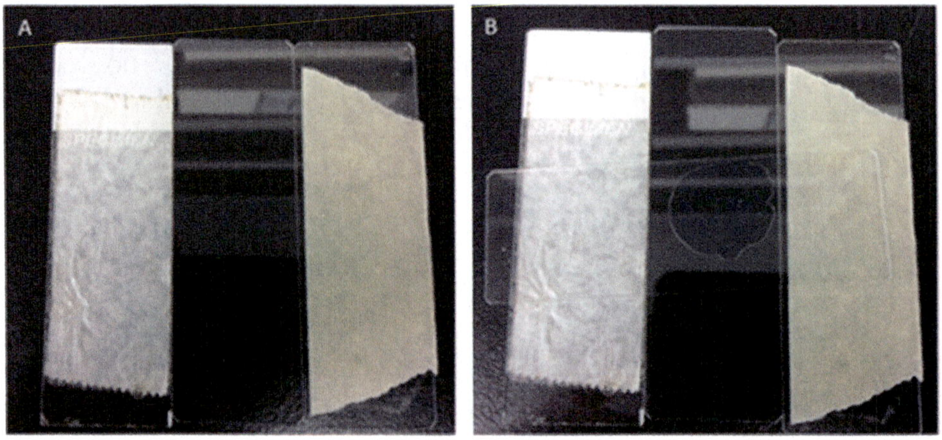

Fig. 2 Agarose pad formation. (*a*) Place one tape-covered glass slide on each side of the glass slide to be used for the sample. (*b*) Once the drop of 2% agarose is placed, quickly cover and flatten it with another glass slide

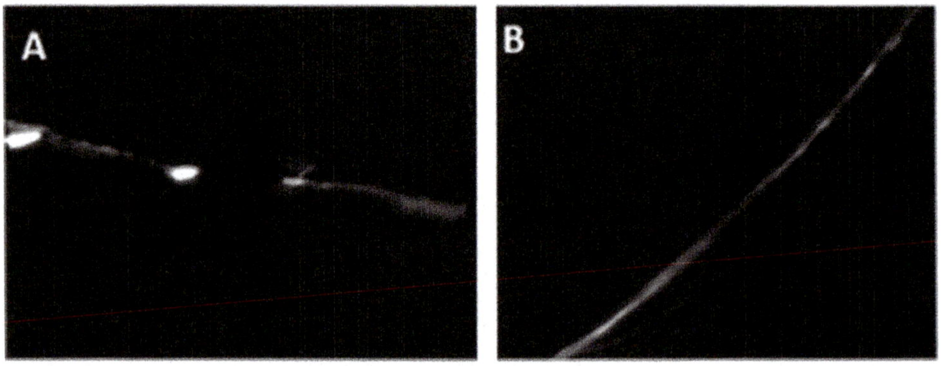

Fig. 3 (**a**) Degenerated *C. elegans* axon represented by a discontinuity of GFP florescent signal. (**b**) Healthy *C. elegans* axon

8. Repeat **steps 1–4** until you have prepared the desired number of pads. We recommend preparing four pads for each worm strain to be assayed.
9. Remove the glass slide used to flatten the pad, and place 10 μL (*see* **Note 12**) of a 60% glycerol solution onto the agarose pad to immobilize the worms (*see* **Note 13**).
10. Place about 25 worms into the glycerol, and gently cover with a cover slip ensuring to omit any air bubbles.
11. Repeat until you have no less than 4 pads with 25 worms on each for each strain (a total of no less than 100 worms per strain).

3.8.3 Visualization of Neuronal Morphology

12. When looking at the worms under the fluorescent microscope, worms with neurodegeneration are counted if the fluorescent signal is discontinuous along the ventral axon (Fig. 3) (*see* **Note 14**).

4 Notes

1. Since dibasic sodium phosphate, Na_2HPO_4, has difficulty dissolving, we recommend preparing this solution a few days before it is needed as it needs to be stirred magnetically overnight for the compiund to properly dissolve.
2. Since the plates for RNAi tests contain IPTG, although not essential, it is good practice to keep them wrapped in aluminum foil to prevent direct exposure to light. Nonetheless, it is better to use freshly made RNAi NGM agar plates when performing tests of any kind, and we advise to not keep plates for more than 2 weeks.
3. Bacterial cultures for RNAi feeding experiments are prepared very similarly, the only difference being the addition of 15 μL of ampicillin in lieu of streptomycin.
4. When placing NGM plates in an incubator, it is good practice to store them in plastic boxes to prevent the plates from drying out. Furthermore, remember to always sterilize the worm pick after every transfer to prevent cross-contamination between strains.
5. If the worms become starved, you will no longer be able to perform tests with them. After transferring starved worms to NGM plates with bacterial food, you will need to wait two generations before being able to start tests anew.
6. Depending on the amount of worms, more or less time will be required. This lysis reaction can be monitored under the dissecting microscope by looking into the Falcon tube: the reaction is considered done when only eggs are visible, and there is the presence of only little worm body debris. It is very important to not keep the worm eggs in bleach for too long as their resistance fades with extended exposure to it.
7. Avoid using micropipettes to transfer the worm larvae as worms will have a tendency to stick to the plastic tips. For this reason, we recommend the use of Pasteur pipettes, or other glass pipettes.
8. Worms in a late-stage paralysis may appear as dead and will not respond to a touch stimuli. If this is the case, you may choose to either heat the pick and then place it near the worm's head and look for a response or look at the worm's pharynx and look to see if it is pumping (alive) or not (dead). For statistical purposes, dead worms are counted as a null event.
9. It is possible that from 1 day to the next you do not have the same number of worms on your plates as they may have crawled up the sides of the plastic petri dish. Usually when

worms do this, they dry out and die very quickly. So, for statistical purposes, remember to count these "lost worms" as null events as well.

10. If you desire to apply this method for a high-throughput purpose, it is recommended to use collect worms with a repeater pipette and to then dispense an equal amount into each well as opposed to picking, by hand, 30 animals into each well. Aside from saving time, by doing so you prevent the early exhaustion of the worms placed in the first well.
11. It is possible to make multiple agarose pads per slide. Instead of placing the agarose drop in the center of the glass slide, you can place one on each end of it to have two pads on the same slide.
12. If you plan on placing more than 25 worms on the pad, we recommend you add around 15 μL of glycerol instead. Having worms that are too closely packed on the pad will result in a more difficult analysis of neuronal health.
13. In our protocol here, we suggest the use of glycerol to immobilize the worms on the pad. However, other compounds for this purpose exist and are often used in *C. elegans* studies. Among these are the paralytic agents sodium azide and levimazole. For the purpose of this neurodegeneration assay, we recommend avoiding such agents as they may affect the worm's neuronal health, potentially leading to higher-than-normal degeneration counts.
14. Since nematodes have their vulva on the ventral side, this results in the appearance of a break in the axon (*see* Fig. 4). It is important to not count this as an axon break as it is simply the result on the *C. elegans* morphology. If there is any doubt

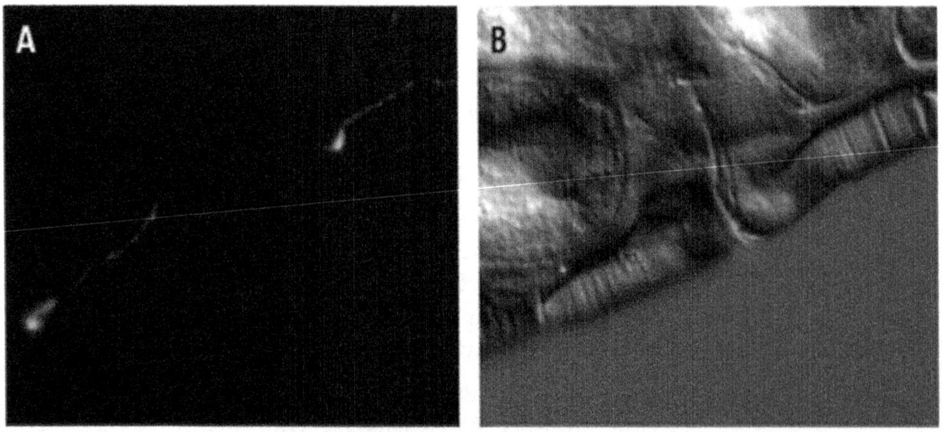

Fig. 4 (**a**) Discontinuity of GFP signal in GABAergic motor neurons in the vicinity of the worm vulva is a regular occurrence, but is not the result of neurodegeneration. (**b**) Differential interference contrast (DIC) image of an adult *C. elegans* vulva

as to whether or not a discontinuity is due to its proximity to the vulva, look at that area in a non-fluorescent manner (e.g., DIC) to see if it is near the vulva.

Acknowledgments

We would like to thank the microscopy platform of the Centre de Recherche du CHUM for allowing usage of their inverted microscope. We also thank the Caenorhabditis Genetic Center (CGC) for providing some strains used during this work. This work was funded by the Canadian Institutes of Health Research (CIHR).

References

1. Therrien M, Parker JA (2014) Worming forward: amyotrophic lateral sclerosis toxicity mechanisms and genetic interactions in Caenorhabditis elegans. Front Genet 5:85
2. Fire A et al (1998) Potent and specific genetic interference by double-stranded RNA in Caenorhabditis elegans. Nature 391(6669): 806–811
3. Timmons L, Fire A (1998) Specific interference by ingested dsRNA. Nature 395(6705):854
4. Kamath RS et al (2003) Systematic functional analysis of the Caenorhabditis elegans genome using RNAi. Nature 421(6920):231–237
5. Winston WM et al (2002) Systemic RNAi in C. elegans requires the putative transmembrane protein SID-1. Science 295(5564): 2456–2459
6. Calixto A et al (2010) Enhanced neuronal RNAi in C. elegans using SID-1. Nat Methods 7(7):554–559
7. Stiernagle T (2006) Maintenance of C. elegans. In: WormBook: the C. elegans research community. WormBook, Pasadena. https://doi.org/10.1895/wormbook.1.101.1. http://www.wormbook.org
8. Vaccaro A et al (2012) Mutant TDP-43 and FUS cause age-dependent paralysis and Neurodegeneration in C. elegans. PLoS One 7(2):e31321
9. Jorgensen EM, GABA (2005) WormBook: the *C. elegans* research community. WormBook 14:1. https://doi.org/10.1895/wormbook.1. http://www.wormbook.org
10. McIntire SL et al (1997) Identification and characterization of the vesicular GABA transporter. Nature 389(6653):870–876.

Suggested Reading

http://www.wormbook.org

Chapter 14

The Use of *Caenorhabditis elegans* to Study Progranulin in the Regulation of Programmed Cell Death and Stress Response

Tsung-Yuan Hsu, Victoria J. Butler, and Aimee W. Kao

Abstract

The nematode *Caenorhabditis elegans* (*C. elegans*) has proven to be a powerful model organism for the study of many biological processes, with major implications for human health and disease. As progranulin is a pleiotropic, secreted protein with both cell autonomous and non-autonomous roles, a multicellular organism such as *C. elegans* is ideal for the investigation of its normal function and pathological effects. The *C. elegans* genome contains a progranulin-like gene known as *pgrn-1*. The nematode *pgrn-1* encodes a protein with three cysteine-rich granulin domains, compared to the seven and a half granulins in the human protein. We have shown that *C. elegans* mutants lacking *pgrn-1* appear grossly normal, but exhibit accelerated apoptotic cell engulfment as well as a stress resistance phenotype (Kao et al., Proc Natl Acad Sci U S A 108:4441–4446, 2011; Judy et al., PLoS Genet 9:e1003714, 2013). In addition, the roles of individual granulins can also be dissected in *C. elegans* (Salazar et al., J Neurosci 35:9315–9328, 2015). Here, we describe methods for studying apoptosis and stress response in *C. elegans*.

Key words *C. elegans*, Progranulin, Stress response, Programmed cell death, Apoptosis

1 Introduction

In 1974, the Nobel Prize winning biologist Sydney Brenner introduced the scientific world to the nematode *C. elegans*. Brenner felt that in addition to yeast and *Drosophila*, a genetically tractable organism was needed for the study of development and neurobiology. In hindsight, Brenner's calculated gamble proved to be a brilliant success as *C. elegans* has become a widely used and highly valuable model organism. Using this small nematode, scientists have invented or reinvented the fields of programmed cell death, genetic regulation of aging, RNA processing, and neuronal development.

Many characteristics render *C. elegans* useful in the lab. Because of its small size, transparent cuticle, and simple anatomy, the entire cell lineage and neuronal circuitry of this nematode have been

Andrew Bateman et al. (eds.), *Progranulin: Methods and Protocols*, Methods in Molecular Biology, vol. 1806, https://doi.org/10.1007/978-1-4939-8559-3_14,

mapped in detail, down to the synapse [1, 2]. This knowledge has made it possible to observe the fate of individual cells which ultimately led to the birth of the programmed cell death field [3]. *C. elegans* exist as males or hermaphrodites; thus isogenic lines can be propagated indefinitely in the lab, allowing for clonal studies as well as those on genetic drift. The genetic tractability of *C. elegans* was greatly enhanced by the full sequencing of its genome in 1998 (the first multicellular organism with a complete genome) which complemented established protocols for gene mutagenesis, transgenesis, and knockdown [4–8]. Compared to other model organisms, *C. elegans* exhibit less genetic redundancy, which can aid in genetic studies. Moreover, mutant strains are developed, collected, and distributed through the centralized *Caenorhabditis* Genetics Center (CGC) repository. As a simple, multicellular organism, *C. elegans* display highly stereotyped behaviors that can be dissected and studied in the lab. For example, sensory processing, which is key to the interaction of any animal with its environment, has been greatly aided by the knowledge gained from *C. elegans* [9]. In addition, were not for pioneering studies in *C. elegans,* the recognition of aging as a genetically regulated process would likely have been delayed for many years [10].

We have taken advantage of the rich genetic and behavioral tools in *C. elegans* to study the normal function of progranulin. The *C. elegans* PGRN-1 protein is less complex than its human counterpart, with three granulin domains rather than seven and a half (Fig. 1). PGRN-1 is secreted from intestinal tissue as well as a subset of head neurons, in a pattern similar to human progranulin [11, 12]. Nematodes lacking the PGRN-1 protein appear grossly normal, and their lifespan is not altered compared to wild-type animals [11]. Using *pgrn-1* mutant and rescue strains, we have identified novel functions for progranulin in programmed cell death and stress response [11, 13]. We also have shown that

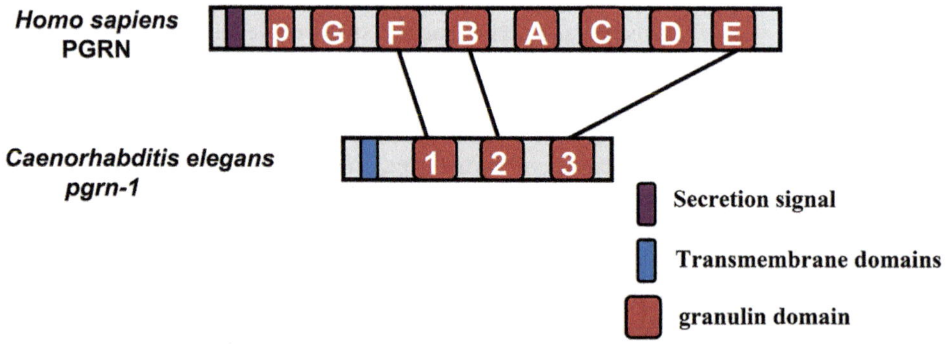

Fig. 1 A comparison of progranulin protein structure between humans and *C. elegans*. Individual granulin domains are indicated as red boxes

individual granulins can contribute to toxicity in a nematode model of neurodegenerative disease [14]. In this chapter, we will describe basic *C. elegans* techniques for investigation in these areas.

1.1 Progranulin in Programmed Cell Death Corpse Clearance

Apoptosis or programmed cell death is a process whereby multicellular organisms maintain cellular homeostasis. If cells are injured, redundant, or no longer necessary, they engage a death program and are eventually cleared by neighboring cells. Metamorphosis and neuronal pruning are two examples of apoptotic programmed cell death. The morphological characteristics of apoptosis include cell shrinkage, membrane blebbing, chromatin condensation, and nuclear fragmentation [15, 16]. Maintaining the balance of cell division and death is important to organismal health. Studies using transgenic and knockout mice provide evidence that disruption of apoptosis can promote tumor development [17].

C. elegans have been invaluable to our understanding of programmed cell death [18]. During the development of an adult hermaphrodite, 131 somatic cells and over 300 germ cells undergo programmed cell death. With differential interference contrast (DIC, Nomarski) microscopy, the timing and location of apoptotic cell death can be followed in the nematode in real time. Therefore, through a process of chemical mutagenesis, identification of mutants lacking apoptotic cells or "corpses," and subsequent mapping and cloning of the responsible genes, the core players in the programmed cell death pathway were first identified in *C. elegans*. Later, conserved homologs of these canonical cell death proteins, EGL-1 (BH3 only), CED-9(Bcl-2), CED-4(Apaf-1), and CED-3(Caspases-3), were found in vertebrates [18–20].

We characterized the process of developmental programmed cell death in *C. elegans pgrn-1* mutants. Using four-dimensional time-lapse Nomarski microscopy to monitor apoptotic corpses in the developing embryo, we have shown that nematodes lacking progranulin display more rapid clearance of apoptotic cells. The time from onset of apoptosis to complete cell clearance in *pgrn-1(−)* animals is almost half that of wild-type animals. Therefore one normal function of progranulin is to slow the rate of cell engulfment [11]. Methods for measuring cell death and clearance in *C. elegans* are described below.

1.2 Progranulin in Stress Response

All organisms must respond appropriately to a wide variety of physiological stressors in order to remain alive. *C. elegans* provides a simple, yet informative system in which to study the genetic, molecular, and physiological basis of organismal stress response [21]. The nematode responds to environmental stressors such as heat shock, oxidative stress, osmotic stress, and hypoxia with stereotyped behaviors [22–24]. The molecular mechanisms underlying these responses are highly conserved between *C. elegans* and mammals [25]. For example, HSF-1 and DAF-16 (the human ortholog FOXO (Forkhead transcription factor)) proteins are part of the heat shock response in both nematodes and mammalian cells, and these genes are implicated in the development of age-related disease in humans [26, 27]. Stress response behaviors can be assayed relatively easily in *C. elegans*. As such, using the nematode to study the effect of mutations in genes that have human homologs provides a powerful genetic system to investigate stress response and its relation to human disease such as cancer and neurodegeneration [28, 29].

Interestingly, loss of *pgrn-1* confers resistance to a variety of external stressors such as heat, osmotic, and ER (endoplasmic reticulum) stress [30]. Stress resistance conferred by *ced* mutations that either block programmed cell death or prevent the engulfment of dying cells is not additive with *pgrn-1* mutations, suggesting that these genes may act in the same pathway to promote organismal-wide stress resistance [30].

Recently, we have utilized a *C. elegans* model of neuronal TAR DNA-binding protein 43 (TDP-43) pathology to investigate the contribution of progranulin and its cleaved granulin products to the neurodegenerative process. Mutations in the human progranulin gene resulting in protein haploinsufficiency cause frontotemporal lobar degeneration (FTLD) with TDP-43 inclusions (FTLD-TDP), but the molecular interactions between these proteins remain unclear. We found that co-expression of TDP-43 with *C. elegans* granulins 2 and 3 (but interestingly not granulin 1) synergistically impairs reproduction, development, growth, and movement [14]. Cholinergic circuits appear particularly susceptible to the co-expression of TDP-43 with individual granulins. These circuits coordinate thrashing, a swim-like locomotion behavior in nematodes, as well as egg laying. These studies suggest that the presence of certain cleaved granulins, rather than or in addition to the loss of full-length progranulin, may contribute to neurodegenerative disease in TDP-43 proteinopathies.

In summary, many human disease proteins and pathways are conserved in invertebrates. This, in combination with the many genetic, molecular, cell biological, and behavioral approaches available in *C. elegans*, makes this organism a powerful system in which to study the function of the progranulin protein and the individual cleaved granulins.

2 Materials

2.1 *C. elegans* Strains

The Bristol strain N2 was used as the standard wild-type strain. *C. elegans* were cultured at 20 °C according to standard procedures on Nematode Growth Medium (NGM) plates with *E. coli* OP50 [31]. Descriptions of strains can be found at www.wormbase.org.

The *pgrn-1(tm985)* strain was provided by the Mitani Laboratory at the Tokyo Women's Medical University (Tokyo, Japan) and out-crossed four times to our laboratory's N2 strain. We generated transgenic strains expressing FLAG-tagged *C. elegans* granulin 1, 2, or 3 behind the endogenous progranulin promoter in a *pgrn-1(-)* background. These strains produce no full-length progranulin, only the individual granulin fragments [14]. Other published strains, including *pmk-1(km25)* and *daf-2(e1370)* control strains, were obtained from the *Caenorhabditis* Genetics Center (CGC) at the University of Minnesota.

2.2 Solution Recipes

1. M9 buffer: 5 g NaCl, 11.32 g Na_2HPO_4, 3 g KH_2PO_4, 200 μL 1 M $MgSO_4$. Add dH_2O to 1 L. Autoclave.
2. 2% agarose solution: prepared by melting 2 g of agarose in 100 mL deionized water using a microwave, until the agarose is completely dissolved. After use, the solution will cool and solidify and can be stored at room temperature for future use.
3. King's Broth: 10 g proteose peptone #2 (DIFCO), 1.5 g anhydrous K_2HPO_4, 15 g glycerol. Add dH_2O to bring volume to 1 L. Autoclave and then add 5 mL of sterile 1 M $MgSO_4$. Add antibiotics (as needed).
4. Juglone stock solution (12 mM): Add 2.1 mg of juglone (Sigma) to 1 mL EtOH. Juglone should be stored dry at −20 °C.

2.3 Preparation of Nematode Growth Medium (NGM) and PGS Petri Plates

C. elegans is maintained on NGM agar poured into petri plates. Several sizes of plates are available (35 mm, 60 mm, 100 mm) and can be bought from companies such as Falcon or Tritech Research. Standard plate size for *C. elegans* growth and maintenance is 60 mm.

1. NGM: 3 g NaCl, 17 g agar, 2.5 g peptone, 975 mL dH_2O. Autoclave for 50 min.
2. Cool flask to 55 °C in a water bath for 15 min.
3. Add 1 mL 1 M $CaCl_2$, 1 mL 5 mg/mL cholesterol in 100% ethanol, 1 mL 1 M $MgSO_4$, and 25 mL 1 M KPO_4 buffer.
4. Follow sterile procedures to dispense the NGM solution into petri plates.
5. Leave plates at room temperature for 1 day before use.
6. Plates stored at 4 °C in an air-tight container can be used for several weeks.
7. For peptone-glucose-sorbitol (PGS) fast-killing plates, make a solution of 1% bactopeptone, 1% NaCl, 1% glucose, 0.15 M sorbitol, and 1.7% bacto-agar. Autoclave and follow sterile procedure to pour into petri dishes at least 1 day prior to use. Store at 4 °C in an air-tight container.

2.4 Preparation of OP50 Bacterial Food Source

C. elegans is grown using the *E. coli* strain OP50 as a food source. OP50 is a uracil auxotroph that displays limited growth on NGM plates. This allows for easier observation and mating of animals.

1. LB Broth: 10 g Bacto-Tryptone, 5 g yeast extract, 5 g NaCl, 10 mL 1 M Tris pH 8; add dH_2O to 1 L. Autoclave.
2. Use a single OP50 colony to inoculate the LB Broth.
3. Grow inoculated cultures overnight at 37 °C.
4. OP50 liquid culture should be stored at 4 °C and can be used for 1 month.
5. Use sterile techniques to seed NGM plates with OP50 (we add 50 μL–35 mm plates and 150 μL–60 mm plates).

2.5 Preparation of Tunicamycin Plates for Assaying Endoplasmic Reticulum (ER) Stress

1. Resuspend tunicamycin (EMD Chemicals) in DMSO to form stock solutions. We make 1 mg/mL, 2 mg/mL, and 5 mg/mL stock concentrations, and store for up to 1 year at −20 °C.
2. Make NGM solution as above. After adding $CaCl_2$, cholesterol, $MgSO_4$, and KPO_4, add tunicamycin stock aliquots to the appropriate final concentrations in the NGM plates, e.g., for a 1 L solution of 5 μg/mL tunicamycin, add 1 mL of 5 mg/mL stock. For 0 μg/mL plates, add 1 mL of DMSO. Pour 35 mm tunicamycin plates.
3. After plates have solidified and cooled, stack and cover with foil (tunicamycin degrades with prolonged light exposure), and allow them to dry overnight at room temperature.
4. Place plates in opaque boxes and store at 4 °C until use. Plates are good for 2–4 months at 4 °C.

2.6 Preparation of Plates for Osmotic Stress Assays

1. Make solution for pouring NGM plates as above with the exception that extra NaCl should be added to generate plates with final salt concentration of 51, 200, 400, and 600 mM (*see* Table 1 for NaCl weights).
2. Add dH_2O to 500 mL. Autoclave and cool to 55 °C.
3. Add the following ingredients in order on a stir plate: 500 μL of 5 mg/mL cholesterol, 500 μL of 1 M $MgSO_4$, 12 mL of 1 M KPO_4.
4. Pour 15 mL of solution into 60 mm petri dishes, and arrange plates in a single layer on lab bench to cool. Once NGM has solidified, turn plates upside down and allow them to dry overnight at room temperature.
5. Wrap stacks of plates in parafilm to preserve moisture, and store at 4 °C for up to 1 week prior to use. Allow inverted plates to equilibrate to room temperature prior to use.

Table 1
Recipe for making NGM-based osmotic stress plates with increasing concentrations of NaCl

NaCl concentration (mM)	51	200	400	500	600
NaCl (g)	1.49	5.85	11.7	14.61	17.53
Agar (g)	10	10	10	10	10
Bactopeptone (g)	1.25	1.25	1.25	1.25	1.25

2.7 Microscopes, Software, and Other Equipment

1. Stereomicroscope for routine *C. elegans* strain maintenance.
2. AxioVision M2 microscope (Zeiss), or similar, fitted with 20×, 63×, and 100× Uplan Apo objectives, DIC microscopy accessories, and AxioCam MRm CCD camera (Zeiss).
3. AxioVision Software Release 4.8.2 SP3, or equivalent.
4. Stratalinker 1800 (Stratagene) or equivalent UV crosslinker.

2.8 Pathogenic Bacteria Strains

1. *Pseudomonas aeruginosa* (*P. aeruginosa*, strain PA14) [32].
2. *Salmonella typhimurium (S. typhimurium*, strain SL1344).

3 Methods

3.1 Cell Corpse Assays

Cells undergoing programmed cell death adopt a refractile and raised, button-like morphology that persists for 10–30 min when observed under DIC optics (Fig. 2). Since cells undergoing programmed cell death die at characteristic times in wild-type animals, mutations that affect cell death can be easily identified and characterized based upon altered morphology or pattern of these cell corpses. As cell death and engulfment are dynamic processes, and because we find that *pgrn-1* null and granulin-expressing animals show delayed development compared to wild-type animals, it is crucial to score the number of cell death corpses at defined developmental stages and regions of the animal. This allows meaningful comparisons between different genetic backgrounds. Here we describe protocols for monitoring the number of apoptotic cells and the kinetics of their clearance in the embryo, larvae, and adult *C. elegans*. Alternative methods of viewing germ cell corpses, which include staining and fluorescent transgene expression, can be found in WormBook [34].

3.1.1 General Protocol for Mounting Animals onto an Agarose Pad

1. Melt the 2% agarose solution by heating in a microwave. Pipet 1–2 mLs of melted agarose into a small glass test tube, and maintain in a heat block set at ~95 °C.
2. Drop ~50 μL of the hot agarose solution onto a microscope slide, and flatten immediately by placing a second microscope slide on top.

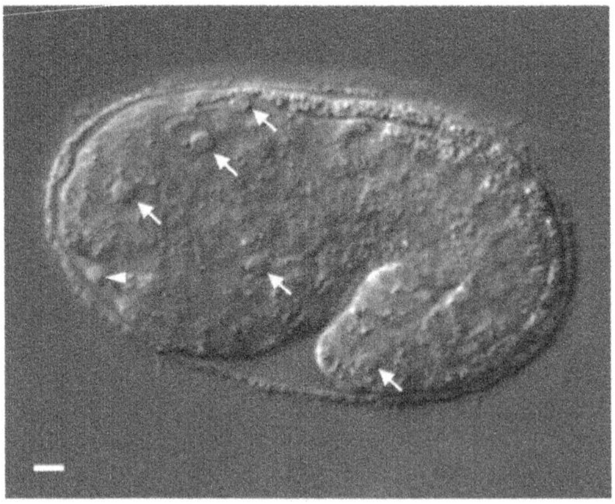

Fig. 2 The morphology of apoptotic cell corpses in a one and a half stage embryo. White arrows indicate the apoptotic cells that have a distinct refractile, raised, button-like structure. Arrowhead indicates a persistent "extruded" corpse, which is not counted in corpse total [33]. Scale bar: 10 μm

3. Once the agarose has solidified, remove the top microscope slide by sliding gently to expose the agarose pad.
4. Cut the agarose pad into ~1 cm^2 with the edge of a glass coverslip.
5. Add a 10 μL drop of 2.5 mM sodium azide (NaN_3) in M9 buffer to the center of the pad.
6. Use a stereomicroscope to pick animals of the appropriate stage (embryo, newly hatched L1 larvae, adult hermaphrodite) to the drop of NaN_3.
7. Gently place a coverslip over the drop of NaN_3, and seal the sides with Vaseline. This prevents the agarose pad from desiccation.

3.1.2 Somatic Cell Corpse Assay During Embryogenesis and Early Larval Stages

The majority of somatic cell deaths (113/131) in a *C. elegans* hermaphrodite occur during embryogenesis, mainly between 250 and 450 min post-fertilization [2, 3]. We commonly score embryonic cell deaths at the comma, 1.5-fold and fourfold stages (*see* **Note 1**). Post-hatching, apoptotic corpses are rarely seen. The exception is in mutants that delay timing of apoptosis or slow engulfment of corpses in which case apoptotic corpses can be quantified in L1 heads.

1. Use the 63× DIC objective to identify animals at the appropriate embryonic stage.
2. Identify and score the number of cell corpses using the 63× or 100× objectives.
3. Scrolling through the focus from the bottom to the top of the animal allows for cell corpses through the whole z-axis to be identified. Alternatively, a z-stack of DIC images can be taken through the animal at 0.3 μm and saved so that cell corpses can be scored at a later time.

3.1.3 Germline Cell Corpse Assay

Like somatic cells, some 300–500 cells in the adult hermaphrodite germline also undergo programmed cell death [35]. The germline is syncytial, such that many germ cell nuclei share a common cytoplasm. However, nuclei that are destined to die become cellularized [35]. Germline cell corpses also show the refractile, raised button-like morphology similar to that of somatic cell corpses, except that germline cell corpses are slightly larger in size than somatic cell corpses (Fig. 3). Cell death in the germline appears to occur in the region where nuclei are arrested in the pachytene stage of meiosis, close to the turn of the gonadal arm (Fig. 3). Occasionally, mature oocytes have been observed to undergo programmed cell death [36].

1. Mount day 1 or 2 adults (24 or 48 h after young adult stage) on a 2% agar pad with a 50 μL drop of 2.5 mM NaN_3 in M9 buffer.
2. *C. elegans* have anterior and posterior gonadal arms that each forms a "U shape" in the body cavity. Identify the gonadal arm that is facing "up" (not obscured by intestinal fat). Count the number of cell corpses in the pachytene region of either the anterior or posterior gonadal arm.

3.2 Four-Dimensional Time-Lapse Nomarski Microscopy to Measure Cell Death Kinetics

To measure the kinetics of cell corpse formation and clearance in the embryo, we use four-dimensional time-lapse Nomarski microscopy as described previously [37].

1. Identify embryos at the appropriate developmental stage, from bean-comma to twofold stage.

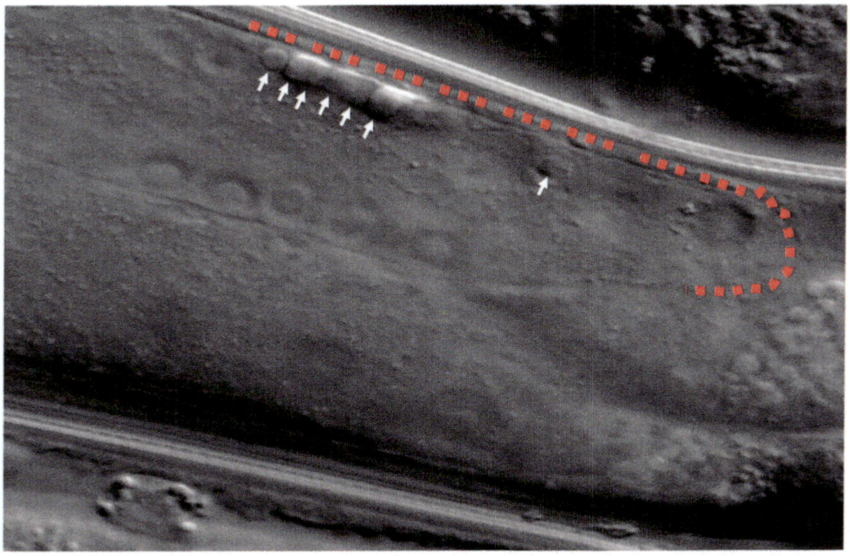

Fig. 3 The morphology of apoptotic cell corpses in the germline of adult hermaphrodite *C. elegans*. Germline apoptosis occurs in the pachytene region at the U-shaped bend of the gonad arm. White arrows indicate the apoptotic cells. Red dotted line indicates the pachytene region

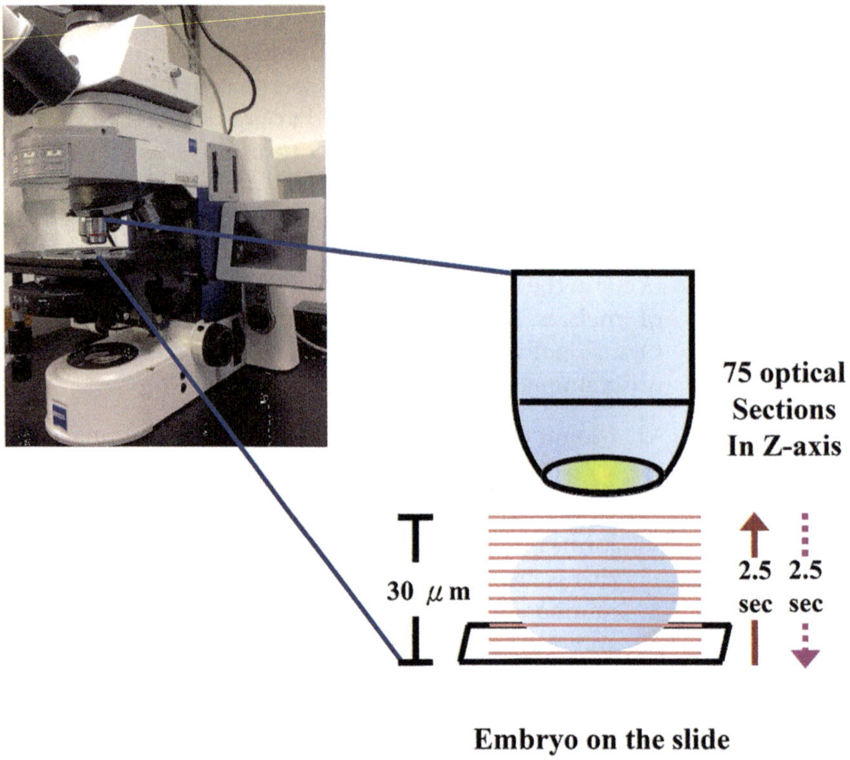

Fig. 4 Microscopy equipment for recording apoptotic cell corpse duration. Serial z-section images over 30 μm will capture the entire thickness of the embryo. We take images at 0.30 μm intervals every 2 min between ~200 and 290 min post-fertilization

2. Using the 63× or 100× DIC objective, set up the microscope parameters for performing serial z-section images. We take images at 0.30 μm intervals every 2 min (Fig. 4). We usually set the white light DIC exposure time to 70 ms.
3. Check images every so often to ensure the focal plane is correct, and adjust during the recording as necessary.
4. Use the recorded images to identify cell corpses based on characteristic morphology, and measure their duration time.

3.3 Stress Assays

We find that *C. elegans* mutations altering the normal course of programmed cell death and apoptotic cell clearance confer animal-wide resistance to a specific set of environmental stressors, namely, ER, heat, and osmotic stress [30]. Below we describe protocols for assaying resistance to common environmental stressors.

3.3.1 The Unfolded Protein Response and ER Stress

1. Pick ~30–40 L4 stage animals to a seeded plate, and let them lay eggs overnight.
2. Transfer 50 unhatched eggs to 35 mm plates containing 0, 1, 2, or 5 mg/mL of tunicamycin. We perform each dose and strain in triplicate (*see* **Note 2**).

3. After 3 days, quantify the number of animals that developed to the L4 stage, which is typically determined by a combination of animal size and maturity of the vulva (*see* **Note 3**).
4. Data can be plotted as the fraction of animals that develop to L4 stage normalized to percent hatching on 0 mg/mL tunicamycin for each strain.

3.3.2 Thermal Stress

1. Pick 50 L4 stage animals per strain and grow overnight at 20 °C on NGM plates.
2. Move day 1 adult worms to 35 °C for 5–12 h (*see* **Note 4**).
3. Score animals for survival. Animals that fail to move in response to a gentle touch with a metal pick are scored as dead.

3.3.3 Osmotic Stress

1. Pick L4 stage animals and grow overnight on OP50 at 20 °C on NGM plates.
2. Expose day 1 adult worms to 51, 200, 400, and 600 mM NaCl added to NGM plates at 20 °C, as previously described [30]. Osmotic stress plates should be seeded with OP50 a few hours prior to use.
3. Score animals for survival after 24 h. Animals that fail to move in response to a gentle touch with a metal pick are scored as dead.

3.3.4 Oxidative Stress

Paraquat

1. Place individual animals in 96-well plates with 100 μL of 250 μM methyl viologen (Paraquat, Sigma-Aldrich) dissolved in M9 [38].
2. Score animals for movement every hour at 25 °C.

Juglone

1. Pick 30 L4 worms into a 30uL drop of M9 containing 250 μM juglone (Sigma) from a 12 mM stock solution [39].
2. A drop of M9 containing an equal amount of ethanol is used as a control.
3. After 20 min, worms are washed off the slide with 100 μL M9 onto seeded NGM plates.
4. Score animals for survival 18 h later.

3.3.5 Genotoxic Stress

1. Transfer day 1 adult animals to unseeded plates and treat with 1200 J/m2 UV light in a Stratalinker 1800 (Stratagene).
2. Score animals for survival every 24 h. Animals that fail to move in response to a gentle touch with a metal pick are scored as dead.

3.3.6 Pathogen Stress

1. Prepare a King's broth culture of *Pseudomonas aeruginosa* (*P. aeruginosa*, strain PA14) [32] or *Salmonella typhimurium* (*S. typhimurium*, strain SL1344) [40] overnight.
2. With a glass rod, spread 10 μL of bacterial culture onto each 35 mm plate (*see* **Note 5**).
3. Incubate seeded plates for 8–24 h (slow killing, on NGM plates) or 8–12 h (fast killing, on PGS plates) at room temperature.
4. Transfer 15–20 L4 stage worms to each plate. Each strain should have 3–5 replicates per trial. Animals can be placed on an unseeded plate to remove residual OP50.
5. *pmk-1(km25)* animals can be used as a short-lived control. *Daf-2(e1370)* animals can be used as a long-lived control.
6. Grow animals at 20 °C and score animals every 4–6 h (fast killing) or once a day (slow-killing) for survival. Animals that fail to move in response to a gentle touch with a metal pick are scored as dead.

4 Notes

1. To enrich for early stage eggs, 25–50 L4 animals can be picked to a seeded NGM plate the day before the assay. The next morning, adult animals can be washed off with M9 buffer, leaving a plate with several hundred eggs for assaying.
2. We find that tunicamycin plates appear to get stronger over time, even with storage at 4 °C after pouring. We wait at least a week before we use newly poured plates.
3. Often we also score development to L4 stage at 4 days after plating eggs. Some slower-growing strains showing stress resistance give significant results at this time point compared to wild-type animals, which is less clear at day 3. ER stress is the most robust stress assay in our hands.
4. Ensure the incubator is maintained at 35 °C for at least 24 h prior to the experiment. We place plates within a plastic box that is kept in the incubator without a lid. Plates are spaced apart and never stacked. We use 35 mm plates for heat stress assays, as the smaller volume will reach assay temperature more quickly.
5. *P. aeruginosa* has two forms of killing: slow killing when grown on minimal NGM and fast killing when grown on peptone-glucose-sorbitol (PGS) media. Animals grown on *S. typhimurium* should be transferred onto new plates every day to prevent overcrowding by progeny.

Acknowledgments

This work was supported by grants from the NIH R21 NS082709, Consortium for Frontotemporal Dementia (CFR), Tau Consortium and Hellman Family Foundation to AWK. For strains, we thank Kaveh Ashrafi (UCSF), Cynthia Kenyon (UCSF), the Mitani Laboratory at the Tokyo Women's Medical University (Tokyo, Japan), and the *C. elegans* Genetic Center (CGC), which is funded by NIH Office of Research Infrastructure Programs (P40 OD010440).

References

1. Conradt B, Xue D (2005) Programmed cell death. WormBook
2. Sulston JE, Schierenberg E, White JG, Thomson JN (1983) The embryonic cell lineage of the nematode Caenorhabditis elegans. Dev Biol 100:64–119
3. Sulston JE, Horvitz HR (1977) Post-embryonic cell lineages of the nematode, Caenorhabditis elegans. Dev Biol 56:110–156
4. C.elegansSequencingConsortium (1998) Genome sequence of the nematode C. elegans: a platform for investigating biology. Science 282:2012–2018
5. Dickinson DJ, Ward JD, Reiner DJ, Goldstein B (2013) Engineering the Caenorhabditis elegans genome using Cas9-triggered homologous recombination. Nat Methods 10:1028–1034
6. Fire A, Xu S, Montgomery MK, Kostas SA, Driver SE, Mello CC (1998) Potent and specific genetic interference by double-stranded RNA in Caenorhabditis elegans. Nature 391:806–811
7. Frokjaer-Jensen C, Davis MW, Hopkins CE, Newman BJ, Thummel JM, Olesen SP, Grunnet M, Jorgensen EM (2008) Single-copy insertion of transgenes in Caenorhabditis elegans. Nat Genet 40:1375–1383
8. Jorgensen EM, Mango SE (2002) The art and design of genetic screens: Caenorhabditis elegans. Nat Rev Genet 3:356–369
9. Bargmann CI (2006) Chemosensation in *C. elegans*. WormBook
10. Kenyon CJ (2010) The genetics of ageing. Nature 464:504–512
11. Kao AW, Eisenhut RJ, Martens LH, Nakamura A, Huang A, Bagley JA, Zhou P, de Luis A, Neukomm LJ, Cabello J et al (2011) A neurodegenerative disease mutation that accelerates the clearance of apoptotic cells. Proc Natl Acad Sci U S A 108:4441–4446
12. Daniel R, He Z, Carmichael KP, Halper J, Bateman A (2000) Cellular localization of gene expression for progranulin. J Histochem Cytochem 48:999–1009
13. Judy ME, Nakamura A, Huang A, Grant H, McCurdy H, Weiberth KF, Gao F, Coppola G, Kenyon C, Kao AW (2013) A shift to organismal stress resistance in programmed cell death mutants. PLoS Genet 9:e1003714. PMCID: PMC24068943
14. Salazar DA, Butler VJ, Argouarch AR, Hsu TY, Mason A, Nakamura A, McCurdy H, Cox D, Ng R, Pan G et al (2015) The Progranulin cleavage products, Granulins, exacerbate TDP-43 toxicity and increase TDP-43 levels. J Neurosci 35:9315–9328
15. Kerr JF, Wyllie AH, Currie AR (1972) Apoptosis: a basic biological phenomenon with wide-ranging implications in tissue kinetics. Br J Cancer 26:239–257
16. Wyllie AH, Kerr JF, Currie AR (1980) Cell death: the significance of apoptosis. Int Rev Cytol 68:251–306
17. Yin C, Knudson CM, Korsmeyer SJ, Van Dyke T (1997) Bax suppresses tumorigenesis and stimulates apoptosis in vivo. Nature 385: 637–640
18. Ellis RE, Jacobson DM, Horvitz HR (1991) Genes required for the engulfment of cell corpses during programmed cell death in Caenorhabditis elegans. Genetics 129:79–94
19. Horvitz HR (1999) Genetic control of programmed cell death in the nematode Caenorhabditis elegans. Cancer Res 59: 1701s–1706s
20. Metzstein MM, Horvitz HR (1999) The C. elegans cell death specification gene ces-1

encodes a snail family zinc finger protein. Mol Cell 4:309–319
21. Rodriguez M, Snoek LB, De Bono M, Kammenga JE (2013) Worms under stress: C. elegans stress response and its relevance to complex human disease and aging. Trends Genet 29:367–374
22. Culotti JG, Russell RL (1978) Osmotic avoidance defective mutants of the nematode Caenorhabditis elegans. Genetics 90:243–256
23. Hedgecock EM, Russell RL (1975) Normal and mutant thermotaxis in the nematode Caenorhabditis elegans. Proc Natl Acad Sci U S A 72:4061–4065
24. Van Voorhies WA, Ward S (2000) Broad oxygen tolerance in the nematode Caenorhabditis elegans. J Exp Biol 203:2467–2478
25. Lant B, Storey KB (2010) An overview of stress response and hypometabolic strategies in Caenorhabditis elegans: conserved and contrasting signals with the mammalian system. Int J Biol Sci 6:9–50
26. Anckar J, Sistonen L (2007) Heat shock factor 1 as a coordinator of stress and developmental pathways. Adv Exp Med Biol 594:78–88
27. Hsu AL, Murphy CT, Kenyon C (2003) Regulation of aging and age-related disease by DAF-16 and heat-shock factor. Science 300:1142–1145
28. Harrington AJ, Hamamichi S, Caldwell GA, Caldwell KA (2010) C. elegans as a model organism to investigate molecular pathways involved with Parkinson's disease. Dev Dyn 239:1282–1295
29. Rea SL, Graham BH, Nakamaru-Ogiso E, Kar A, Falk MJ (2010) Bacteria, yeast, worms, and flies: exploiting simple model organisms to investigate human mitochondrial diseases. Dev Disabil Res Rev 16:200–218
30. Judy ME, Nakamura A, Huang A, Grant H, McCurdy H, Weiberth KF, Gao F, Coppola G, Kenyon C, Kao AW (2013) A shift to organismal stress resistance in programmed cell death mutants. PLoS Genet 9:e1003714
31. Brenner S (1974) The genetics of Caenorhabditis elegans. Genetics 77:71–94
32. Tan MW, Mahajan-Miklos S, Ausubel FM (1999) Killing of Caenorhabditis elegans by Pseudomonas aeruginosa used to model mammalian bacterial pathogenesis. Proc Natl Acad Sci U S A 96:715–720
33. Denning DP, Hatch V, Horvitz HR (2012) Programmed elimination of cells by caspase-independent cell extrusion in C. elegans. Nature 488:226–230
34. Gartner (2008) Germline survival and apoptosis. WormBook
35. Gumienny TL, Lambie E, Hartwieg E, Horvitz HR, Hengartner MO (1999) Genetic control of programmed cell death in the Caenorhabditis elegans hermaphrodite germline. Development 126:1011–1022
36. Riddle DL, Blumenthal T, Meyer BJ, et al (1997) Programmed cell death in *C. elegans. C. elegans* II
37. Schnabel R, Hutter H, Moerman D, Schnabel H (1997) Assessing normal embryogenesis in Caenorhabditis elegans using a 4D microscope: variability of development and regional specification. Dev Biol 184:234–265
38. Possik E, Pause A (2015) Measuring oxidative stress resistance of caenorhabditis elegans in 96-well microtiter plates. J Vis Exp: JoVE
39. Gomez F, Monsalve GC, Tse V, Saiki R, Weng E, Lee L, Srinivasan C, Frand AR, Clarke CF (2012) Delayed accumulation of intestinal coliform bacteria enhances life span and stress resistance in Caenorhabditis elegans fed respiratory deficient E. coli. BMC Microbiol 12:300
40. Aballay A, Yorgey P, Ausubel FM (2000) Salmonella typhimurium proliferates and establishes a persistent infection in the intestine of Caenorhabditis elegans. Curr Biol 10: 1539–1542

Chapter 15

Application of Zebrafish and Knockdown Technology to Define Progranulin Neuronal Function

Babykumari P. Chitramuthu and Hugh P. J. Bennett

Abstract

The zebrafish (*Danio rerio*), a small tropical fish, has become a powerful model for the study of early vertebrate development, human diseases, and drug screening. Zebrafish provides large numbers of optically clear embryos, and its development is very rapid. Overexpression or under-expression of proteins can be effectively achieved by microinjection of mRNA or morpholino antisense oligonucleotides (MOs), respectively, into developing embryos at the 1–2 cell stage. The function of a particular protein can be revealed by correlating gene expression patterns with the phenotypes observed from over- or under-expression. We defined the expression pattern of zebrafish progranulin A (zfPGRN-A), an orthologue to the single human PGRN by whole-mount in situ hybridization (ISH) and immunofluorescence (IF). The MO-mediated knockdown of zfPGRN-A expression generated embryos that display abnormal motor neuron development resulting in touch-evoked swimming deficits.

Key words Zebrafish, Microinjection, Morpholino, PGRN, Motor neuron, In situ hybridization, Immunofluorescence

1 Introduction

The zebrafish, a small tropical fish, has become a popular model for the study of gene function during vertebrate development and disease [1, 2]. It has the ability to regenerate damaged tissues [3, 4] and is being used in the identification of potential therapeutics [5, 6]. Zebrafish are maintained at 28.5°C in our laboratory on a 14-h/10-h light/dark cycle [7]. The temporal and spatial expression of target genes can be determined by in situ hybridization and immunofluorescence analyses. Morpholino antisense oligonucleotide (MO)-mediated knockdown technology coupled with microinjection is used to define gene function [8, 9]. The optical clarity of embryos facilitates unhindered real-time observation during organogenesis and assessment of gene expression. Protein knockdown or overexpression can be effectively achieved by microinjection of MOs or mRNA, respectively, into developing embryos at

Andrew Bateman et al. (eds.), *Progranulin: Methods and Protocols*, Methods in Molecular Biology, vol. 1806,
https://doi.org/10.1007/978-1-4939-8559-3_15,

the 1–2 cell stage. Analyzing phenotypes resulting from over- or under-expression can be correlated with gene expression patterns and thus can reveal the function of a particular protein.

To define zfPGRN-A function during development, we performed whole-mount in situ hybridization (ISH) and immunofluorescence (IF) analyses to first characterize its expression pattern. The temporal and spatial expression of zfPGRN-A mRNA are determined by ISH technique, and that of zfPGRN-A protein are achieved by IF. The ISH of whole-mount zebrafish embryos was assessed using antisense RNA probes labeled with digoxigenin, which were synthesized by in vitro transcription of cloned and linearized template of zfPGRN-A with digoxigenin-linked nucleotides. After being dechorionated, fixed, and permeabilized, the postfixed developmentally staged embryos are incubated with zfPGRN-A sense and antisense probes. After washing away the excess probes, the embryos are incubated with alkaline phosphatase-conjugated anti-digoxigenin antibody. PGRN expression is detected using a chromogenic substrate for alkaline phosphatase. Using the IF technique, the dechorionated, fixed, permeabilized, and postfixed embryos were incubated with zfPGRN-A antibody. After extensive washing in phosphate buffered saline Tween (PBST) solution, the embryos were incubated with the appropriate secondary antibody conjugated with Alexa-488. The zfPGRN-A immunoreactivity was observed using a GFP filter-equipped microscope.

In order to define the biological function of PGRN, we knocked down zfPGRN-A expression by employing a microinjection technique [10]. In our protocol, MO directed against zfPGRN-A is filled within a micro-capillary filament and attached to the holder of the microinjector. Air pressure forces the solution out of the filament into the cytoplasm of 1–2 cell-staged embryos. The injected embryos are incubated at 28.5 °C to develop further. The injected MO can be monitored inside all cells of the developing embryo. Blocking of translation initiation is achieved by the MO binding to target mRNA within each cell with high affinity and specificity. Marker gene approaches following over- or under-expression reveal the functional roles of the protein products by the morphological, physiological, or molecular changes.

2 Materials

2.1 Whole-Mount In Situ Hybridization

2.1.1 Digoxigenin-Labeled RNA Probe Synthesis

1. Agarose.
2. Buffer-saturated phenol (ultrapure): Store at 4 °C. This is an eye, skin, and respiratory tract irritant and suspected carcinogen. Care should be taken while handling.
3. DNA templates: cDNA or genomic DNA from zebrafish embryos or adult tissue to amplify PCR product for probe.

4. Gene-specific PCR primers to make sense and antisense riboprobes for *GRN* and antisense riboprobes for *SHH*.
5. Chloroform: Toxic and suspected carcinogen. Work under fume hood.
6. DIG-RNA Labeling Mix with Sp6, T7, and T3 RNA polymerase (Roche, Laval, QC).
7. Diethyl pyrocarbonate (DEPC): Add 1 mL of DEPC in 1 L of water, incubate at 50 °C overnight, autoclave for 20 min, and bring to room temperature before use. DEPC is an eye, skin, and respiratory irritant. Avoid contact with skin and eyes. Use safety glasses, gloves, and mask.
8. EDTA 0.5 M (pH 8.0): Combine 186.1 g of EDTA in 800 mL of water. Add 15 g of NaOH pellets and make up to 1 L by adding water when all pellets are dissolved.
9. HQ Mini Plasmid Purification Kit (Invitrogen).
10. NaOH.
11. *Not*I and *Hin*dIII restriction enzymes (Invitrogen).
12. RNase-free DNAse I (Roche): Prepare small aliquots, and store at −20 °C.
13. RNase inhibitor (Roche).
14. *Taq* polymerase (Invitrogen).
15. TOPO TA Cloning Kit-Dual Promoter pCRII-TOPO (Invitrogen).
16. 3.0 M sodium acetate (pH 5.5).
17. 100% ethanol.
18. 75% ethanol.
19. Agarose gel electrophoresis equipment.
20. One Shot competent cells.
21. 4 M LiCl.

2.1.2 Embryo Preparation

1. Dumont (Watchmaker's) Forceps pattern no. 5 (Fine Science Tools, North Vancouver, B.C.).
2. 100× egg water: Prepare by dissolving 6 g of Instant Ocean salt (Aquarium Systems Inc., Mentor, OH) in 1 L of water. Add one part of salt with 99 parts of water to make a working solution.
3. Methanol.
4. Paraformaldehyde (PFA): Make fresh 4% (w/v) solution in PBS for each experiment. The solution needs to be carefully heated using a stirring hot plate in a fume hood to dissolve, then cool to room temperature, and store at −20 °C as 50 mL aliquots for later use. PFA is toxic. Use gloves, safety glasses, and dust mask.

5. Phenylthiocarbamide (PTC): 0.3% (w/v) in egg water. PTC is highly toxic. Use gloves, safety glasses, and dust mask.
6. Phosphate buffered saline (PBS, Fisher): Reconstitution in 1 L of water results in 10× PBS. Autoclave before storage at room temperature. Prepare working solution by dilution of one part with nine parts sterile distilled water.
7. Liquid nitrogen for freezing the embryos to be used for RNA analysis.
8. cDNA synthesis kit.

2.1.3 Whole-Mount In Situ Hybridization

1. Six-well cell culture cluster (Costar).
2. Acetic anhydride.
3. Acetylation mix: combine 125 μL of triethanolamine and 27 μL of acetic anhydride in 10 mL sterile water.
4. AG501-X8 Resin (Bio Rad, Mississauga, ON).
5. Bovine serum albumin (purified fraction V) (BSA): Prepare 100 mg per mL and store at −20 °C.
6. Alkaline Tris (pre-staining) buffer: 100 mM Tris HCl, pH 9.5, 50 mM $MgCl_2$, 100 mM NaCl, 0.1% Tween 20 (v/v), and 1 mM levamisole.
7. Baskets are prepared using nylon mesh and Eppendorf tubes (*see* Fig. 1). To make the baskets, trim Eppendorf tubes (1.5 mL) with or without rims to remove the conical end. Cut nylon mesh into small pieces corresponding to the size of the cut ends of the Eppendorf tubes. In the fume hood, positon tubes with the nylon mesh covering the cut end on an electrical hot plate until both the tube and the nylon mesh stick together. Break off the excess mesh from the baskets and preserve them in 100% methanol until used.
8. Blocking buffer: 1× PBST, 2% calf serum (vol/vol), 2 mg/mL BSA.
9. Citric acid monohydrate: Make 1 M solution. Filter and preserve at 4 °C.
10. Deionized formamide: Add formamide to 10 g /L of AG501-X8 and mix slowly for 30 min and repeat once. Filter the solution to take away the resin, aliquot into 20 mL and preserve in the dark at −20 °C.
11. Formamide, high-purity grade: Formamide is highly corrosive on contact with the skin or eyes. It is also a teratogen. Handle with proper safety attire including gloves and goggles.
12. Glycerol.
13. Heparin sodium salt: Make stock solution as 50 mg/mL and store at 4 °C.

Fig. 1 *Eppendorf-nylon mesh baskets used to hold staged embryos.* Staged embryos from different experimental conditions are placed in baskets with or without rims. This permits easy orientation of the order of the baskets while performing various steps involved in the protocol

14. Hybridization mix (HM): 50% deionized formamide, 5× SSC, 0.1% Tween 20, 50 μg/mL of heparin, 500 μg/mL of RNase-free tRNA; adjust to pH 6.0 by adding citric acid (460 μL of 1 M citric acid solution for 50 mL of HM). Prepare mix without tRNA, and then add tRNA only to the volume needed for pre-hybridization and hybridization.
15. Levamisole hydrochloride: 100 mM, prepare fresh when needed.
16. Methanol.
17. $MgCl_2$: Prepare 1 M stock and store at room temperature.
18. NaCl: Make 5 M stock solution and save at room temperature.
19. Nitro blue tetrazolium (NBT). Dissolve 50 mg of NBT in 0.7 mL of *N*, *N*-dimethylformamide and 0.3 mL of sterile water and store at −20 °C. Protect this solution from light.
20. N,N-Dimethylformamide anhydrous (DMF): Irritant, toxic, combustible, and suspected teratogen. Handle with proper safety attire including gloves and goggles.
21. Paraformaldehyde (PFA): prepare solution in PBS as outlined in Subheading 2.1.2.
22. Phosphate buffered saline (PBS) as outlined in Subheading 2.1.2.

23. Phosphate buffered saline Tween (PBST): 10% (v/v). Polyoxyethylene sorbitan monolaurate (Tween 20) (Sigma) stock solution is prepared by diluting 10 mL of Tween 20 with 90 mL of sterile water. After complete homogenization of the 10% Tween 20 solution, it is stored at room temperature. Guard this solution from light. Combine one part of 10× PBS, one part of 10% Tween 20 with eight part of sterile water to make 1× PBST.
24. Proteinase K.
25. Saline-sodium citrate buffer (SSC) (Sigma): Dissolve powdered blend in 1 L of water to make 20× SSC stock solution and store at room temperature.
26. Sheep anti-digoxigenin-AP Fab fragments (Roche Diagnostics, Laval, QC): Store at 4 °C.
27. Staining buffer: Add 23 μL of 50 mg/mL NBT and 13 μL of 50 mg/mL BCIP in 4 mL of alkaline Tris (pre-staining buffer). Prepare fresh when needed.
28. Stop solution: 1× PBS, pH 5.5, 1 mM EDTA, 0.1% Tween 20 (v/v).
29. Triethanolamine.
30. Tris–HCl: 1 M stock solution Tris–HCl pH 9.5. Dissolve 157.1 g of Tris base in 900 mL of water, adjust pH to 9.5 by adding HCl, and then adjust the volume to 1 L by adding distilled water.
31. tRNA from baker's yeast type X, lyophilized powder (Sigma): Resuspend in water at a concentration of 50 mg/mL and preserve at −20 °C as small aliquots.
32. 5-Bromo 4-chloro 3-indolyl phosphate (BCIP). Dissolve 50 mg of BCIP in 1 mL of *N*, *N*-dimethylformamide anhydrous and store at −20 °C. Protect this solution from light.

2.2 Whole-Mount Immunofluorescence

2.2.1 Embryo Preparation

Same as listed for in situ hybridization (*see* Subheading 2.1.2 for in situ technique).

2.2.2 Whole-Mount Immunofluorescence

1. Blocking buffer: 1× PBST, 5% calf serum (v/v), 1% DMSO.
2. Dimethyl sulfoxide (DMSO).
3. Fetal calf serum (Invitrogen).
4. Methanol.
5. Paraformaldehyde (PFA): prepare solution in PBS as outlined in Subheading 2.1.2.
6. Phosphate buffered saline (PBS).

7. Phosphate buffered saline Tween (PBST): prepare as outlined in Subheading 2.1.3.
8. Proteinase K.
9. Primary antibody: The rabbit polyclonal anti-zfPGRN-A was raised against synthetic peptide corresponding to residues 242–256 of zfPGRN-A (RAEWEDHKQKKPETQC) by solid-phase chemistry, and polyclonal antibodies were raised in rabbits by immunization with the peptides conjugated to keyhole limpet hemocyanin via the carboxyl-terminal cysteine. Antisera were purified on peptide affinity columns prior to use. Peptide synthesis, antigen conjugation, antibody production, and purification were carried out at the Sheldon Biotechnology Centre, McGill University, Montreal, Quebec, Canada.
10. Secondary antibody: Alexa Fluor 594 donkey anti-rabbit IgG antibody.
11. znp1 monoclonal antibody (1:200), a marker for motor neurons (MNs).
12. Secondary antibody: Alexa Fluor 488 goat anti-mouse.

2.3 Microinjection

2.3.1 Embryo Preparation

1. Wild-type zebrafish: purchased from Aquatica Tropicals (Florida).
2. Egg water: Prepare 100× egg water as described in Subheading 2.1.2. Combine one part of salt with 99 parts of distilled water to prepare a working solution.
3. Nylon netting to catch and transfer fish between tanks.
4. Petri dishes.
5. Zebrafish housing system (Aquaneering Inc. San Diego, CA).
6. Laboratory aquarium (Allentown Caging Equipment Co).

2.3.2 Microinjection

1. Dissecting microscope (Leica, MS 5).
2. Dumont #5 watchmaker forceps.
3. Fluorescein isothiocyanate (FITC).
4. Incubator (28.5 °C).
5. Microinjection pipettes: Microinjection pipettes are prepared from standard-walled glass capillaries with an inner filament (1.0 mm × 0.58 mm). Microinjection pipettes (needle) are drawn using a needle puller (Sutter Instrument Company, Novato, CA). Needle can be broken under the dissecting microscope using the highest magnification by gently scraping the tip of the needle with clean forceps immediately prior to use to produce a sharp pipette tip to penetrate the egg chorion during microinjection.

6. Microinjection station to hold embryos during microinjection. Pour about 40 mL of 2% agarose in a Petri dish (150 mm × 15 mm) and let it solidify. Prepare a furrow by making a wedge-shaped cut using a sterile scalpel. Two furrows can be made per dish.
7. Microinjection system (PLI-100 microinjection system, Harvard Apparatus, Saint-Laurent, QC). This consists of a micropipette holder to hold the micropipette, dissecting microscope, nitrogen gas cylinder connected to a microinjector (to apply pressure to deliver volumes from femtoliters to microliters), and a foot switch. Microinjection of FITC tagged antisense MO into 1–2 cell-staged embryo using microinjector and uniform distribution of morpholino antisense oligonucleotide (Fig. 2).
8. Morpholino antisense oligonucleotide (MO) are directed against the initiator ATG sequence and 5' UTR of the zfP-GRN-A gene (Gene tools, Philomath, OR). MOs are diluted in 1× Danieau buffer with 0.1% phenol red and 0.05% FITC-dextran. Phenol red facilitates the estimation of the volume of DNA injected. FITC also indicates the efficiency of incorporation of MO inside the developing embryo.
9. 0.5% phenol red in DPBS.
10. 1× Danieau buffer: 58 mM NaCl, 0.7 mM KCl, 0.4 mM $MgSO_4$, 0.6 mM $Ca(NO_3)2$, and 5.0 mM HEPES pH 7.6.

3 Methods

3.1 Whole-Mount In Situ Hybridization

In order to gain greater insights into the relative abundance and specific location of the *GRN* gene within the embryo during development, the analysis of the spatiotemporal expression was carried out by in situ hybridization. For all the stages examined, the sense riboprobe of the corresponding *GRN* gene was used as a negative control. Because of the discrete and well-documented expression pattern known for sonic hedgehog (*SHH*), an antisense riboprobe for this gene was used as a positive control [11]. Embryo preparation and probe synthesis for whole-mount in situ hybridization are illustrated in Fig. 3. Diagrammatic representation of the steps involved in the whole-mount in situ hybridization technique is emphasized in Fig. 4.

3.1.1 Digoxigenin-Labeled RNA Probe Synthesis

PCR Amplification and Gene Cloning for Probe Template

1. The following PCR reaction (50 μL) is set up with genomic DNA or cDNA as DNA templates (Table 1; *see* **Note 1**).
2. The size of the PCR product is confirmed by agarose gel electrophoresis (*see* **Note 2**).
3. The TOPO cloning reaction is set up as the options shown in Table 2 depending on whether chemically competent or electro-competent *E. coli* cells is used in downstream transfor-

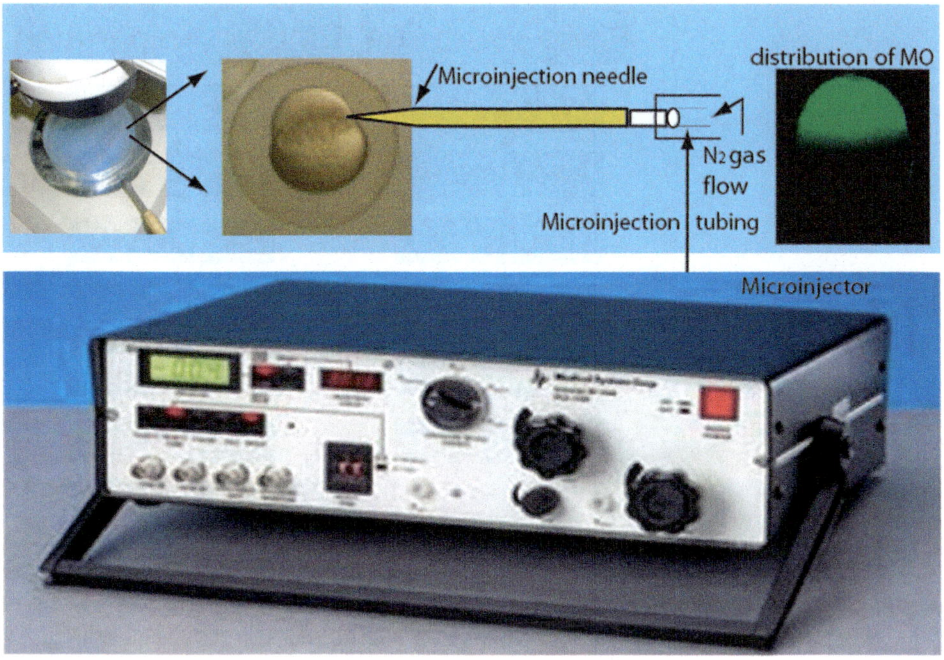

Fig. 2 *Microinjection system for zebrafish embryos used in zebrafish gene knockdown technology.* The microinjection system consists of a micropipette holder to hold a micropipette with its tips cut to form the microinjection needle, dissecting microscope, nitrogen gas cylinder connected to a microinjector (to apply pressure to deliver volumes from femtoliters to microliters), and footswitch. Embryos are lined in the furrow of microinjection chamber (A), and the microinjection needle is loaded with fluorescein-tagged morpholino antisense oligonucleotide (MO). The MO is injected into the cytoplasm of the 1–2 cell-staged embryo (B), and the distribution of MO (C) is monitored by fluorescence microscopy, while the embryos are maintained at 28.5 ° C until collected

mation. PCR product and TOPO vector from TOPO TA Cloning Kit are mixed together and incubated for 5 min at room temperature.

4. The TOPO cloning product is transformed into One Shot competent cells (chemically competent *E. coli*).
5. The presence of insert is analyzed and selected based on white (with cloned insert sequence) or light blue (without insert) colonies.

Preparing the DNA Template from Plasmid Clones

1. Following plasmid purification 5 μg of plasmid is linearized for each probe by digesting with the appropriate restriction enzyme that has a unique site located 5′ (for antisense probes) or 3′ (for sense probes) to the insert. zfPGRN-A antisense riboprobe was transcribed using SP6 polymerase with *Not*I-linearized templates, and the corresponding sense riboprobes were transcribed using T7 polymerase with *Hin*dIII-linearized templates.
2. Linearized DNA is purified by phenol/chloroform extraction.

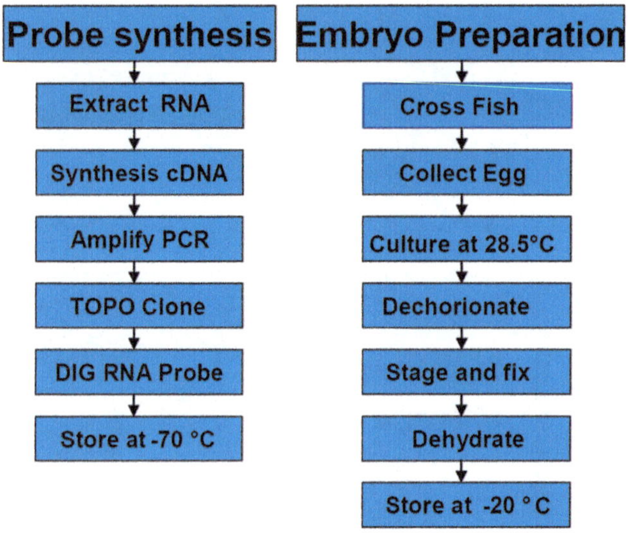

Fig. 3 *Workflow scheme for the whole-mount in situ hybridization procedure.* (1) Probe synthesis and embryo preparation. Probe synthesis can be done in advance, and the probes can be kept at −80 °C in accordance with Fig. 4 for long periods. Staged, postfixed, and dehydrated embryos can be stored at −20 °C until used

3. DNA is precipitated with sodium acetate and ethanol (*see* **Note 3**), centrifuged, and washed with RNase-free 70% ethanol.
4. DNA is resuspended in 20 μL DEPC water.
5. An aliquot is confirmed for purity and recovery by agarose gel electrophoresis.

RNA Labeling Reaction

1. 1–5 μg purified template DNA is pipetted to a sterile, RNase-free reaction vial. Then, enough sterile, RNase-free, DMPC-treated, double-distilled water is added to the vial to make the total sample volume 13 μL. Three riboprobes are labeled with DIG for the experiment: zfPGRN-A antisense probe (prepared using RNA Polymerase SP6), zfPGRN-A sense probe for negative control (prepared using RNA polymerase T7), and ssh (*sonic hedgehog*) antisense probe for positive control (prepared using RNA polymerase SP6).
2. The reaction vial is placed on ice, and then the following reagents are pipetted (Table 3).

Fig. 4 (continued) advance and stored at −80 °C as small aliquots for long periods. (P2) Staged embryos after fixing can be stored in methanol at −20 °C until used. (P3) Most whole-mount in situ hybridization experiments take 3–4 days to complete the procedure. For weakly expressed genes, (P4) staining reaction will take up to 16 h at room temperature or more if the staining reaction is carried out at 4 ° C. (P5) Images are acquired under microscope and analyzed for expression pattern

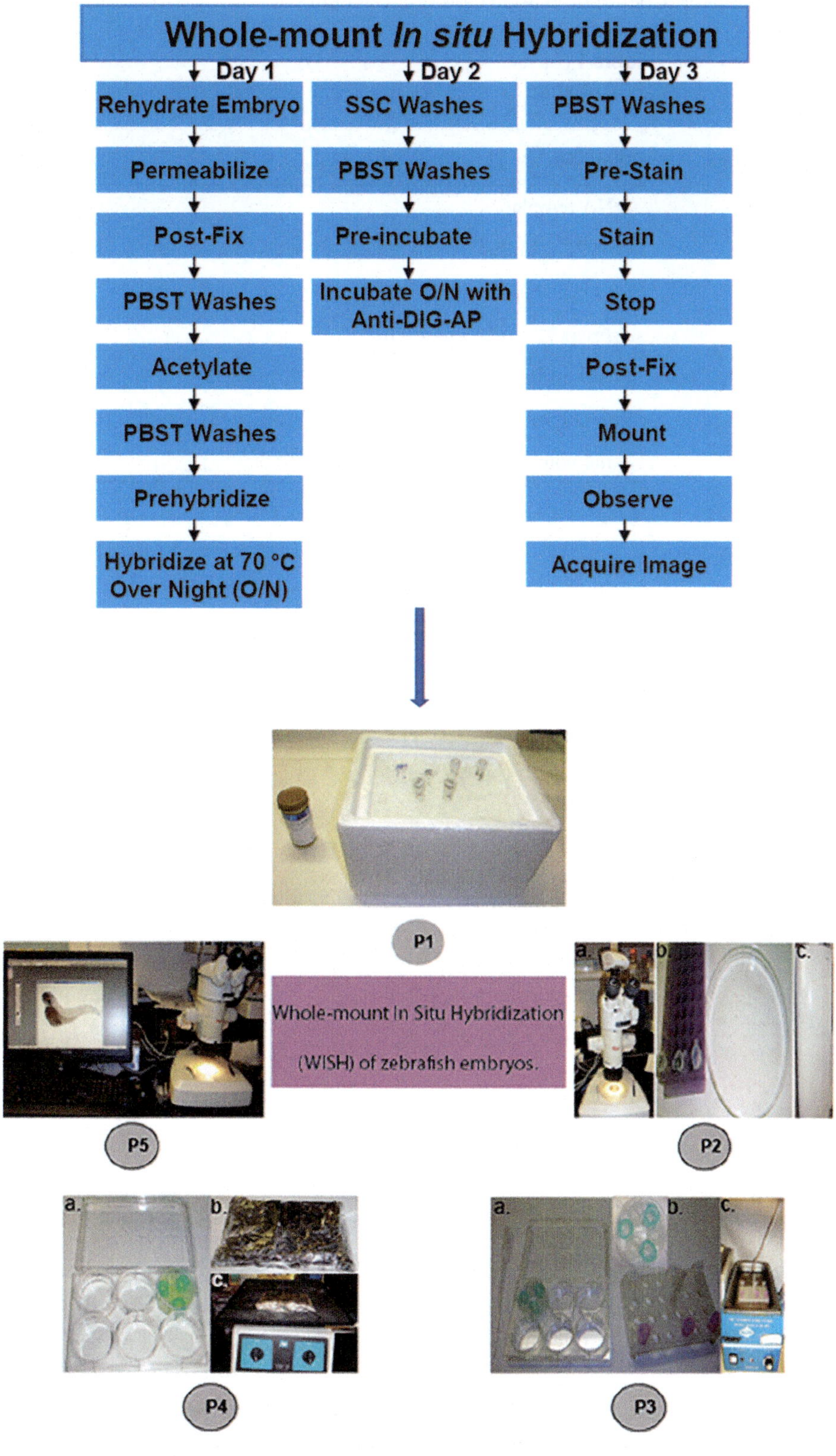

Fig. 4 *Workflow and diagrammatic representation of the steps involved in the whole-mount in situ hybridization technique.* (Top panel) Whole-mount in situ hybridization experiments generally take 3 days to complete the entire procedure. For weakly expressed genes, the staining reaction will take up to 16 h at room temperature or more if the staining reaction is carried out at 4 °C. (Bottom panel) (P1) Probe synthesis can be done in

Table 1
PCR Mixture

Reagent	Volume
DNA template	10–100 ng
10× PCR buffer	5 μL
50 mM dNTPs	0.5 μL
Primers (100–200 ng each)	1 μM each
Water add to a final volume of	49 μL
Taq polymerase (1 unit/μL)	1 μL
Total volume	50 μL

Table 2
TOPO Cloning Reaction Mixture

Reagent	Chemically competent *E. coli*	Electro-competent *E. coli*
Fresh PCR product	0.5–4 μL	0.5–4 μL
Salt solution	1 μL	–
Dilute salt solution	–	1 μL
Water	Add to total volume of 5 μL	Add to total volume of 5 μL
TOPO vector	1 μL	1 μL
Final volume	6 μL	6 μL

3. The reaction mixture is agitated gently and centrifuged briefly.
4. The reaction mixture is incubated for 2 h at 37 °C.
5. 2 μL RNase-free DNase I is added to remove template DNA.
6. The reaction mixture is incubated for 30 min at 37 °C.
7. The reaction is stopped by adding 2 μL 0.2 M EDTA (pH 8.0).
8. The reaction products are precipitated with 2.5 μL 4 M LiCl and 75 μL 100% ethanol (*see* **Note 3**).
9. The precipitating solution is mixed well and stored at −20 °C overnight.

Table 3
RNA Labelling Reaction Mixture

Reagent	Volume (μL)
10× NTP DIG-labeling mixture	2
10× transcription buffer	2
Protector RNase inhibitor	1
RNA polymerase SP6 or RNA polymerase T7	2
Final volume	20

10. The suspension is spun for 30 min in microfuge. The pellet is washed with 50 μL 70% ethanol, spun 3 min, dried, and resuspended in 50 μL for 30 min at 37 °C.
11. An aliquot of 1–2 μL is run on an agarose gel to verify the probe quality.
12. This reaction should yield about 10 μg of digoxigenin-labeled RNA probe (100–200 ng/μL).
13. RNA transcripts are analyzed by agarose gel electrophoresis.

3.1.2 Embryo Preparation

1. Zebrafish are purchased from Aquatica Tropicals.
2. Zebrafish are maintained on a 14-h/10-h light/dark cycle at 28.5 °C in a laboratory aquarium.
3. Embryos are collected at the required developmental stages (Day 1, Day 2, and Day 3 postfertilization). Staged according to Kimmel et al., 1995 [12].
4. Chorions are separated manually with Dumont Watchmaker's Forceps no. 5.
5. Embryos are fixed in 1 mL4% paraformaldehyde (PFA) in 1× PBS for 2 h at room temperature or overnight at 4 °C.
6. The embryos are washed with 1 mL PBS three times, then transferred into 1 mL of 100% methanol (MeOH), and kept at −20 °C until used for whole-mount in situ hybridization or immunofluorescence analysis.
7. For RNA extraction, staged embryos are frozen in liquid nitrogen and kept at −80 °C.
8. RNA is extracted from staged embryos, and cDNA is synthesized for PCR amplification and subsequent cloning.

3.1.3 Whole-Mount In Situ Hybridization of Zebrafish Embryos

Day 1

Rehydration

1. Six-well plates are prepared with 4 mL/well successive dilutions of methanol in 1× PBS as 75%, 50%, and 25% (v/v) methanol (*see* **Note 4**).
2. Dehydrated embryos of the same developmental stage in 100% MeOH are placed into the baskets prepared using nylon mesh (*see* Fig. 5, **Note 5**).
3. Embryos are rehydrated by moving the basket from one well to the next well of the plate containing 75%, 50%, and 25% MeOH for 5 min each.
4. Embryos are rinsed three times, 5 min per wash, in 100% PBST.

Permeabilization

1. The rehydrated embryos are permeabilized by digestion in 4 mL of 1× PBST buffer with proteinase K (10 μg/mL) at room temperature for the indicated period in Table 4.

Postfixation and PBST Washes

1. Embryos are incubated for 20 min in 4 mL 4% (wt/vol) paraformaldehyde in 1× PBS.
2. Embryos are rinsed three times, 5 min per wash, in 4 mL1× PBST to remove residual paraformaldehyde.

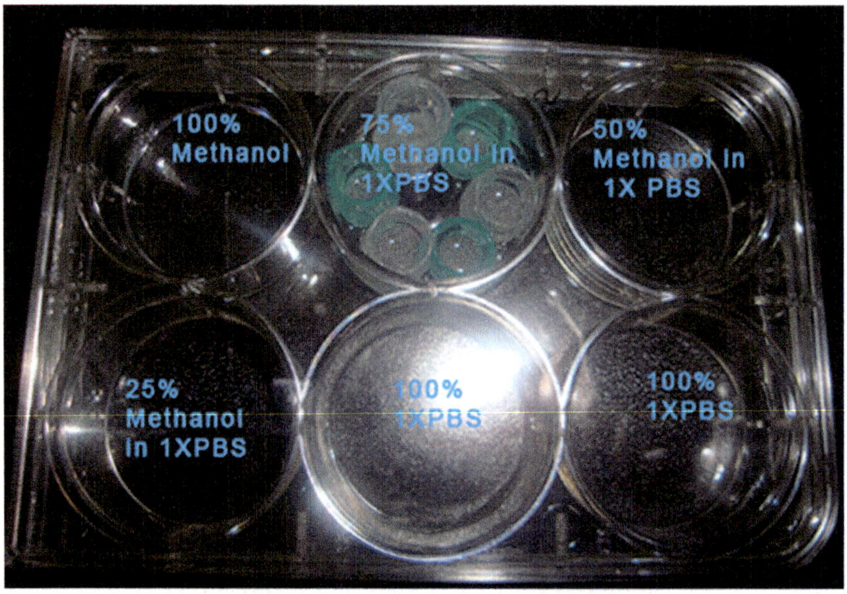

Fig. 5 Successive dilutions of methanol in 1× PBS. Six-well plates are prepared with successive dilutions of methanol in 1× PBS as 75%, 50%, and 25% (v/v) methanol. Embryos are rehydrated by moving the basket from one well to the next well of the plate

Table 4
Developmental Stage and Digestion Time

Developmental stage (in hours)	Proteinase K digestion
0–6	15 s
6–12	30 s
12–18	3 min
24	12 min
48	25 min
72	40 min
96–120	50 min

Acetylation

1. Embryos are treated twice for 10 min each in 4 mL acetylation mix (125 μL triethanolamine and 27 μL acetic anhydride in 10 mL ddH_2O) made immediately prior to use to minimize endogenous phosphatize activity.
2. Embryos are rinsed twice, 10 min per wash in 4 mL 1× PBST.

Pre-hybridization

1. Approximately 20 embryos per basket are moved from their baskets to 1.5 mL sterile Eppendorf tubes and pre-hybridized with 200 μL hybridization mix for 2–4 h in a 70 °C water bath.

Hybridization

100–200 ng of the probe in nuclease-free water is pipetted and is added to the pre-hybridized embryos. The solution is mixed and hybridized overnight at 70 °C in a water bath (*see* **Note 6**).

Preadsorption of Embryos with Antibody

1. About 30–50 embryos are moved from their baskets to 1.5 mL sterile Eppendorf tubes.
2. Embryos are incubated with 2 μL anti-DIG-AP (1:500) antibody in 1 mL blocking buffer.

Embryos are kept at room temperature for 2–4 h and then stored at 4 °C overnight. Remove them from the fridge first thing in the morning to let them reach room temperature.

Day 2

SSC Washes

1. Six-well plates are prepared with successive dilutions of 4 mL pre-warmed HM in 2× SSC as follows: 75% (v/v) HM, 50% (v/v) HM, and 25% (v/v) HM (*see* **Note7**).
2. The hybridized embryos are transferred to baskets.
3. Hybridized embryos are washed by moving the basket from one well to the next well of the plate for 15 min each in a 70 °C water bath.

4. Washed twice, for 15 min per wash, in 4 mL 2× SSC at 70 °C.
5. Washed twice, for 30 min per wash, in 4 mL 0.2× SSC at 65 °C.

PBST Washes

1. Six-well plates are prepared with successive dilutions of 4 mL 2× SSC in 1× PBST as follows: 75% (v/v) 2× SSC, 50% (v/v) 2× SSC, and 25% (v/v) 2× SSC).
2. Hybridized embryos are washed by moving the basket from one well to the next well of the plate for 5 min at room temperature on a rocker at minimum speed.

Preincubation

1. The embryos are incubated for 3–4 h at room temperature in 4 mL blocking buffer.

Incubation with Anti-DIG Antibody

1. Embryos are incubated with anti-DIG-AP antibody solution (1:5000) with 4 mL blocking buffer overnight at 4 °C with gentle agitation.

Day 3

PBST Washes

1. The embryos are washed briefly in 4 mL PBST.
2. Rinse six times, 15 min per wash, in 4 mL PBST at room temperature with gentle agitation.

Pre-staining

1. The embryos are rinsed two times for 15 min with 4 mL alkaline Tris buffer (pre-staining buffer) at room temperature with gentle agitation (*see* **Note 8**).

Staining

1. Embryos are subjected to staining solution at room temperature in the dark.
2. The staining reaction is monitored with a stereomicroscope every 30 min. Reaction time varies from 15 min for highly expressed genes to 16 h for weakly expressed genes. For short period, staining is carried out at room temperature for longer incubation is done at 4 °C (*see* **Note 9**).

Stop Reaction

1. The staining reaction is halted when the desired staining intensity is reached by moving the embryos containing baskets to the well containing 4 mL stop solution.
2. The embryos are rinsed two times in 4 mL stop solution for 10 min at room temperature on a rocker with gentle agitation (*see* **Note 10**).

Postfixation

1. The embryos are incubated for 20 min in 4 mL 4% (wt/vol) paraformaldehyde in 1× PBS.
2. The embryos are rinsed three times, 5 min per wash, in 4 mL 1× PBST to remove residual paraformaldehyde (*see* **Note 11**).

Mounting, Observation, and Image Acquisition

1. Embryos are transferred in the minimum possible volume of PBS to a six-well plate containing 4 mL 100% glycerol, placed on a rocker, and agitated gently overnight at room temperature in the dark.
2. The following day, the embryos are mounted in 100% glycerol and observed under microscope.
3. Images are acquired and saved in a file format that allows improving the quality through image software such as Photoshop. Examples of these images are shown in Fig. 6.

3.2 Whole-Mount Immunofluorescence

To assess the spatial and temporal expressions of zfPGRN-A protein levels, immunofluorescence was performed by monitoring antigen-antibody interactions using antigen-specific antibodies. The amino-terminal sequences of mature PGRN-A was used to raise site-specific antibodies. For this reason, while the in situ hybridization technique detects the message localization, immunofluorescence analysis labels only the mature zfPGRN-A protein. The rabbit polyclonal anti-zfPGRN-A was raised against synthetic peptide corresponding to residues 242–256 of zfPGRN-A (RAEWEDHKQKKPETQC). For the preadsorption negative control, the zfPGRN-A antibody was preincubated for 2 h at room temperature with an excess of the purified peptide to which the antibody was raised. This peptide/antibody mixture was then diluted in blocking solution to replace the primary antibody solution used to incubate with the embryos in the negative control. Steps involved in whole-mount immunofluorescence are illustrated in Fig. 7. The expression of zfPGRN-A within the MNs was further confirmed by the co-localization with znp1, a marker for motor neurons.

3.2.1 Embryo Preparation

Same as mentioned for in situ hybridization (*see* Subheading 3.1.2 for in situ technique).

3.2.2 Whole-Mount Immunofluorescence

Day 1

Rehydration, Permeabilization, Postfixation, and PBST Washes

Same as described for in situ hybridization technique (*see* Subheadings “Rehydration”, “Permeabilization”, “Postfixation” and “PBST Washes”)

Blocking

Embryos are incubated for 3–4 h at room temperature in 1 mL blocking buffer.

Incubation with Anti-zfPGRN-A Antibody

Embryos are incubated with 1 mL anti-zfPGRN-A (1:500) antibody solution in blocking buffer or peptide: antibody mixture diluted with blocking buffer which serves as control for non-specific binding overnight at 4 °C with gentle agitation.

Day 2

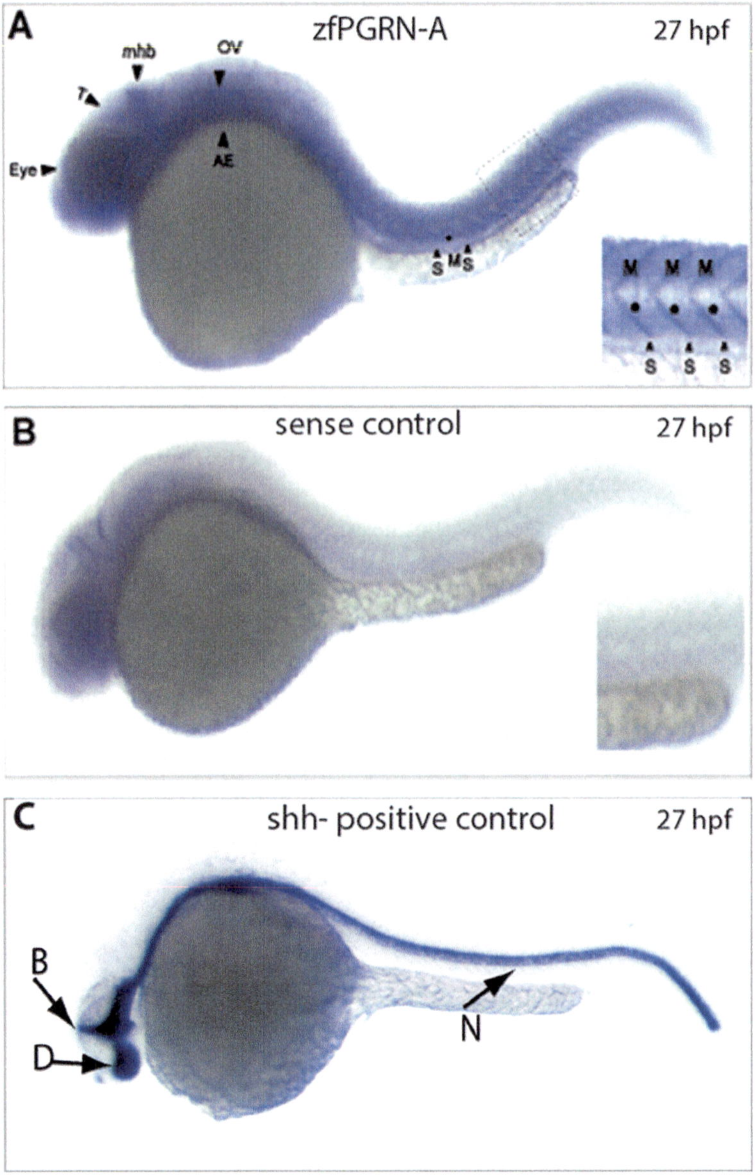

Fig. 6 *Developmental expression of zebrafish PGRN-A*. Whole-mount in situ hybridization analysis of *zfPGRN-A* expression revealed by (**a**) Antisense riboprobe, (**b**) Sense riboprobe negative control and (**c**) similar analysis of expression of *sonic hedgehog* revealed by *ssh* antisense riboprobe as positive control. Lateral view of 27 hpf (hours postfertilization) embryos revealed *zfPGRN-A* expression within the peripheral and central nervous systems. Panel A: In the head region, *zfPGRN-A* is strongly expressed in the eyes and brain particularly in the tectum (T), midbrain-hindbrain boundary (mhb), otic vesicle (ov), and anterior endoderm (AE). zfPGRN-A is also expressed robustly in the somite borders (s) and weakly in the motor neurons (M) and the spinal cord. Panel B: Sense riboprobe detected weak signal for the antisense transcript. Panel C: Shh antisense riboprobe revealed the expression of *Shh* [11] within the diencephalon (D), brain (B), and notochord (N) served as a positive control of the experiment. Images were captured at 6× magnification, and hatched boxes represent 2–3× zoom

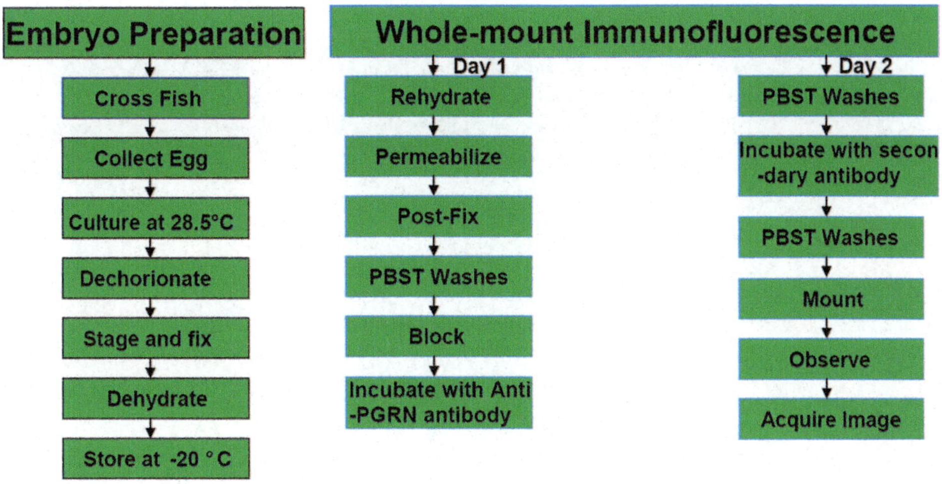

Fig. 7 *Workflow scheme for whole-mount immunofluorescence analysis.* The procedure includes two parts: (1) embryo preparation and (2) whole-mount immunofluorescence. The whole process takes 2–3 days using the staged and stored embryos

Preadsorption of Embryos with Antibody

1. About 30–50 embryos are rehydrated quickly in 1.5 mL sterile Eppendorf tubes containing 1 mL volume of blocking buffer.
2. Embryos are incubated with 5 μL goat anti-rabbit-Alexa Fluor 488 (1:200) antibody in 1 mL blocking buffer.
3. Left at room temperature for 2–4 h.

PBST Washes

1. Embryos incubated with anti-zfPGRN-A antibody are rinsed briefly in 1 mL PBST.
2. Rinse six times, 15 min per wash, in 1 mL PBST at room temperature with gentle agitation.

Incubation with Anti-Rabbit-Alexa Fluor 488

Embryos are incubated with goat anti-rabbit-Alexa Fluor 488 (1:200) antibody in 1 mL blocking buffer for 2 h.

PBST Washes

1. Embryos incubated with secondary antibody are washed briefly in 1 mL PBST.
2. Rinse four times, 15 min per wash, in 1 mL PBST at room temperature with gentle agitation.

Mounting, Observation, and Image Acquisition

1. Embryos are transferred in the minimum possible volume of PBS to a six-well plate containing 4 mL 100% glycerol. Place on a rocker and agitate gently overnight at room temperature in the dark.
2. Next day, the embryos are mounted in 100% glycerol and observed under the microscope.

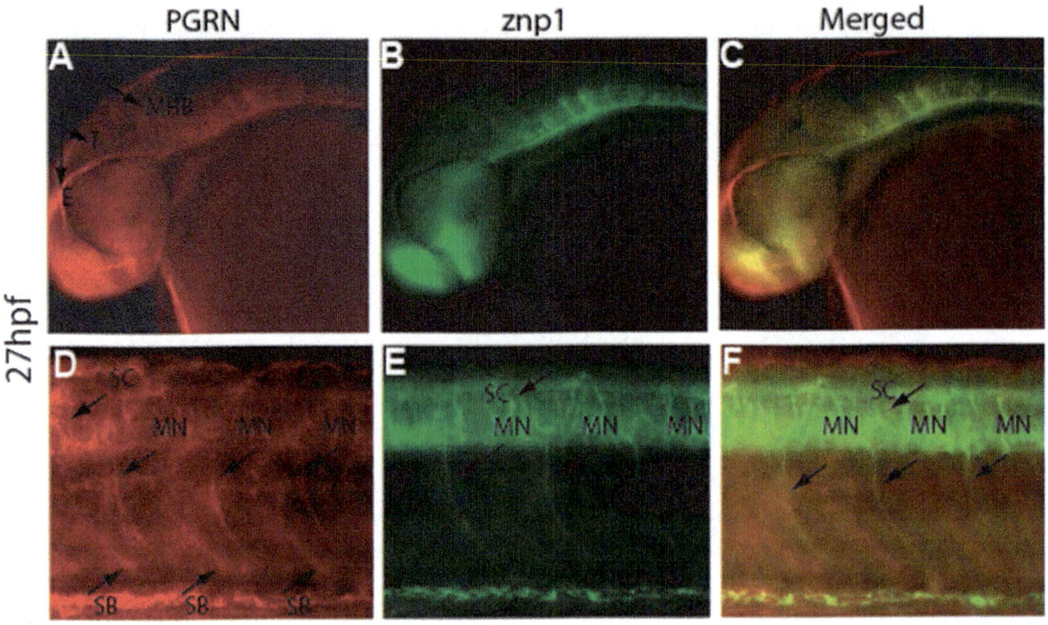

Fig. 8 Whole-mount immunolocalization of zfPGRN-A in primary motor neurons at 27 hpf (hours postfertilization). Lateral views (anterior, left; dorsal, top) of whole-mount embryos labeled with zfPGRN-A polyclonal antibodies (**a**, **d**) and znp1 monoclonal antibodies (**b**, **e**) at 27 hpf. Embryos show zfPGRN-A expression in the developing eye, tectum, and hindbrain (**a**); spinal cord, somite boundaries, and motor neurons (**d**); znp1 expression in the eye, hindbrain, spinal cord, and motor neurons (**b** and **e**). Merged images show co-localization of zfPGRN-A and znp1 (**c**, **f**), [9]. Abbreviations: T, tectum; E, eye; MHB, midbrain-hindbrain boundary; SB, somite boundary; MN, motor neurons; SC, spinal cord. Images were captured at 20× magnification. Scale Bar = 50 μm [9]

3. Images are captured and saved in a file format that allows improvement of the image quality using software such as Adobe Photoshop 7.0. Examples of these images are shown in Fig. 8.

3.3 Microinjection

Microinjection has been widely used for analyzing gene function [10] and generating transgenic fish [13].To uncover functional roles of zfPGRN-A, we knocked down its protein level by microinjection of antisense MO oligonucleotides directed against the starter methionine (MO-1) and 5' UTR (MO-2) of zfPGRN-A. The MOs bind to target mRNA with high affinity and specificity, thereby blocking the initiation of zfPGRN-A translation. Steps involved in the microinjection procedure are illustrated in the following diagram (Fig. 9).

3.3.1 Embryo Preparation

1. Wild-type zebrafish purchased from Aquatica Tropicals (Florida) are maintained on a 14-h/10-h light/dark cycle at 28.5 °C in a laboratory aquarium. A zebrafish facility of approximately 200 fish produces hundreds of embryos daily by natural spontaneous spawning.
2. In the late afternoon of the day before eggs are required, fish were transferred to a breeding tank. In the morning, after the

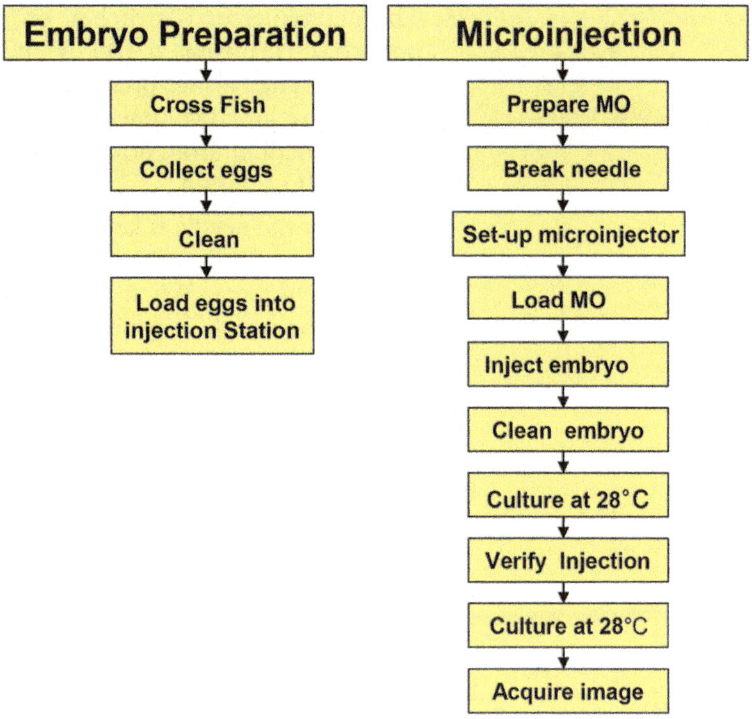

Fig. 9 *Workflow scheme for embryo preparation and microinjection.* Injection of embryos with antisense morpholinos (or mRNA) should be carried out immediately after collection by loading them in the injection chamber using the needle filled with MO solution

light cycle begins and 10–15 min after spawning has stopped, the eggs that have fallen through the divider are collected from the bottom of the tank and are checked under a dissecting microscope. Embryos at the one- to two-cell stage are selected for microinjection.

3. Embryos are cleaned by changing fresh egg water three a few times. This has to be done quickly since the eggs remain in the one-cell stage only for 30 min after fertilization. The embryos are then collected using transfer pipette and carefully arranged in the furrows of the injection station. In order to facilitate microinjection, only a little egg water is used to ensure that the embryos are held firmly within the furrow to prevent them from floating away.

3.3.2 Microinjection

1. The tube containing 5 μL morpholino antisense oligonucleotide (MO) solution is centrifuged to separate small particles before loading the supernatant into the microinjection needle to avoid these particles from the MO solution clogging the needle (*see* **Notes 12** and **13**).
2. Under the highest magnification view of the dissecting microscope, the tip of the microinjection needle is cracked gently with clean forceps (*see* **Note 13**).

3. The microinjector is switched on, and the pressure valve is opened. P-clear is set at 100 psi and the injection pressure set at 15 psi. The balance pressure is monitored and regulated to expel the required volume (2 nL) of MO solution and to prevent the medium from flowing back into the needle and diluting the MO solution.
4. The tip of the needle is kept inside the tube that contains the MO solution to allow the solution to migrate inside the needle by capillary force. When the MO solution has migrated to the tip, the needle is attached into the needle holder on the microinjector. Alternatively, MO can be aspirated into the needle using the fill button after introducing the needle into the needle holder of the injector.
5. The flow of MO through the tip of the cracked needle into injection chamber is verified by pressing the foot pedal. For the best distribution of MO in the 1-cell embryo, a volume of 2 nL is used.
6. Magnification of the microscope is set to 1.6× for microinjection.
7. The needle tip is brought close to the target embryo and focused on the cells to be injected.
8. Pressure is applied using foot pedal ensuring the flow of MO through the needle tip.
9. Injection time and injection pressure balance are calibrated for appropriate volume of MO discharge.
10. The tip of the needle is pierced through the chorion directly into the cytoplasm of the 1–2 cell-staged embryo although injection can also be done through the yolk and then into the cytoplasm of the dividing cell.
11. MO is discharged by applying pressure via the foot pedal, and the injected MO is visible as a red spot in the injected embryo due to the presence of phenol red in the solution.
12. The needle tip is ejected slowly from the injected embryo by gently lifting the needle.
13. The needle is then placed slowly to the adjacent embryo to be injected. Injection is repeated for all of the embryos in the furrow.
14. All dead and uninjected embryos are collected immediately from the Petri dishes and discarded.
15. The injected embryos are transferred to a dish with egg water and incubated at 28.5 °C.
16. The embryos are checked again generally 3 h after injection under the microscope through GFP filter. Dead or abnormal embryos were removed. The successfully injected embryos are

incubated at 28.5 °C until the appropriate developmental stage is reached.

17. The embryos are dechorionated manually using Dumont #5 Watchmaker's Forceps after 1 day. Morphological defects associated with the MO were photographed and documented. Examples are shown in Fig. 10. The embryos are then fixed for 2 h at room temperature and kept at −20 °C for staining.

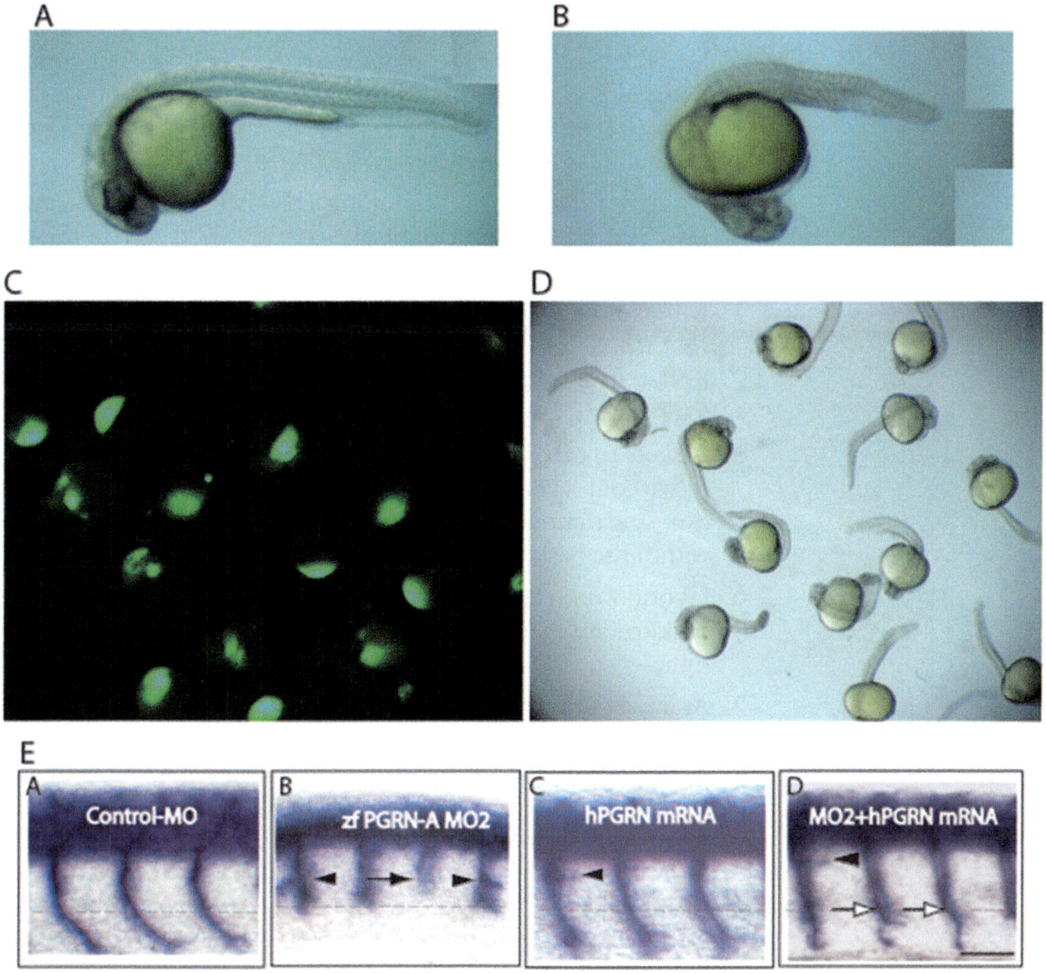

Fig. 10 *Antisense morpholino (AMO)-based translation inhibition of zebrafish PGRN-A.* Lateral view of 27hpf (hours postfertilization) (**a**, **b**, **d**, and **e**) and 4hpf embryos (**c**) revealed phenotypic consequences of zfPGRN-A knockdown. Panel A: WT embryo at 27 hpf showing normal development. Panel B and D: zfPGRN-A knockdown embryos at 27 hpf showing compromised development. Panel C: zfPGRN-A knockdown embryos at 4hpf showing distribution of injected AMO directed against zfPGRN-A. Panel E: Lateral views (anterior to the left; dorsal to the top) of whole-mount embryos labeled with znp1 antibodies at 27 hpf (**a**–**d**): in wild type (**a**), embryo injected with zfPGRN-A MO (**b**), embryo injected with 100 pg hPGRN (human progranulin) mRNA (**c**), and Pgrn-A MO + 100 pg hPGRN mRNA (**d**). Normal motor axon development (**a**), increase in branched axons (**b**, **c**; black arrowheads), truncated motor axons (**b**, black arrows), branched motor axons (**d**, black arrowheads), and partial rescue of motor axons (**d**, white arrow) occur. (**d**) Scale bar = 50 μm [9]

4 Notes

1. Use less plasmid DNA and more genomic DNA as a template for the PCR reaction. Make sure to include a 10 min extension at 72 °C after the last cycle to ensure that all PCR products are full length and 3′ adenylated to facilitate TOPO cloning.
2. There should be a single, discrete band on agarose gel from the PCR reaction. In the absence of a single, discrete band, gel-purify the fragment before using the TOPO TA Cloning Kit. Take special care to avoid sources of nuclease contamination. Alternatively, the PCR reaction can be optimized to eliminate multiple products and smearing caused by non-specific PCR amplification.
3. We have found that precipitating DNA with sodium acetate/LiCl and ethanol overnight at −20 °C works better than at −80 °C for 40 min.
4. Unless otherwise stated, the in situ hybridization procedures are carried out with embryos in baskets as shown in Fig. 1 and RNase-free solutions in 6-well plates (4 mL/well) at room temperature under sterile conditions.
5. Baskets to carry embryos are prepared in different colors with or without a rim on the top as shown in Fig. 5. This permits easy orientation of the order of the baskets and the embryos within the basket to correspond to a particular probe.
6. Hybridization at 70 °C with 50% deionized formamide reduces background staining and provides better hybridization signals.
7. At day 2, pre-warm hybridization mix, 2× SSC and 0.2× SSC, by placing them in a beaker with water and keep it in a 70 °C water bath for at least 20 min. Withdraw hybridization mix and probe from the tubes with hybridized embryos, substitute with 1 mL of pre-warmed hybridization mix, and incubate for 15 min at 70 °C. During incubation, prepare six-well plates with baskets for 2× SSC washes and place them at 70 °C. After 15 min transfer the embryos to basket and continue with 2× SSC washes. Maintain another water bath at 65 °C. Place six-well plates with 0.2× SSC for high-stringency washes to prevent non-specific hybridization of the probe.
8. After 1× PBST washes, washing the embryos in Alkaline Tris buffer twice for 15 min prior to staining with BCIP and NBT will help to provide clean signals.
9. If the gene of interest possesses antisense (AS) transcripts, the sense probe will detect a signal for the AS transcript. For instance, we have found AS transcripts to zebrafish *Progranulin-1* and *Progranulin-2* genes, members of the granulin family of secreted growth factors [14].

10. Subjecting embryos in methanol for 30 min or longer after the staining reaction is completed converts the signal to a true purple color and washes away non-specific staining.
11. Fixed embryos can be maintained in the dark in stop solution at 4 °C for several months.
12. In order to avoid generating non-specific effects, the MO injection solution and dilution should be prepared using pure distilled water. Pilot experiments should be carried out using various dilutions of MO to eliminate non-specific toxicity phenotypes caused by high concentrations of MO.
13. Prepare the injection solutions and needles in advance to facilitate microinjection of embryos at the one-cell stage.

References

1. Zon LI (1999) Zebrafish: a new model for human disease. Genome Res 9(2):99–100
2. Lieschke GJ, Currie PD (2007) Animal models of human disease: zebrafish swim into view. Nat Rev Genet 8(5):353–367
3. Ausoni S, Sartore S (2009) From fish to amphibians to mammals: in search of novel strategies to optimize cardiac regeneration. J Cell Biol 184(3):357–364
4. Major RJ, Poss KD (2007) Zebrafish heart regeneration as a model for cardiac tissue repair. Drug Discov Today Dis Models 4(4):219–225
5. Love DR et al (2004) Technology for high-throughput screens: the present and future using zebrafish. Curr Opin Biotechnol 15(6):564–571
6. Guo S (2009) Using zebrafish to assess the impact of drugs on neural development and function. Expert Opin Drug Discov 4(7):715–726
7. Westerfield M (2007) The Zebrafish book. In: A guide for the laboratory use of zebrafish (Danio rerio), 5th edn. University of Oregon Press, Eugene. (4th Edition available online)
8. Chitramuthu BP et al (2010) Molecular cloning and embryonic expression of zebrafish PCSK5 co-orthologues: functional assessment during lateral line development. Dev Dyn 239(11):2933–2946
9. Chitramuthu BP et al (2010) Progranulin modulates zebrafish motoneuron development in vivo and rescues truncation defects associated with knockdown of survival motor neuron 1. Mol Neurodegener 5:41
10. Nasevicius A, Ekker SC (2000) Effective targeted gene 'knockdown' in zebrafish. Nat Genet 26(2):216–220
11. Krauss S, Concordet JP, Ingham PW (1993) A functionally conserved homolog of the Drosophila segment polarity gene hh is expressed in tissues with polarizing activity in zebrafish embryos. Cell 75(7):1431–1444
12. Kimmel CB et al (1995) Stages of embryonic development of the zebrafish. Dev Dyn 203(3):253–310
13. Culp P, Nusslein-Volhard C, Hopkins N (1991) High-frequency germ-line transmission of plasmid DNA sequences injected into fertilized zebrafish eggs. Proc Natl Acad Sci U S A 88(18):7953–7957
14. Cadieux B et al (2005) The zebrafish progranulin gene family and antisense transcripts. BMC Genomics 6:156

Chapter 16

Methods to Investigate the Molecular Basis of Progranulin Actions on Brain and Behavior In Vivo Using Knockout Mice

Jan Boddaert, Hans Wils, and Samir Kumar-Singh

Abstract

Currently one of the few molecules that equally excites a neuroscientist, a cancer biologist, an immunologist, and a developmental biologist is progranulin (GRN/Grn)—a pluripotent growth factor that plays key roles in cell survival, proliferation, development, tissue regeneration, inflammation, wound healing, and angiogenesis. However, the molecular pathways associated with GRN signaling involved in these varied physiological processes are not understood. Gene inactivation has been considered as one of the best methods to delineate the biological role of a protein, and gene targeting is a direct means to disrupt a gene's open reading frame and block its expression, for instance, in a mouse. Such a gene knockout animal model also served as an in vivo disease model where loss of gene or its function is thought to be the primary disease mechanism, as is the case with progranulin loss of function in frontotemporal lobar degeneration (FTLD). It is estimated that up to half of the cases of familial, dominant FTLD might be due to GRN haploinsufficiency. To understand the molecular pathways associated with GRN loss, constitutive and conditional progranulin knockout ($Grn^{-/-}$) mice have also been constructed in several laboratories, including ours. These mice show several disease-characteristic features and suggest that continued studies on the $Grn^{-/-}$ mice would be instructive in the understanding of complex GRN biology in health and disease.

Key words Progranulin, Gene knockout, Mouse model, Preclinical model, Methods

1 Introduction

Loss-of-function mutations in the *progranulin* (*GRN*) gene have been identified as one of the most common causes of frontotemporal lobar degeneration (FTLD), the second most common cause of presenile dementia [1, 2] where intraneuronal deposits of TDP-43 occur in the frontal and temporal regions of the brain [3]. Progranulin also has unparalleled neuroprotective capabilities that are able to diminish the effect of ischemic stroke and is suggested as a potential therapeutic for cerebrovascular accidents as well as Alzheimer's disease. The versatility of progranulin is striking and is

Andrew Bateman et al. (eds.), *Progranulin: Methods and Protocols*, Methods in Molecular Biology, vol. 1806,
https://doi.org/10.1007/978-1-4939-8559-3_16, © Springer Science+Business Media, LLC, part of Springer Nature 2018

reflected by its crucial role in many physiological functions in the body [4, 5] and also being highly expressed in many cancers [6, 7].

Fully matured progranulin (GRN) is an 88-kDa secreted precursor glycoprotein and contains 7½ repeats of a cysteine-rich motif. GRN is also proteolytically cleaved by many proteinases to form 6-kDa GRN peptides that often have opposing functions to GRN. GRN binds to nucleic acid and has important functions in the regulation of transcription and splicing [8]. However, the molecular pathways associated with GRN and GRN-peptide signaling are currently not understood, which is a prerequisite to any drug development program.

To study the mechanisms involved in GRN deficiency-mediated neurodegeneration, several *Grn* knockout mouse models have been developed, leading to the elucidation of many aspects of the FTLD pathology [9–12]. For instance, $Grn^{-/-}$ mouse models showed significant astro- and microgliosis and accelerated aging, accompanied by disease-characteristic brain protein accumulations, all characteristic features of FTLD-TDP [12]. Interestingly, $Grn^{-/-}$ mice also have increased brain levels of detergent-insoluble hyperphosphorylated TDP-43 comparable to those observed in TDP-43 mutant and wild-type-overexpressing mice [13, 14]. These data suggest that $Grn^{-/-}$ mice are indeed a bona fide model for FTLD-TDP pathology [12] and can be successfully utilized for many GRN-related studies including, but not limited to, validation of in vitro cell signaling data, studies of cell non-autonomous injury to neurons by the glial compartment, and increased clearance of apoptotic cells by increased macrophage activity as occurs in *C. elegans* development [15]. Importantly, $Grn^{-/-}$ mice are also a preclinical model to test future FTLD therapies. This chapter would discuss how these mice could be bred and characterized for such studies.

2 Materials

2.1 Genotyping

1. Progranulin knockout mice at 2–3 weeks of age.
2. Mouse restrainer.
3. Scalpel blade (N° 10).
4. Proteinase K (20 mg/mL).
5. PCR tail lysis buffer (for 1 L): 10 mM Tris (1.21 g), 100 mM NaCl (5.84 g), 10 mM EDTA (3.72 g), 0.5% SDS (5.00 g) in 1 L double-distilled water (ddH_2O). Mix the solution and adjust to pH 8 and filter-sterilized using 45 μm filter. Alternatively, commercial pre-mixed lysis buffer can be used.
6. PCR Master Mix containing Taq polymerase 1 U/μL, 20 mM $MgCl_2$, 50 mM dNTPs, 10× PCR buffer. Premade commercial

Table 1
PCR reaction components

Reagent	Volume
Master Mix (nDTPs, Taq polymerase, $MgCl_2$)	6 μL
Primers	1 μL each
DNA template	0.5 μL
Molecular grade water	1.5 μL
Total volume	12 μL

Table 2
Primers used for genotyping

Gene symbol	GenBank accession	Sequence
Grn-Fw	NM_008175.4	TCACTGTGTCTGGGACTTCCA
Grn-rev	NM_008175.4	GCAGCAGTGGTAGCCATCA

PCR mixes are available such as EconoTaq green PCR Master Mix (Table 1).

7. Primers: working dilution at 0.1 μM (*see* Table 2).
8. 0.5× TBE buffer: dissolve 7.44 g of EDTA, 108 g of Tris, and 55 g of boric acid in 1 L of water to obtain 10 × TBE buffer as stock solution. Dilute to 0.5× in distilled water (dH_2O).
9. Detection marker such as GelRed.
10. 2% agarose gel: dissolve 1 g of agarose in 50 mL of 0.5 × TBE buffer. Add 5 μL of GelRed and let the gel solidify for 30 min.
11. Gel marker such as Orange G.
12. 50 bp DNA ladder.
13. PCR thermal cycler such as the Biorad Thermal cycler.
14. 1.5 mL plastic microcentrifuge tube with lid (e.g., Eppendorf standard microcentrifuge tube; cat. nr.: 3810×).
15. Block heater or water bath at 56 °C and 85 °C.
16. Microcentrifuge (e.g., Eppendorf 5417R).
17. Agarose gel electrophoresis apparatus such as the Gel XL Ultra V2 (Labnet International).
18. UV light and photographic equipment such as GeneScan G:Box (Syngene) and the GeneSnap image acquisition software.

2.2 Behavioral Analysis

2.2.1 Morris Water Maze

1. Circular pool with a diameter of 150 cm and height of 30 cm.
2. Fixed room cues to aid orientation.
3. 25.8 °C opacified water: various opacifiers exist such as nontoxic opacifiers (e.g., Lytron 621) or nontoxic resin-based white paint.
4. Round platform 1 cm below surface in one of the quadrants.
5. Video tracking system. Many commercial tracking systems exist such as Chromotrack, San Diego Instruments, USA. At the probe trials, time spent in each quadrant and crossing of the previous platform location can also be measured manually.

2.2.2 Passive Avoidance

1. Light-dark transition box: two-compartment box containing one small, brightly lit compartment and one larger, dark compartment equipped with a 0.3 mA electric shock grid. Both compartments are connected with a sliding door.

2.2.3 Open Field

1. Brightly lit 2500 cm^2 (50 cm × 50 cm) arena with center circle marking and 7 × 7 cm corner markings.
2. Video tracking system. Many commercial tracking systems exist such as Chromotrack, San Diego Instruments, USA.

2.2.4 Y-Maze

1. Y-maze composed of three equally spaced arms (120°, 41 cm long and 15 cm high). The floor of each arm is made of Perspex and is 5 cm wide.
2. Moderate lighting (100 lux at center platform).
3. Video tracking system. Many commercial tracking systems exist such as Chromotrack, San Diego Instruments, USA.

2.3 Histology and Immunohistochemistry

2.3.1 Tissue Harvesting, Fixation, Embedding, and Preparation

1. Ketamine (100 mg/kg bodyweight) and xylazine (16 mg/kg bodyweight) combination for anesthetizing the mice through intraperitoneal injection.
2. Kreb's solution: dissolve 6.8 g NaCl, 0.37 g KCl, 0.37 g $CaCl_2.H_2O$, 0.3 g $MgSO_4.7H_2O$, 0.19 g $NaH_2PO_4.2H_2O$, 2.1 g $NaHCO_3$, and 1.98 g glucose in 1 L of dH_2O.
3. Phosphate-buffered saline (PBS): dissolve 1.53 g $NaH_2PO_4.2H_2O$, 14.4 g $NA_2HPO_4.12H_2O$, and 45 g NaCl in 1 L dH_2O to obtain a 5× PBS stock solution. Dilute five times with dH_2O.
4. 23 gauge winged infusion set.
5. Two burets, frame stand, and clamps.
6. Three-way stopcock.
7. 2% (w/v) paraformaldehyde (PFA): prepare a 2% (w/v) PFA in PBS or any buffered saline. Carefully heat the solution up to 60 ° C while stirring. Make sure the solution does not boil.

Cool to room temperature afterward. PFA is toxic: always work under a hood and wear safety glasses, gloves, and a dust mask. Prepare fresh PFA solution for each experiment.

8. 100%, 96%, 90%, 80%, 70% ethanol in water.
9. Methyl-benzoic acid.
10. Xylene.
11. Acetone.
12. Paraffin.
13. Optional: tissue processor for paraffin embedding (e.g., Citadel 2000). Contact the institute histology department for usage of this equipment.
14. Microtome.
15. Poly-L-lysine-coated slides.

2.3.2 Histology and Immunohistochemistry

1. Citric acid antigen retrieval buffer (pH 6): sodium citrate 0.1 M stock solution, dissolve 29.41 g of 1 L H_2O. Citric acid 0.1 M stock solution: dissolve 21.01 g citric acid. H_2O in 1 L H_2O. Store both at 4 °C. Prepare retrieval buffer by mixing 41 mL of 0.1 M sodium citrate with 9 mL of citric acid and adjust to a final volume of 500 mL with dH_2O. Adjust to pH 6.
2. H_2O_2: prepare a 0.3% solution of hydrogen peroxide in dH_2O. Never exceed 1% of H_2O_2 for immunohistochemical purposes. H_2O_2 is highly corrosive in high concentrations: always wear a lab coat, safety mask, and gloves.
3. Phosphate-buffered saline (PBS) 0.01 M.
4. Bovine serum albumin (BSA): prepare a 1% BSA solution with PBS fresh for each experiment.
5. Normal horse serum (NHS): prepare a 20% NHS solution in PBS containing 1% BSA.
6. Primary antibodies diluted in PBS with 1% BSA (*see* Table 3).
7. Biotin-labeled secondary antibody diluted in PBS with 1% BSA (*see* Table 3).
8. Extravidin with conjugated horseradish peroxidase (HRP): dilute in PBS with 1% BSA.
9. Diaminobenzidine (DAB): dissolve 10 mg DAB in 40 mL PBS, and mix for 5 min at room temperature. Add 1 μL H_2O_2 per 2 mL of DAB. Adjust pH to 7.2. Filter and aliquot. Commercial kits are available such as DAB+ (DAKO).
10. DAPI (4′,6-diamidino-2-phenylindole): make a 5 mg/mL DAPI stock solution by dissolving one 10 mg vial of 14.3 mM DAPI dihydrochloride in 2 mL dH_2O. Dilute to 300 nM in PBS. Protect from light. Aliquots of the stock solution can be

Table 3
Antibodies used for immunohistochemistry

Antibody	Epitope/marker	Spp.	Clonality[1]	Dilution	Source
Antihuman GRN	Progranulin	Goat	pAb	1/250	R&D Systems
Anti-Iba1	Ionized calcium binding adaptor molecule 1	Rabbit	pAb	1/1000	Wako Chemicals
Anti-GFAP	Glial fibrillary acidic protein	Mouse	mAb	1/10,000	Dako
Anti-TDP-43	TAR DNA-binding protein 43	Rabbit	pAb	1/2000	Proteintech
Anti- p409/410 TDP-43	TAR DNA-binding protein 43	Rabbit	pAb	1/3000	CosmoBio
Anti- p403/404 TDP-43	TAR DNA-binding protein 43	Rabbit	pAb	1/3000	CosmoBio
Anti-ubiquitin	Ubiquitin	Rabbit	pAb	1/2000	Dako
Anti-p62	p62 protein	Guinea pig	pAb	1/2000	Progen Biotechnik
Anti-cleaved caspase 3	Caspase-3	Rabbit	pAb	1/2000	Cell Signaling Technology
Anti-FUS	Fused in sarcoma	Rabbit	pAb	1/100	Bethyl Laboratories

[1]mAb = monoclonal antibody; pAb = polyclonal antibody

stored at −20 °C, where they are stable for at least 6 months. DAPI is a known mutagen and should be handled with care. The dye must be disposed of safely and in accordance with applicable local regulations.

11. Ehrlich's hematoxylin (i.e., from Klinipath): For immunohistochemical counterstaining, prepare 1:10 diluted hematoxylin.
12. 1% eosin: dissolve 1% eosin in 100 mL 70% EtOH. Add 1 drop of acetic acid per 100 mL to obtain a better staining.
13. 5% acetic acid: dilute 5 mL glacial acetic acid in distilled water.
14. Cresyl violet: prepare 1% (w/v) cresyl violet solution in dH_2O.
15. Xylene-based mounting medium (e.g., Entellan).
16. Fluorescent-compatible mounting medium (e.g., Citifluor).
17. Hydrophobic marker pen: the hydrophobic marker provides a barrier to liquids such as antibody solutions applied to the sections resulting in a more uniform immunohistochemical staining reducing the amount of reagents needed.
18. Plastic slide container (Thermo Scientific).

19. Plastic tray with black bottom.
20. Water bath: heat the water bath to 100 °C.
21. Vacuum pump.
22. Bright-field microscope allowing up to 600× magnification.
23. Image analysis and quantification software such as ImageScope v.8 (Aperio Technologies) or ImageJ IHC profiler.

2.3.3 Western Blotting/TDP-43 Aggregation

1. Standard protein concentration measurement assay such as the bicinchoninic acid (BCA) assay (Pierce; cat. nr. 23225).
2. Low-salt (LS) buffer (for 1 L): 10 mM Tris, 5 mM EDTA, 1 mM dithiothreitol (DTT), 10% sucrose, protease inhibitor cocktail (e.g., complete mixture, Roche), phosphatase inhibitor (e.g., PhosphoStop, Roche) pH 7.5.
3. Triton-X (TX) buffer: LS buffer, 1% Triton X-100, 0.5 mM NaCl.
4. TX buffer with 30% sucrose.
5. Sarkosyl (SARK) buffer: LS, 1% N-lauroyl sarcosine, 0.5 M NaCl.
6. SDS buffer: 2% SDS, 50 mM Tris, cocktail of protease inhibitors, pH, phosphatase inhibitor, 7.6.
7. Urea buffer: 7 M urea, 2 M thiourea, 4% 3-[(3-cholamidopropyl) dimethylammonio]-1-propanesulfonate (CHAPS), 30 mM Tris–HCl, pH 8.5.
8. Polyacrylamide gel. Commercial, precast gels are available such as NuPage 4–12% Bis-Tris gel (Life Technologies).
9. Polyvinylidene difluoride membrane.
10. Methanol (MeOH).
11. 10× transfer buffer: add 24.2 g Tris (Ultrapure) and 112.6 g glycine to 1 L dH_2O.
12. Blotting buffer: mix 400 mL methanol (MeOH) with 200 mL 10× transfer buffer. Add 2 mL 10% SDS to obtain 0.01% SDS and increase volume to 2 L with dH_2O.
13. Wash buffer: add 2 mL 0.1% Tween to 8 mL 10× PBS. Increase volume to 2 L with dH_2O.
14. 5% skimmed milk blocking buffer: add 10 g skimmed milk to 200 mL wash buffer.
15. Sonicator, e.g., UP50H ultrasonic processor (Hielscher, Germany).
16. Electro-blotting apparatus, e.g., iBlot (Invitrogen).
17. ECL Plus chemiluminescence detection system.
18. Primary and secondary antibodies: (*see* Table 4).

Table 4
Antibodies used for Western blotting

Antibody	Epitope/marker	Species	Clonality [1]	Dilution	Source
Anti-TDP-43	TAR DNA-binding protein 43	Rabbit	pAb	1/2000	Proteintech
Anti- p409/410 TDP-43	TAR DNA-binding protein 43	Rabbit	pAb	1/500	CosmoBio
Anti- p403/404 TDP-43	TAR DNA-binding protein 43	Rabbit	pAb	1/500	CosmoBio
Anti-GAPDH	Glyceraldehyde-3-phosphate dehydrogenase	Mouse	mAb	1/20000	Meridian life science

[1]mAb = monoclonal antibody; pAb = polyclonal antibody

2.4 Cell Biology

2.4.1 Neuronal Cell Cultures

Cell Treatment

1. Dimethyl sulfoxide (DMSO).
2. MG132: dissolve in dimethyl sulfoxide (DMSO) to a concentration of 10 μM.
3. Clasto-lactacystin b-lactone: dissolve in DMSO to a concentration of 10 μM.
4. Pan-caspase inhibitor Z-VAD-FMK (pan-caspase inhibitor): dissolve in DMSO to a concentration of 50 μM.
5. Neurobasal medium.
6. Recombinant Grn.
7. 10% horse serum in PBS.

Cell Counting

1. PBS.
2. 6-well plate.
3. Countess Automated Cell Counter (Invitrogen).
4. Dulbecco's minimal eagle medium.
5. BrdU.
6. Spectrophotometer.

2.4.2 Viability Assay

1. Cytotoxicity assay such as CytoTox-Glo cytotoxicity assay (Promega).
2. 96-well, white-walled tissue culture plates compatible with luminometer and fluorometer (clear or solid bottom).
3. Luminescence plate reader or multimode reader.
4. Positive control cytotoxic compound.

2.4.3 Cytology/Immunocytochemistry

1. Glass coverslips.
2. 3% PFA in PBS.

3. 0.2% Triton X-100 in PBS.
4. 0.75% glycine in PBS.
5. 1% BSA in PBS.
6. DAPI (4′,6-diamidino-2-phenylindole, dihydrochloride).

3 Methods

3.1 $Grn^{-/-}$ Mice

3.4.1 Creation and Genotyping

Several *Grn* knockout mice exist where one or more of the initial coding exons have been targeted to result in very short *Grn* transcripts that are destroyed by nonsense-mediated decay of the truncated transcripts. For example, a knockout mouse was generated where exons 2–4 were targeted with Cre-LoxP system [12]. In this mouse model, Cre expression driven by an actin ACTB promotor resulted in *Grn* constitutive knockout mice (Fig. 1), although tissue-specific promoters can also be used to generate conditional knockout mice. The *Grn* knockout mice were generated on a mixed Bl6/129Sv background. At weaning mice are genotyped.

Tail Biopsy and DNA Extraction

1. Mix 200 μL lysis buffer +12.5 μL Proteinase K for each reaction.
2. Place the animal in a restrainer. Collect tail biopsy by cutting 1–2 mm of the tail with a N° 10 scalpel blade and transfer to an micro test tube.
3. Add 200 μL of lysis buffer with Proteinase K and incubate overnight at 56 °C.

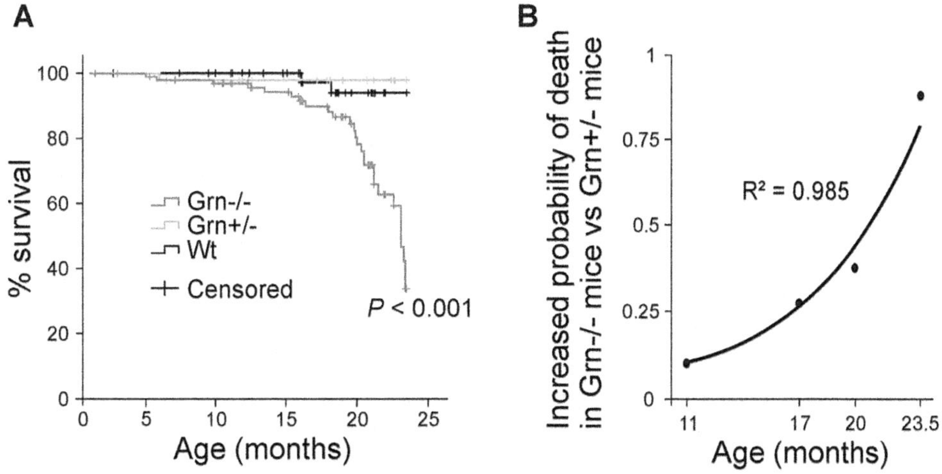

Fig. 1 Aged $Grn^{-/-}$ mice generally show decreased survival as exemplified here in a strain described in ref. [12]. (**a**) After 10 months of age, an increased mortality is observed in the $Grn^{-/-}$ strain compared to WT and $Grn^{+/-}$ mice. (**b**) The mortality observed in these $Grn^{-/-}$ strain mice is exponential in nature compared to $Grn^{+/-}$ littermates (reproduced from ref. [12] with permission from John Wiley & Sons, Inc.)

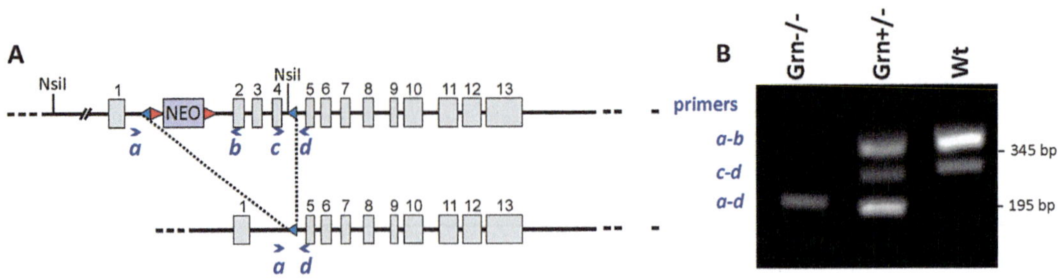

Fig. 2 Generation of Grn knockout (Grn$^{-/-}$) in strain developed in ref. [12]. (**a**) Schematic representation of the targeted deletion strategy for the *GRN* locus: the upper line drawing represents the WT *GRN* locus, while the lower one represents the *Grn* locus, which has exons 2–4 deleted. A multiplex PCR assay using PCR primers a, b, c, and d was designed to differentiate between Grn$^{-/-}$, Grn$^{+/-}$, and nontransgenic (WT) mice that could result from breeding of the Grn$^{+/-}$ mice. The locations of the PCR primers a, b, c, and d are indicated in the WT and Grn- loci. (**b**) PCR products of the multiplex PCR assay using PCR primers a, b, c, and d analyzed on agarose gel show that the genotypes of Grn$^{-/-}$, Grn$^{+/-}$, and WT mice could be easily differentiated (reproduced from ref. [12] with permission from John Wiley & Sons, Inc.)

4. Incubate 45 min at 85 °C to stop the enzymatic reaction.
5. Centrifuge 10 min at 14,000 rpm (app. 18,000 × *g*) and collect supernatant.

PCR Reaction

1. Set up the PCR reaction (total volume = 12 μL) (*see* also Table 1 and Subheading 2.2.1).

Visualization of PCR Products

1. Verify the size of the PCR product by 2% agarose gel electrophoresis in 0.5 × TBE buffer. Load 5 μL PCR product +5 μL marker on the gel. Load 1 μL 50 bp DNA marker +5 μL marker.
2. Run at 100 V for 30 min.
3. Visualize under UV light (aperture 5.6; 500 ms illumination) (*see* Fig. 2).

3.1.1 Husbandry

All animal experiments should be conducted in accordance with institutional guidelines, and ethical permission should be acquired before proceeding with any animal experimentation. Animals are housed in groups in a 12-h day/12-h night cycle. Animals are kept on fresh wood shavings, and food and water were ad libitum available (*see* **Note 1**).

3.2 Behavior Analysis

Grn$^{-/-}$ animals are routinely tested on an experimental battery containing a Morris water maze test, open field and Y-maze exploratory behavior test, and fear-conditioning test such as one testing for passive avoidance. On the Morris water maze, Grn$^{-/-}$ do not show significant differences in target quadrant preference compared to Grn$^{+/+}$ animals. However, Grn$^{-/-}$ animals (generally) show an impaired acquisition performance and a decline in retention performance in terms of platform crossings [12, 16, 17]. This reflects an impaired spatial learning ability and memory deficit. FTLD patients also often show emotional impairment. This is reflected in the impaired fear-conditioning in $Grn^{-/-}$ animals [18]. In an open field test, $Grn^{-/-}$ animals perform less center-crossings compared to wild-type animals indicating increased anxiety [11, 19, 20]. Animals are tested at 13–15 months and 20–22 months of age.

3.2.1 Morris Water Maze Test

Acquisition Training

1. Train the animals during eight trial blocks over 4 days with a 15 min intertrial interval. The visual cues in the room should remain consistent. Be aware that the position of the experimenter in the room is also a cue and should be constant in all experiments.
2. Place the animal in one of four different starting points in semi-random order at every training point.
3. Let the mouse swim freely for 120 s.
4. If the animal reaches the platform, leave it there for 15 s. If it does not reach the platform within 120 s, guide it or place it on the platform and leave it there for 15 s before returning it to its cage. Extremely nervous or anxious animals might have to be gently restrained on the platform to avoid them jumping back into the water.
5. Measure the escape latency and path length.

Probe Trial

1. Remove the platform.
2. Let the animals swim freely for 100 s.
3. Record the trajectory and measure percentage of time spent in each quadrant and number of crossings through the position where the platform used to be.

3.2.2 Passive Avoidance/Fear Conditioning

1. Perform the experiment during dark phase of the day/night cycle of the animal.
2. Place the animal in the brightly lit compartment of the step-through box.
3. After 5 s, open the door to the dark compartment.
4. Upon complete entry in the dark compartment, apply a 1 s 0.3 mA shock.
5. Repeat exactly 24 h later and time escape latency up to 300 s.

3.2.3 Open Field

1. Perform the experiment during dark phase of the day/night cycle of the animal.
2. Place the animals in the arena. Allow for 1 min adaptation. Always let animals start from the same corner.
3. Record trajectories for 10 min.
4. Calculate the path length and number of entries in the center circle or the corners of the open field.

3.2.4 Y-Maze

1. Place the animal in one arm of the Y-maze facing away from the center platform and toward the wall.
2. Allow the animal to explore for approx. 8 min.
3. Record the sequence of arm entries of the animal. An animal makes a complete entry when the whole tail is in the arm.
4. Clean the apparatus with 5% ethanol and let dry after each trial.
5. Exclude animals that make less than five entries.
6. Calculate the number of correctly alternated triplets upon total visits.

The number of maximum spontaneous alterations is then the total number of arms entered minus 2 and the percentage alteration is calculated as (actual alternations/maximum alternations)*100.

3.3 Neuropathological Methods

3.3.1 Tissue Harvesting

1. Euthanize the animal using CO_2 overdose and collect the desired tissue.
2. Rinse tissue in PBS. Tissue can either be snap-frozen in liquid nitrogen for RNA or protein analysis or further processed as described in Subheading 3.3.2.
3. Alternatively, when no further RNA or protein analysis is required, animals can be perfused. For this, animals are deeply anesthetized by intraperitoneal injection with a ketamine/xylazine combination immediately before perfusion (*see* **Note 2**). Open the chest cavity and expose the heart. Insert a needle into the left ventricle and cut open the atria. Perfuse the animal with 10 mL of Krebs solution followed by perfusion with 10 mL of 2% PFA (*see* **Note 3**). Collect the desired tissue and further fixate for min. 2 h.

3.3.2 Fixation and Embedding

1. After dissection, rinse the tissue in PBS.
2. Fix in 2% PFA for 18 h at room temperature (RT). Keep at 4 °C in 70% ethanol for maximum 72 h.
3. Embed the tissue in paraffin using the gradient listed in Table 5.
4. Make paraffin blocks of the tissue and store the blocks at room temperature.

Sectioning and Slide Preparation for Histology

1. Cut 5 μm thick sections on a microtome.
2. Capture sections and carefully transfer to 70% ethanol and 45 °C heated water bath subsequently (*see* **Note 4**).
3. Capture the sections on freshly coated poly-L-lysine-coated slides or superfrost slides (ThermoFischer) and incubate at 37 °C overnight. Store at room temperature until needed.
4. Before the start of any histological procedure, remove the paraffin in the sections by dipping the slide sequentially in xylene (two times 5 min), acetone (1 min), and then hydrate in 100%,

Table 5
Paraffin embedding gradient

	Solution	Time
Step 1	70% ethanol	1 h
Step 2	80% ethanol	1 h
Steps 3–4	90% ethanol	2 × 1 h
Steps 5–7	100% ethanol	3 × 1 h
Step 8	Methyl-benzoic acid	1.5 h
Steps 9–10	Xylene	2 × 1 h
Steps 11–12	Paraffin	2 × 1.5 h

90%, and 80% EtOH (20 dips each). Wash 1 min for two times in dH_2O. Optionally, after the acetone wash and before the hydration in EtOH, a circle can be drawn around the tissue with a hydrophobic marker (Dako) to minimize the antibody use.

3.3.3 Histology for Gross Neuropathology (Table 6)

Hematoxylin/Eosin Stain

1. After rehydration, stain the sections in hematoxylin for 8 min.
2. Rinse in dH_2O.
3. Differentiate in 5% acetic acid (20 dips).
4. Put sections under running tap water for 10 min for "bluing" reaction.
5. Stain in eosin for 6 min.
6. Dehydrate in 80% EtOH and 96% EtOH (10 dips each); 100% EtOH (2 × 1; 10 dips each), methyl-benzoic acid (10 dips), and xylene (two times, 2 min each).
7. Coverslip using a xylene-based mounting medium (e.g., Entellan).
8. Capture images on a bright-field microscope (e.g., a Zeis Axioskop AC1 equipped with an Olympus UC30 camera) using image capturing software (e.g., analySIS getIT).

Cresyl Violet Stain

1. Stain tissue in a 1% cresyl violet solution for 1 min.
2. Differentiate in dH_2O for 5 min.
3. Dehydrate as described above and coverslip using a xylene-based mounting medium (e.g., Entellan).
4. Capture images on a light microscope as described above.

Gross Neuropathology

For the pathological analyses, the investigator is blinded to the genotypes. To determine the anatomical distribution of pathology, images of stained sections were captured on different anatomical locations to include the cerebellum, cerebral cortex, hippocampus, thalamus, and

Table 6
Histological stains

Stain	Marker
Cresyl violet	Stains perikarya deep blue
Hematoxylin/eosin	Stains nuclei purple/stains cytoplasm red

hypothalamus. Neuronal loss on sections stained with hematoxylin/eosin, cresyl violet, or neu-N immunostaining as well as TDP-43 immunopathology is assessed qualitatively, but semiquantitative image analysis approaches could also be applied. Development of gliosis is quantitatively analyzed through Iba1 and GFAP staining. For more details on the procedure of pathological evaluation, readers are referred to the work of J. Lewis [21] and N.J. Cairns [20].

Quantitative Analysis

1. Convert the immunostained sections into high-resolution digital images.
2. The positively stained area is analyzed using quantification software such as ImageScope v.8 (Aperio Technologies) or ImageJ IHC profiler.
3. Measures are expressed as a percentage of the total area analyzed (% burden).

3.3.4 Immunohistochemistry and Antibodies

To estimate and understand the development of neuropathology reflected by processes such as progressive gliosis and TDP-43 aggregate deposition, immunohistochemical analysis is performed.

Antigen Retrieval and Blocking of Endogen Peroxidases Nonspecific Antigens

1. Fill plastic slide container with citric acid antigen retrieval buffer and add slides. Heat water bath to 99 °C. Heat slides for 20–30 min (*see* **Note 5**) and let cool in container for 45 min before transferring the slides back to the glass jar. Wash two times in dH_2O (*see* **Note 6**).
2. Add 100 mL of 0.3% hydrogen peroxide; incubate for 20 min (max 30 min). Wash for 5 min under running tap water, 1 min in dH_2O, and 5 min in PBS.
3. Transfer slides to plastic tray and add 100–200 μL (depending on tissue size) of NHS; incubate for minimum 30 min. Remove the NHS with a vacuum pump immediately before applying the primary antibody.

Fluorescent Staining: TDP-43, Phosphorylated TDP-43

Redistribution of TDP-43 and accumulation of phosphorylated TDP-43 can also be visualized with by double labeling of TDP-43 (Proteintech Group) and phosphorylated TDP-43 (Cosmo Bio) antibodies and detected with Alexa Fluor 488 and 594 conjugated

Table 7
Incubation parameters for immunohistochemistry

	Duration	Temperature	Remarks
Primary antibody	Overnight[a]	4 °C	
Secondary antibody	30 min	Room temperature	Protect from light
Extravidin[b]	30 min	Room temperature	
DAB[b]	3–10 min	Room temperature	

[a]Alternatively, sections can also be incubated for 1–3 h at room temperature
[b]Only for chromogenic stains. Other chromogenic stains such as nitro blue tetrazolium chloride/5-bromo-4-chloro-3-indolyl phosphate (NBT/BCIP; HRP based) or 3-amino-9-ethylcarbazole (AEC; alkaline phosphatase based) can also be used

secondary antibodies or a similar set of labeled antibodies compatible with the filters available on investigators' fluorescent microscope. All antibodies are diluted in PBS/BSA.

1. Add the antibodies and incubate as shown in Table 7. Wash three times in PBS after each step (*see* **Note 7**).
2. Counterstain with DAPI for 2–5 min at room temperature.
3. Coverslip with a fluorescent-compatible mounting medium (e.g., Citifluor).
4. Capture images using a fluorescence microscope (e.g., Axiophot, Zeiss).

Non-fluorescent Staining: Iba1 and GFAP

Development of gliosis was studied by visualization of microglia (Iba1) and astroglia (GFAP) (Fig. 3).

1. Add the antibodies and incubate as shown in Table 5. Wash three times in PBS after each step as described before.
2. Add DAB solution and watch color development.
3. Stop reaction in PBS.
4. Wash slides in tap water and dH_2O, each for 5 min.

Counterstain and Coverslip

1. Counterstain with 10% hematoxylin for 5 min. Wash in tap H_2O and dH_2O.
2. Dehydrate as described in Subheading 3.3.3.
3. Coverslip using a xylene-based mounting medium (e.g., Entellan).
4. Capture images on a light microscope as described above.
5. Quantify using software such as ImageScope v.8 (Aperio Technologies) or ImageJ IHC profiler.
6. Measures are expressed as a percentage of the total area analyzed (% burden).

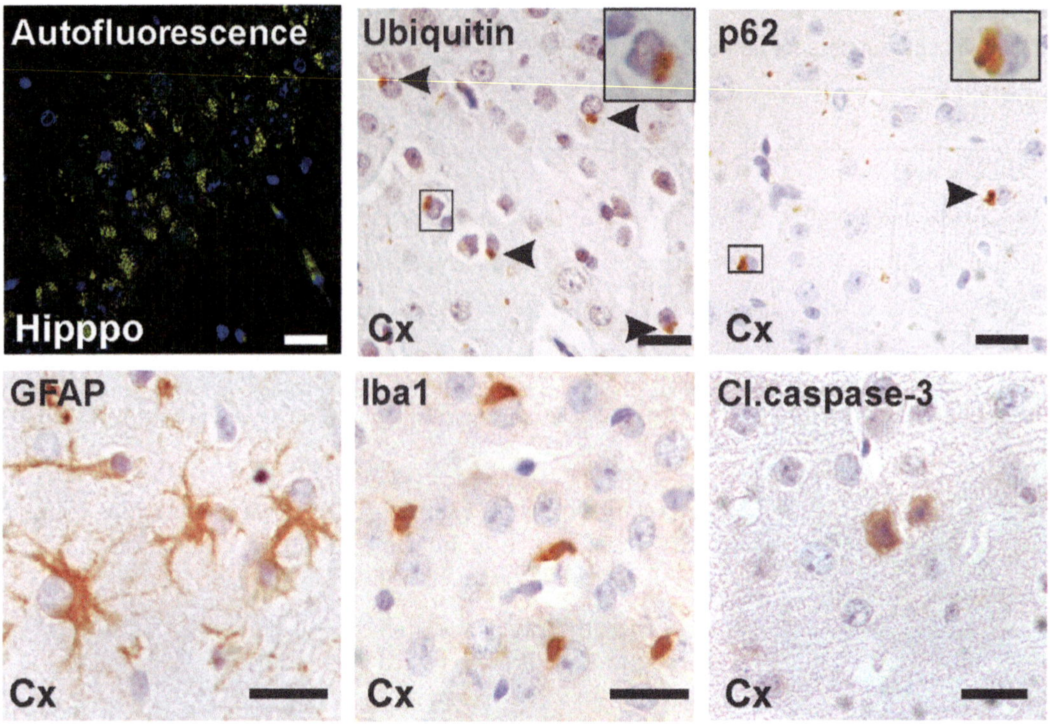

Fig. 3 Examples of pathology in Grn knockout mice. The upper panel shows aging and distribution of UPS (ubiquitin-proteasome system) pathology. The lower panel shows development of astrogliosis (GFAP) and microgliosis (Iba1) and occasional cleaved caspase-3-positive apoptotic neurons (cl. caspase-3) (reproduced from ref. [12] with permission from John Wiley & Sons, Inc.)

3.3.5 Western Blotting

Protein Extraction

TDP-43 accumulation is analyzed by Western blot (Fig. 4). Proteins are extracted in a serial solubilization assay with buffers of increasing detergent strength.

1. Homogenize the tissue in LS buffer, 5× volume/weight (V/wt) by sonication (50–60 Hz, 0.5 pulse/sec, 5 min, keep on ice). Centrifuge at 25,000 × *g* for 30 min at 4 °C. Save supernatants as LS fraction. Wash by resuspending the pellet in LS buffer and centrifuge as before.
2. Dissolve the pellet in TX buffer (5× V/wt) and sediment at 180,000 × *g* for 30 min at 4 °C. Save the supernatants as the TX fraction.
3. Remove myelin by homogenization and vortexing in TX buffer with 30% sucrose (5× V/wt). Centrifuge at 180,000 × *g* for 30 min at 4 °C.
4. Homogenize the pellet in SARK buffer (5× V/wt) and incubate at 22 °C on a shaker for 1 h at 150 rpm. Centrifuge at 180,000 × *g* for 30 min at 22 °C. Store supernatants as the SARK fraction.

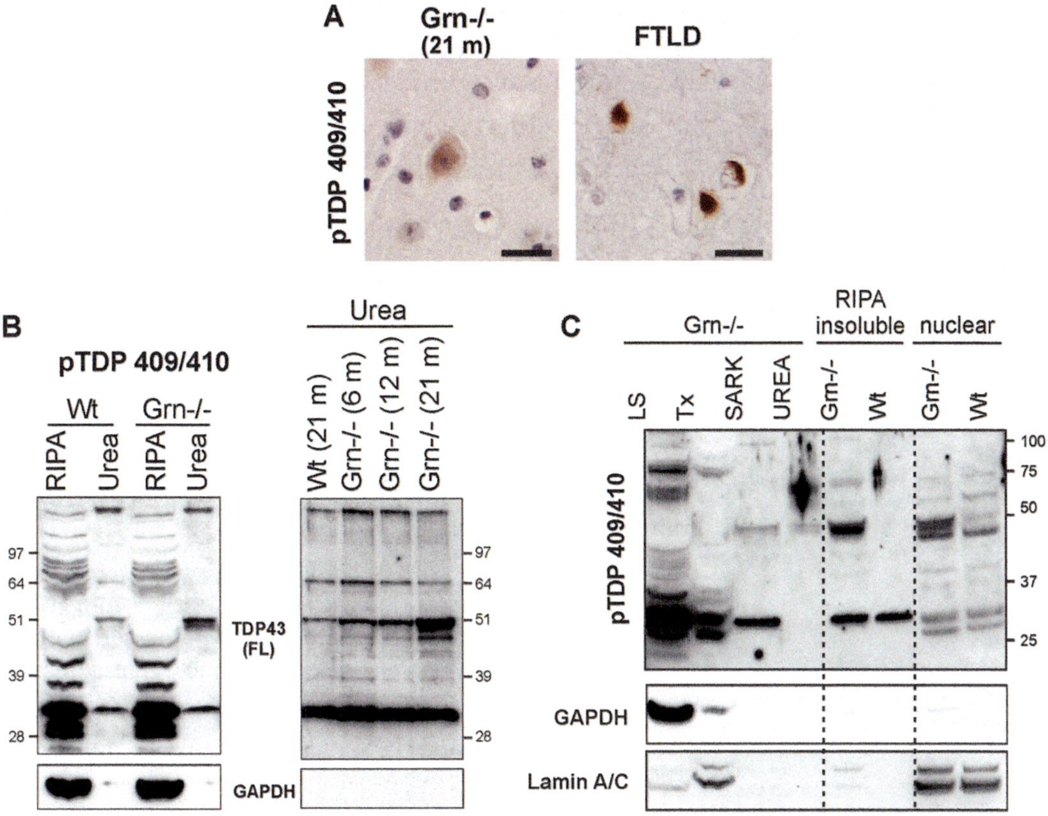

Fig. 4 TDP-43 pathology illustrated by anti-phosphorylated TDP-43 409/410 antibody. (**a**) Example of rare TDP-43-positive neurons in $Grn^{-/-}$ mice, which were never observed in WT mice (Wt). FTLD control is shown. (**b**) Western blots of RIPA and urea fractions. Full-length (FL) TDP-43 is approximately 50 kD and was observed in the urea fraction of $Grn^{-/-}$ mice, but not in WT mice. Immunoblot analysis in the urea fraction of animals at 6-, 12-, and 21 months of age shows an increase of TDP-43 in aged $Grn^{-/-}$ mice, while TDP-43 was not observed in the WT mice at the same age. (**c**) Western blot shows predominant recovery of full-length pTDP-43 sarkosyl-soluble and -insoluble (urea) fractions in a sequential extraction of $Grn^{-/-}$ mouse brain lysates (lanes 1–4) and in the urea fractions of $Grn^{-/-}$ brain after RIPA/urea protein extraction (lanes 5 and 6) (reproduced from ref. [12] with permission from John Wiley & Sons, Inc.)

5. Resuspend the pellet in either SDS (1× V/wt) or Urea buffer (1× V/wt) and sediment at 25,000 × *g* for 30 min at 22 °C. Save the supernatants as the SDS and urea fractions, respectively.

Western Blotting

1. Determine protein concentration with a standard bicinchoninic acid (BCA) colorimetric assay.
2. Load equal amounts of protein of each sample and the protein marker on the NuPage® Bis-Tris gradient gel in a total volume of 10 μL and electrophoresed at 180 V for 45 min in a compatible buffer (see vendor guides for more details).
3. Cut the polyvinylidene difluoride membrane to an appropriate size (e.g., 10 × 10 cm), wash in MeOH for 30 sec, and incubate

in transfer buffer for 10 min to make the membrane hydrophilic. Do not touch the membrane with anything other than forceps. Mark the protein side.

4. Wash the gel (10 min) and wet filter papers and sponges in blotting buffer and assemble the blotting sandwich. Avoid bubbles.
5. Blot for 1 h at 25 V/50 Watt at room temperature.
6. Block membranes in PBS with 5% skimmed milk and 0.1% Tween for 1 h at room temperature.
7. Add primary antibodies and incubate for 1 h at room temperature on shaker. Wash five times and each for 5 min with wash buffer (*see* **Note 8**).
8. Add HRP-conjugated secondary antibodies and incubate 1 h at room temperature on shaker. Wash five times and each for 5 min with wash buffer.
9. Cover the membrane with the ECL Plus reagent (or other HRP-based similar detection), and incubate for 5 min in the dark at room temperature. Measure chemiluminescence on an ECL Plus™ chemiluminescent or other comparable detection system.
10. Normalize data to GAPDH (low-salt fractions) or lamin A/C for 2% SDS fractions.

3.4 Cell Biology

3.4.1 Neuronal Methods (Counting, Preparation, and Viability Assay)

Primary neuronal cell (PNC) cultures derived from $Grn^{-/-}$ knockout mice allow to test various different challenges in a Grn-null background with a relatively short-term readout of the results [22]. As such, important data can rapidly be obtained through PNC cultures. Primary neuronal cell cultures are prepared according to *Neuronal Cell Culture: Methods and Protocols* [22].

Cell Treatment

Neuronal cell cultures are challenged by proteasomal inhibition for exacerbation of ubiquitin proteasome system (UPS) pathology and cell survival challenge.

1. Add 10 μ M MG132/clasto-lactacystin b-lactone to challenge the cell into depositing protein aggregates.
2. Add 50 μM Z-VAD-FMK (pan-caspase inhibitor) to avoid preemptive apoptosis.
3. For pathway analysis, grow neurons in neurobasal medium for 2 days.
4. Stimulate for 10 min with 30 nM recombinant GRN or 10% horse serum as control.

Cell Counting

Automated Cell Counter

1. Seed 5×10^4 cells in 6-well plates.
2. Culture for 14–16 h under basal conditions.

3. Wash two times with PBS.
4. Culture in medium containing 2% FCS.
5. Count with the Countess automated cell counter (Invitrogen).

DNA Synthesis Measurement

1. Seed 2×10^4 cells.
2. Culture 14 h under basal conditions.
3. Wash two times in PBS.
4. Incubate in DMEM containing 10 μM BrdU for 24 h.
5. Record absorbance at 450 nm. Use 260 nm as a reference.

Viability Assay

The relative number of viable neurons was measured using the CytoTox-Glo cytotoxicity commercial kit (Promega)—used following manufacturer's instructions.

1. Transfer cells in culture medium to a 96 well plate.
2. Add test compounds and vehicle controls to a final volume of 100 μL per well.
3. Incubate for 24 h.
4. Add 50 μL of CytoTox-Glo™ Cytotoxicity assay reagent to all wells.
5. Mix briefly and incubate for 15 min at room temperature.
6. Measure luminescence.
7. Add 50 μL of lysis reagent.
8. Mix and incubate at room temperature for 15 min.
9. Measure luminescence.
10. Calculate the luminescent contribution of previously viable cells (after lysis) by subtracting the luminescent signal resulting from experimental cell death from total luminescence death (viable cell luminescence = total luminescence minus experimental dead cell luminescence).

3.4.2 Cytology/ Immunocytochemistry

1. Grow cells on glass coverslips.
2. Fix in 3% PFA.
3. Permeabilize with 0.2% Triton X-100 in PBS for 10 min.
4. Quench the aldehyde groups using 0.75% glycine in PBS for 20 min.
5. Block with 1% BSA in PBS for 30 min.
6. Incubate with the primary antibody overnight at 4 °C.
7. Incubate with the secondary antibody/antibodies for 30 min at room temperature and visualize with Alexa Fluor 488 or Alexa Fluor 594.
8. Counterstain the nuclei with DAPI.

9. Images are taken with Axiovert fluorescence microscope equipped with an Apotome or on a LSM 510 confocal microscope (Zeiss).
10. Perform the quantification by either manual counting of cortical neurons with cytoplasmic pTDP-43 accumulation or by measuring optical density of total pTDP-43 immunoreactivity in mouse embryonic fibroblasts (MEF) or cortical neurons treated with MG132 [13, 23].

4 Notes

1. For aging of the animals, keep the animals in a SPF facility (to reduce risk of infection), and be gentle while changing cages. $Grn^{-/-}$ animals show decreased survival at postnatal stage [24] and after 15 months of age [12]. This was correlated to stress in one study, which suggested reducing the frequency of animal handling and cage changing in the breeding of *Grn* knockout mice [24].
2. When performing perfusion, do not apply cervical dislocation as this will rupture the blood vessels in the neck and impair the perfusion of the brain and hampering neuropathological analysis.
3. For optimal perfusion, the fluids should enter the heart at a trickling rate and not a steady stream: individual drops should just be visible.
4. Be careful when floating the tissue in the water bath. If tissue sections disintegrate, then try to reduce the amount of alcohol on the sections and/or lower the temperature of the water bath.
5. For antigen-retrieval, instead of using a water bath, it is possible to use a microwave and heat for 4 + 3 + 3 + 3 + 3 min at 800 W. Check buffer level at each interval to avoid tissue not being scorched dry.
6. After antigen retrieval, leave the sections in the antigen retrieval buffer until 5 degrees higher than room temperature. Also take care that the sections never dry.
7. The volumes and time of washing steps increase in each step to increase washing efficiency. First, apply a small volume of PBS with a pipette and immediately take it off. Repeat this step but now wash for 1 min. For the last washing step, fill a jar with PBS and wash the sections for 5 min.
8. To improve signal/noise ratio, increase the amount of wash steps but not the time of each step. For example, 3 × 4 min washing will result improve signal/noise better than 2 × 6 min of washing. Use an excess of washing buffer wherever possible.

References

1. Baker M et al (2006) Mutations in progranulin cause tau-negative frontotemporal dementia linked to chromosome 17. Nature 442(7105):916–919
2. Cruts M et al (2006) Null mutations in progranulin cause ubiquitin-positive frontotemporal dementia linked to chromosome 17q21. Nature 442(7105):920–924
3. Neumann M et al (2006) Ubiquitinated TDP-43 in frontotemporal lobar degeneration and amyotrophic lateral sclerosis. Science 314(5796):130–133
4. Bateman A, Bennett HP (1998) Granulins: the structure and function of an emerging family of growth factors. J Endocrinol 158(2):145–151
5. Toh H et al (2013) Expression of the growth factor progranulin in endothelial cells influences growth and development of blood vessels: a novel mouse model. PLoS One 8(5):e64989
6. Liau LM et al (2000) Identification of a human glioma-associated growth factor gene, granulin, using differential immuno-absorption. Cancer Res 60(5):1353–1360
7. Kumar-Singh S (2011) Progranulin and TDP-43: mechanistic links and future directions. J Mol Neurosci 45(3):561–573
8. Buratti E, Baralle FE (2008) Multiple roles of TDP-43 in gene expression, splicing regulation, and human disease. Front Biosci 13:867–878
9. American Thoracic Society; Infectious Diseases Society of America (2005) Guidelines for the management of adults with hospital-acquired, ventilator-associated, and healthcare-associated pneumonia. Am J RespirCrit Care Med 171(4):388–416
10. Martens LH et al (2012) Progranulin deficiency promotes neuroinflammation and neuron loss following toxin-induced injury. J Clin Invest 122(11):3955–3959
11. Petkau TL et al (2012) Synaptic dysfunction in progranulin-deficient mice. Neurobiol Dis 45(2):711–722
12. Wils H et al (2012) Cellular ageing, increased mortality and FTLD-TDP-associated neuropathology in progranulin knockout mice. J Pathol 228(1):67–76
13. Wils H et al (2010) TDP-43 transgenic mice develop spastic paralysis and neuronal inclusions characteristic of ALS and frontotemporal lobar degeneration. Proc Natl Acad Sci U S A 107(8):3858–3863
14. Janssens J et al (2013) Overexpression of ALS-associated p.M337V human TDP-43 in mice worsens disease features compared to wild-type human TDP-43 mice. Mol Neurobiol 48(1):22–35
15. Kao AW et al (2011) A neurodegenerative disease mutation that accelerates the clearance of apoptotic cells. Proc Natl Acad Sci U S A 108(11):4441–4446
16. Yin F et al (2010) Behavioral deficits and progressive neuropathology in progranulin-deficient mice: a mouse model of frontotemporal dementia. FASEB J 24(12):4639–4647
17. Minami SS et al (2014) Progranulin protects against amyloid beta deposition and toxicity in Alzheimer's disease mouse models. Nat Med 20(10):1157–1164
18. Filiano AJ et al (2013) Dissociation of Frontotemporal dementia-related deficits and Neuroinflammation in Progranulin Haploinsufficient mice. J Neurosci 33(12):5352–5361
19. Kayasuga Y et al (2007) Alteration of behavioural phenotype in mice by targeted disruption of the progranulin gene. Behav Brain Res 185(2):110–118
20. Ghoshal N et al (2012) Core features of frontotemporal dementia recapitulated in progranulin knockout mice. Neurobiol Dis 45(1):395–408
21. Ahmed Z et al (2010) Accelerated lipofuscinosis and ubiquitination in granulin knockout mice suggest a role for progranulin in successful aging. Am J Pathol 177(1):311–324
22. Darbinyan A et al (2013) Isolation and propagation of primary human and rodent embryonic neural progenitor cells and cortical neurons. Methods Mol Biol 1078:45–54
23. Kleinberger G et al (2010) Increased caspase activation and decreased TDP-43 solubility in progranulin knockout cortical cultures. J Neurochem 115(3):735–747
24. Ahmed Z et al (2007) Progranulin in frontotemporal lobar degeneration and neuroinflammation. J Neuroinflammation 4:7

Chapter 17

Methods to Investigate the Protection Against Neurodegenerative Disorders Provided by Progranulin Gene Transfer in the Brain

Jackalina M. Van Kampen

Abstract

Progranulin (PGRN) is a multifunctional protein that is widely expressed throughout the brain, where it has been shown to be a critical regulator of CNS inflammation (Ahmed et al., J Neuroinflammation 4:7, 2007; Yin et al., J Exp Med 207:117–128, 2010; Martens et al., J Clin Investig 122:3955–3959; Inestrosa and Arenas). PGRN functions as an autocrine neuronal growth factor, important for long-term neuronal survival (Ahmed et al., J Neuroinflammation 4:7, 2007; Nat Rev Neurosci 11:77–86, 2009). Together, these critical roles in the CNS suggest that enhancing PGRN expression may provide neuronal support and protection for neurodegenerative disorders, such as Parkinson's disease (PD). Here, we describe the application of PGRN gene transfer using in vivo delivery of lentiviral expression vectors in a rodent model of PD.

Key words Gene therapy, Lentiviral vectors, Neuroprotection, Progranulin, *Substantia nigra*, Parkinson's disease, Immunohistochemistry

1 Introduction

Progranulin (PGRN) is a 68.5 kDa multifunctional glycoprotein. It is widely expressed throughout the periphery where it has been implicated in the regulation of tumorigenesis, wound repair, development, and inflammation [1]. In the central nervous system, PGRN is expressed primarily in neurons and microglia and serves as an autocrine neurotrophic factor and potent regulator of neuroinflammation [2–7]. Indeed, PGRN deficiency has been shown to result in dysregulated neuroinflammation, elevated indices of cellular aging, synaptic dysfunction, disrupted lysosomal processing, and potentiated responses to neurotoxic insults [2, 6, 8–12]. Progranulin activates cell survival signaling pathway [13] and protects neurons against various insults [13–16]. By virtue of its neurotrophic and anti-inflammatory properties, PGRN has become a target of interest for the support of neuronal health and prevention of neurodegeneration. Therapies aimed

Andrew Bateman et al. (eds.), *Progranulin: Methods and Protocols*, Methods in Molecular Biology, vol. 1806,
https://doi.org/10.1007/978-1-4939-8559-3_17,

at elevating expression levels of PGRN may be of benefit to a variety of brain disorders.

Unfortunately, there is currently no pharmacological means of regulating PGRN levels in the brain, and local delivery of recombinant PGRN to the CNS poses a challenge, owing to its large size and blood–brain barrier limitations. However, these obstacles can be overcome through the use of in vivo gene transfer using recombinant viral vectors derived from retroviruses, adeno-associated virus (AAV), adenovirus, or herpes simplex virus (HSV). Of these, HIV-1-based (lentiviral) vectors, a form of retrovirus, are ideally suited for the targeted delivery of PGRN to the CNS. While AAV is one of the safest and most commonly used vectors for gene delivery, it is a very small virion, limited to a total packaging capacity of approximately 4.7 kb. This seriously limits the size of the transgene that can be packaged and makes AAV not suitable for larger genes. While adenoviral and HSV vectors have large packaging capabilities, they tend to induce an inflammatory response [17]. Lentiviral vectors, however, can integrate into the host chromosomes without triggering cytotoxicity or a significant inflammatory response [18]. The size of the wild-type HIV genome is approximately 10 kb making it better suited for delivery of larger genetic payloads. A lentiviral vector can provide stable long-term transgene expression and is capable of efficiently transfecting both dividing and nondividing cells [19, 20], including neurons of rodent and primate brain [20–22]. Thus, lentiviral gene transduction represents an effective tool for targeted overexpression of PGRN in the brain.

Here, we describe the methods utilized to enhance PGRN expression in the *substantia nigra* (SN) of C57BL/6 mice. In these animals, lentiviral gene delivery has been shown to effectively elevate PGRN expression in the SN resulting in neuroprotection against the parkinsonism-inducing neurotoxin, 1-methyl-4-phenyl-1,2,3,6-tetrahydropyridine (MPTP) [16].

2 Materials

2.1 Lentiviral Stock Storage/Preparation

1. Progranulin- and GFP-expressing lentiviral vectors (Fig. 1) can be generated using ViraPower™ HiPerform™ Promoterless Gateway® Vector Kit obtained from Invitrogen Corporation (Carlsbad, CA) using the manufacturer's instructions. Briefly, it is first necessary to generate entry clones containing the promotor of interest (CMV) and the gene of interest (mPGRN). The CMV immediate-early promoter allows high-level, constitutive expression of the gene of interest in mammalian cells. A Gateway LR Recombination is then performed to simultaneously transfer the two DNA fragments into the pLenti6.4/R4R2/V5-DEST vector using purified plasmid DNA from

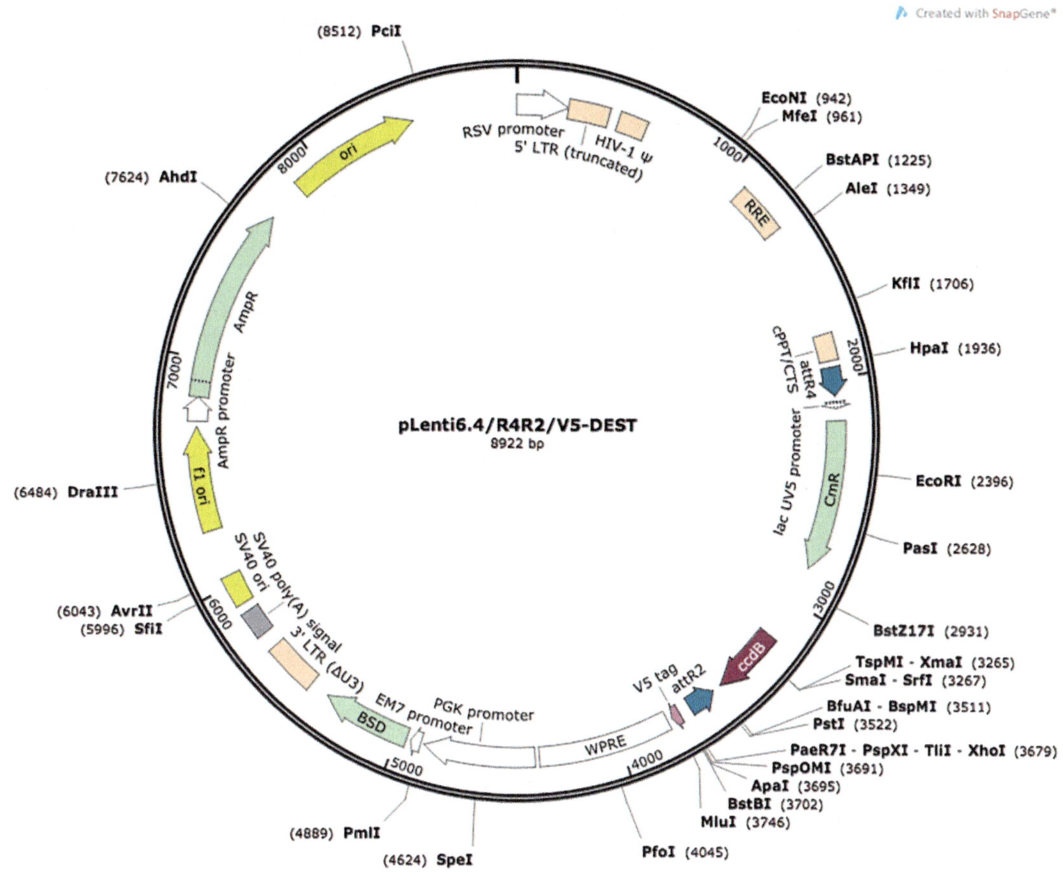

Fig. 1 Lentiviral Gateway® destination vector map

each entry clone to transform *E. coli* cells, creating an expression clone. The expression vector should be verified by transfecting the plasmid directly into a mammalian cell line, such as HEK 293 cells, and assaying for mPGRN or GFP expression. The 293FT cell line is then used to produce lentiviral stocks. The titer (concentration) should be at least 1×10^8 TU/mL, as determined by a blasticidin resistance assay.

2. PCR microtubes with a conical bottom (approximately 20 μL volume).
3. Crushed wet ice.
4. Ice bucket.
5. Mini centrifuge.

2.2 Infusion Cannula

1. Glass Hamilton micro-syringe and needle (10–50 μL).
2. Distilled water.
3. 1 mL syringe and 26-gauge (Ga) needle.
4. 50 PE polyethylene tubing.

5. A 31 Ga stainless steel infusion cannula, approximately 20 mm in length. This can be purchased from Plastics One or can be made from pieces of stainless steel tubing. To make an infusion cannula, a length of 31 Ga stainless steel tubing can be placed within a shorter length of 26 Ga stainless steel tubing and fixed in place with a watertight glue or dental cement.
6. Micro-syringe pump (Harvard Instruments).
7. Small frozen gel packs (*see* **Note 1**).

2.3 Intracerebral Infusion

1. Anesthesia vaporizer and corresponding inhalant anesthetic.
2. Medical grade oxygen.
3. Stereotaxic frame with holder arm and mouse adaptor and gas anesthesia mask.
4. Shaver.
5. Three-step prep: Betadine scrub, 70% ethanol, Betadine solution. Distilled water and sterile gauze will also be needed for preparation of the incision site.
6. Ophthalmic ointment (e.g., Duratears).
7. Sterile surgical instruments: a #10 scalpel and blade, retractors, bone scraper, forceps.
8. Drill with a drill bit approximately 0.6 –0.8 mm. This can be a micro-drill from a surgical supply store, a dental drill, or a handheld rotary tool from the local hardware store. In any case, drill bits can be purchased from any supplier of surgical instruments. The bit should create an opening in the skull just large enough to allow the infusion cannula to pass through. Drill bits should be sterilized prior to surgery.
9. Marking pencil.
10. Cleaning wires. These should be small enough to fit inside the tip of the infusion cannula, should it need cleaning. Hamilton syringes often come with cleaning wires, which are well-suited for this purpose.
11. Timer.
12. Suture thread or adhesive for wound closure.
13. Saline (0.9%), an analgesic, and an antibiotic for postsurgical care.
14. Warming pad.

2.4 Histological Confirmation

1. For tissue harvesting, prepare 100–150 mL/animal of 4% paraformaldehyde (400–500 mL for rats). To make paraformaldehyde solution, add distilled water to paraformaldehyde at a concentration of 10% (w/v). This takes a significant volume of paraformaldehyde. Heat to 65 °C while stirring. Slowly

add a few drops of NaOH at a time until the solution becomes clarified. Allow the solution to cool to room temperature. Filter. Dilute the paraformaldehyde solution to 4% using half distilled water and half potassium phosphate buffer. To prepare the potassium phosphate buffer, add 2.65 g NaH_2PO_4 and 14 g K_2HPO_4 to each L of distilled water. If necessary, adjust the pH to approximately 7.2–7.5.

2. Approximately 50–100 mL/animal of phosphate-buffered saline (0.1 M): 11.36 g Na_2HPO_4, 3.5 g $NaH_2PO_4 \cdot H_2O$, 9 g NaCl, 1 L distilled water (pH: 7.4).
3. Peristaltic pump.
4. Necropsy instruments: rongeurs, #10 scalpel and blade, forceps, small scissors, micro spatula.
5. Centrifuge tubes or glass vials.
6. 30% sucrose in 0.1 M PB. To each 100 mL of 0.1 M PB, add 30 g of sucrose. Heat slightly to dissolve.
7. To prepare 0.1 M PB: Prepare sodium phosphate dibasic stock (0.5 M) by dissolving 35.5 g of Na_2HPO_4 in a final volume of 500 mL of distilled water. Some crystallization will occur when the solution is stored at 4 °C. Warm on a hot plate and stir until the crystals dissolve. Prepare sodium phosphate monobasic stock (0.5 M) by dissolving 30 g of $NaH_2PO_4 \cdot H_2O$ in a final volume of 500 mL of distilled water. Prepare 0.1 M sodium phosphate dibasic: Put 80 mL of sodium phosphate dibasic stock (0.5 M) in a beaker and add distilled water to give a final volume of 400 mL. Prepare 0.1 M sodium phosphate monobasic: Put 30 mL of sodium phosphate monobasic stock (0.5 M) in a beaker and add distilled water to give a final volume of 150 mL. Bring the 0.1 M sodium phosphate dibasic solution to pH 7.2 by adding as much as needed of the 0.1 M sodium phosphate monobasic. The resulting solution is 0.1 M phosphate buffer pH 7.2.
8. Freezing microtome or cryostat.
9. Tissue culture plates (24 well for storage and 12 well for IHC).
10. Millonigs (1.2 M) stock solution: To 400 mL of distilled water, add 84.4 g $NaH_2PO_4 \cdot H_2O$ and 19.3 g NaOH and warm slightly while stirring. Once completely dissolved, adjust the volume to a final volume of 500 mL. Store stock solution in a warm location.
11. Millonigs working solution: Add 100 mL of Millonigs stock solution to 890 mL of distilled water. Warm slightly while stirring. Add 10 mL of 6% sodium azide (pH: 7.2–7.4).
12. Tris-buffered saline (TBS): To each L of distilled water, add 13.22 g Trizma (Tris) HCl, 1.94 g Tris base, and 14 g NaCl (pH: 7.4).

13. TBS Triton: To each L of TBS, add 3 mL of Triton-X. This makes a final solution of TBS with 0.3% Triton.
14. Shaker.
15. Sodium chloride sodium citrate buffer (SSC): To prepare 20× SSC stock solution, dissolve 175.3 g NaCl and 88.2 g sodium citrate in 800 mL of distilled water. Adjust the pH to 7.0 with a few drops of HCl. Adjust the volume to 1 L with distilled water. For a 2× SSC working solution, dilute the stock with distilled water.
16. Hybridization oven.
17. Goat serum.
18. Donkey serum
19. Primary antibodies: anti-GFP (Millipore polyclonal chicken anti-GFP; AB1690), anti-TH (Millipore monoclonal mouse anti-TH; MAB 318), anti-mouse PGRN (R&D Systems polyclonal sheep anti-progranulin, AF2557).
20. Fluorescent secondary antibodies: Alexa Fluor 488 IgG (H + L) goat anti-chicken, Alexa Fluor 594 IgG (H + L) goat anti-mouse, Alexa Fluor® donkey anti-sheep 594 (H + L).
21. Clean glass microscope slides.
22. Glass coverslips (No. 1).
23. Mounting medium for fluorescence (e.g., Fluoromount-G).
24. Nail polish.
25. Drying rack.
26. Aluminum foil.
27. Fluorescent microscope with imaging FITC filter, Texas Red filter.

3 Methods

Lentiviral vectors are based on the human immunodeficiency virus (HIV) which is the virus responsible for the development of acquired immunodeficiency syndrome (AIDS). Although lentiviral vectors have been modified to provide a safer version of the HIV, in which the viral replication genes have been removed, they can still pose some biohazardous risk since they retain the ability to transduce primary human cells. For this reason, it is recommended that you treat these lentiviral vectors as Biosafety Level 2 (BL-2) organisms and strictly follow all published BL-2 guidelines with proper waste decontamination. Handle all lentiviruses in compliance with established institutional guidelines. Since safety requirements for use and handling of lentiviruses may vary at individual institutions, we recommend consulting the health and safety guidelines and/or officers at your institution prior to their use.

3.1 Lentiviral Stock Storage/Preparation

1. Viral titers can decrease as much as 10% with each freeze/thaw cycle. Therefore, lentiviral stocks should be aliquoted and stored at −80 °C. Do not freeze/thaw more than three times. The volume of each aliquot should be determined by the number of infusions to be performed within no more than a 2-h period. Allow a few microliters of added volume, as the infusion cannula requires a lead volume of approximately 1 μL and a small quantity is invariably lost when initially testing the cannula for proper flow. For example, if four infusions of each vector are to be performed each day, a total of 4 μL should be infused into the brain. In this case, an aliquot of 8 μL is recommended. Small, clear PCR tubes with a conical bottom are recommended (*see* **Note 2**). Stored in this way, lentiviral stocks are good for up to 1 year.
2. Immediately prior to surgery, defrost one aliquot of lentiviral stock on wet ice.
3. Once defrosted, spin the aliquot in a mini centrifuge for 3–5 s to ensure the entire contents are at the bottom of the tube.
4. Keep the aliquot of lentiviral stock on ice for the duration of the procedure.

3.2 Preparation of the Infusion Cannula

1. Fill a 10–50 μL Hamilton syringe with 1–2 μL of distilled water. Push down the plunger slightly until a small drop forms at the tip.
2. Fill a 1 mL syringe, containing a 26-gauge needle, with distilled water. With the infusion cannula at one end, attach the needle to the other end of the polyethylene tubing and fill the tubing and infusion cannula with distilled water until a small drop forms at the cannula tip.
3. While pinching the end of the tubing to prevent escaping water, remove the 1 mL syringe and attach the Hamilton syringe. Once the needle of the Hamilton syringe is securely in place, release the pinch. It is important to minimize any potential air bubbles inside the syringe.
4. Pull back the plunger of the Hamilton syringe until an air bubble, approximately 1 cm in length, is visible in the tubing just above the cannula (*see* **Note 3**).
5. You are now prepared to load the tubing with the lentivirus. Place the tip of the cannula into the aliquot and slowly withdraw the plunger. Watch the air bubble in the tubing move and make sure no other air bubbles arise. Draw up the contents of the aliquot.

6. Place the Hamilton syringe in a Harvard syringe pump. Adjust the diameter settings according to the diameter of your syringe. Set the flow rate at 0.2 μL/min.
7. Place the infusion cannula securely into the holder arm of the stereotaxic frame. Wrap the arm with frozen flexible gel packs to keep the lentivirus cool.

3.3 Intracerebral Infusion of Lentiviral Vector

1. Anesthetize an adult C57Bl6 mouse (*see* **Note 4**) in an induction chamber using 5% isoflurane carried by medical grade oxygen.
2. Test the depth of anesthesia by moderately pinching the toe and/or tail. Once nonresponsive, place the animal in a stereotaxic frame and maintain anesthesia at 1–2% isoflurane for the duration of the surgery.
3. Shave the fur from the top of the head and prepare the incision site with a standard three-step prep, which includes Betadine scrub, followed by 70% ethanol, then followed by Betadine solution.
4. Administer a lubricating eye ointment, such as Duratears, to prevent the eyes from drying out during surgery.
5. With a scalpel, create an incision down the length of the scalp. Using a small bone scraper, clear remaining tissue away from the skull. Use retractors to keep the incision site open during surgery.
6. Locate the bregma. This is the anatomical point on the skull at which the coronal suture is intersected perpendicularly by the sagittal suture.
7. Center the infusion cannula using the bregma, making sure to be flush with the skull, and mark the anterior–posterior, medial–lateral, and dorsal–ventral coordinates.
8. Locate the proper coordinates for your infusion site using a brain atlas [23]. For the SN_C, a target of interest in PD, position the infusion cannula at the following coordinates, relative to the bregma: A.P. −2.9 mm, M.L. +1.30 mm, D.V. −4.60 mm (*see* **Note 5**). Mark the site with a small dot.
9. Using a dental- or rotary-style drill, make a hole in the skull at the marked coordinate. Be careful not to drill into the cortical tissue.
10. Confirm that the flow by starting the pump and checking the cannula tip for fluid (*see* **Note 6**). Once the flow is confirmed, stop the pump.
11. Reposition the infusion cannula to the designated coordinates and lower the cannula to the appropriate depth. Lower the cannula approximately 0.1 mm beyond the target and draw back, again to the target depth. This will create a small pocket

that will allow the fluid to flow more readily and prevent backflow.

12. Begin pumping at 0.2 μL/min for 5 min (*see* **Note 7**). Leave the cannula in place for an additional 4 min to allow the fluid to diffuse away from the cannula tip and prevent it from being pulled back up upon removal of the cannula.
13. Slowly withdraw the cannula. Swab the tip of the cannula with distilled water to clear any residual tissue and prevent the tip from getting blocked.
14. Close the incision site with individual subcutaneous sutures, surgical staples, or tissue adhesive.
15. As standard postoperative care, administer 1 mL of subcutaneous saline (0.9%) for rehydration, an analgesic for pain management, and an antibiotic. Allow the animal to recover in a quiet, isolated cage with a warming pad before transferring to a clean home cage.
16. Allow a minimum of 1–2 weeks for full infectivity.

3.4 Confirmation of Infectivity

Following in vivo transduction of your lentiviral construct, you will need to assay for proper expression of your reporter protein at the target site. Immunohistochemistry can be used to determine the extent and localization of infectivity through examination of GFP immunolabeling.

1. Sacrifice animals by cardiac perfusion using 4% paraformaldehyde (approximately 100 mL) preceded by either PBS or 0.9% saline.
2. Extract the brain and place it in a vial filled with 4% paraformaldehyde overnight. Do not exceed 24 h, as this will over-fix the tissue, creating excessive protein cross-links, which may interfere with the performance of certain antibodies.
3. Transfer the brains to 30% sucrose for cryoprotection and place on a shaker at room temperature overnight. Use a low setting.
4. Adequate absorption of the sucrose will be indicated by the sinking of the brains in the vial. Once the brains have sunk, they are ready to be sectioned using a freezing microtome or cryostat. For immunohistochemistry using free-floating sections, as described below, a freezing microtome is preferable. Collect every 12th coronal section (20 μm) through the brain and store sections free-floating in 24-well tissue culture plates filled with Millonigs working solution.
5. For immunohistochemical detection of the reporter protein, eGFP, transfer one well of tissue from each animal to 12-well tissue culture plates filled with TBS (3–4 mL/well). All washes should be done on a shaker at a low setting.

6. Heat the tissue at 65 °C in 2× SSC for 30 min.
7. Wash in TBS Triton three times for 10 min each.
8. Incubate the tissue in 5% goat serum in TBS Triton for 1 h.
9. Incubate the tissue in the primary antibody (Millipore polyclonal chicken anti-GFP, AB1690; 1:1000) in TBS Triton with 5% serum at 4 °C overnight. To localize GFP expression relative to dopaminergic neurons of the SN_C (Fig. 2), the tissue may be co-incubated with a TH primary antibody (Millipore monoclonal mouse anti-TH, MAB 318; 1:10,000).
10. The following day, wash in TBS three times for 10 min each.
11. Incubate the tissue in the appropriate secondary conjugated antibodies. For detection of GFP, Alexa Fluor® 488 goat anti-chicken IgG (H + L) (1:500) and, for TH, Alexa Fluor® 594 goat anti-mouse IgG (H + L) (1:500) plus 2% goat serum in TBS for 3–4 h. From this point onward, treat the tissue as light-sensitive and protect it from light. This can be achieved by simply wrapping plates with aluminum foil.
12. Wash the tissue in TBS three times for 10 min each.
13. Transfer the tissue to Millonigs for storage until mounting.
14. Mount the sections onto clean glass slides using 0.1 M PB. Dry slides upright on a drying rack in the dark. Drying time will vary anywhere from 30 min to 24 h, depending on the environmental conditions. The appropriate drying time should be determined in-house).
15. Cover the slides using glass coverslips (No. 1–1.5) and a mounting medium designed for fluorescently labeled tissue (e.g., Fluoromount-G).
16. Use clear nail polish around the edges of the coverslip to seal them in place. Allow the slides to dry flat overnight.
17. Allow at least 1 week for the slides to cure before imaging on a fluorescent microscope. Alexa Fluor® 488 has a peak excitation wavelength of 490 nm and peak emission wavelength at 525 nm. A FITC filter set generates the best imaging results for this fluorescent dye. Alexa Fluor® 594 has a peak excitation wavelength of 590 nm and a peak emission wavelength of 617 nm. A Texas Red filter set generates the best imaging results for this fluorescent dye.
18. For immunolabeling of PGRN, repeat **steps 5–17** using a primary antibody raised against mouse PGRN (R&D Systems polyclonal sheep anti-progranulin, 1:1000; AF2557), donkey serum, and Alexa Fluor® donkey anti-sheep 594 (H + L) (1:500) as the secondary antibody.

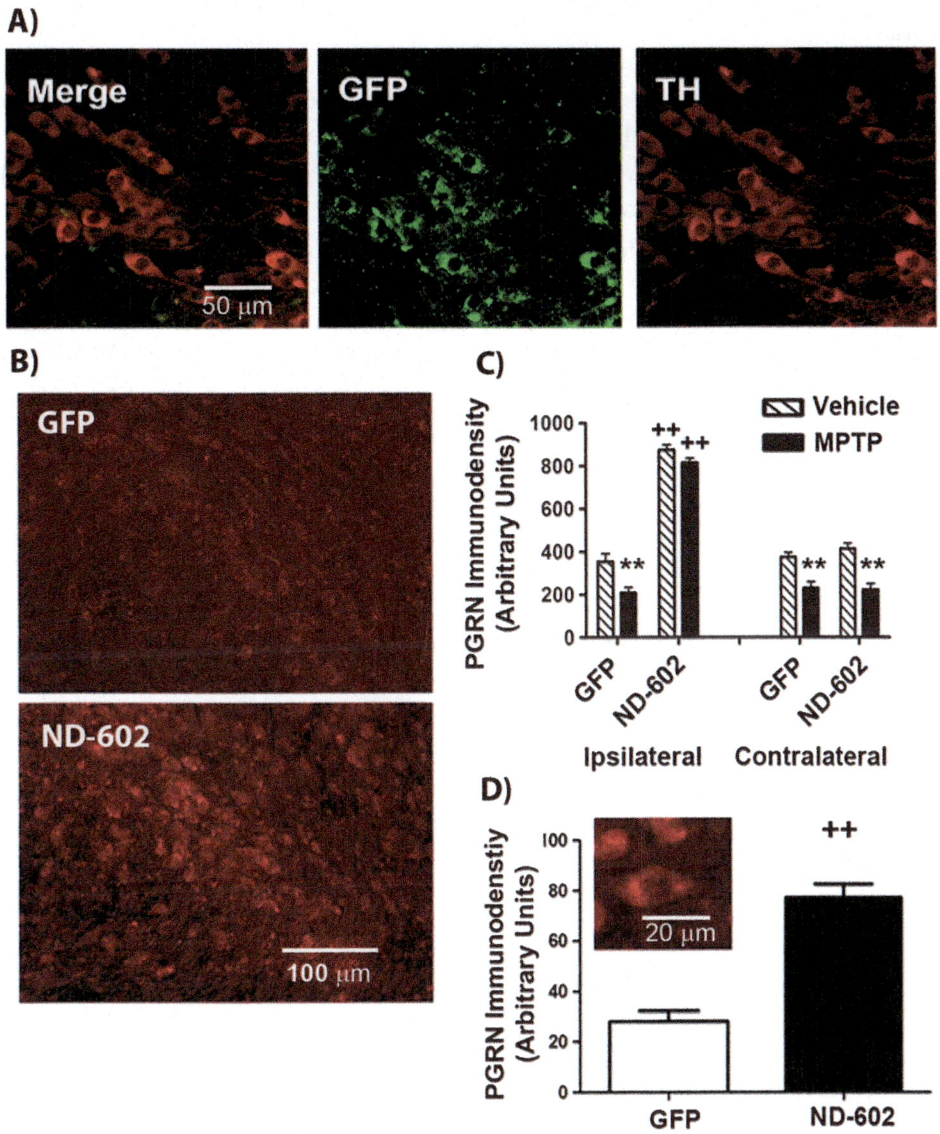

Fig. 2 Immunohistochemical analysis of enhanced PGRN expression in the SN_C following lentiviral gene transfer. (**a**) Fluorescent photomicrographs depicting GFP expression in TH+ neurons of the ipsilateral SN_C 4 weeks following unilateral intranigral ND-602 infusion. (**b**) Fluorescent photomicrographs depicting PGRN immunolabeling in the ipsilateral SN_C 4 weeks following unilateral intranigral infusion of either pLenti6/R4R2/V5-mPGRN or pLenti6/R4R2/V5-eGFP. (**c**) In GFP controls, MPTP produced a significant reduction in PGRN immunolabeling within the *substantia nigra* of both hemispheres. Unilateral intranigral delivery of pLenti6/R4R2/V5-mPGRN significantly elevated PGRN immunolabeling in the ipsilateral SN_C of both vehicle- and MPTP-treated animals. No effect of pLenti6/R4R2/V5-mPGRN was seen in the contralateral nigra. Each bar represents the mean (±S.E.M., $n = 5–9$) optical density (arbitrary units) averaged across four coronal sections through the SN_C. (**d**) The density of PGRN immunolabeling within TH+ neurons of the ipsilateral SN_C was significantly elevated in those animals treated with pLenti6/R4R2/V5-mPGRN. Inset fluorescent photomicrograph depicts PGRN immunolabeling in dopaminergic neurons of the ipsilateral SN_C following pLenti6/R4R2/V5-mPGRN treatment. Each bar represents the mean (±S.E.M., $n = 5–9$) optical density (arbitrary units) measured in ten cells in each of four coronal sections through the ipsilateral SN_C. Figure derived from [16]. **sig. diff. from vehicle control, $p < 0.001$. $^{++}$sig. diff. from GFP control, $p < 0.001$

4 Notes

1. An alternative to gel packs is to freeze moistened paper towels and wrap them around the tubing using aluminum foil.
2. Tubes with a conical bottom are required to permit efficient uptake by the cannula.
3. Before the surgery, it is advisable to calibrate the tubing by filling it with distilled water and pushing the plunger in 1 μL increments, marking the tubing along the way. This will give you a visual confirmation of the approximate volumes being infused.
4. C57Bl6 mice are used here as part of the MPTP model of PD, but this gene delivery approach can also be applied to rats.
5. These coordinates target the left *substantia nigra pars compacta*, a key region of interest in PD. However, other regions of the brain have also been targeted with this PGRN gene transfer approach.
6. The tip of the cannula should be wiped clean after each infusion, but the tip can still get blocked. If a blockage is noted, cleaning wires, supplied with the Hamilton syringe, can be used to clear the cannula.
7. It is a good idea to peek at the air bubble to provide visual confirmation that things are flowing appropriately.

References

1. He Z, Bateman A (2003) Progranulin (granulin-epithelin precursor, PC-cell-derived growth factor, acrogranin) mediates tissue repair and tumorigenesis. J Mol Med 81:600–612
2. Yin F, Banerjee R, Thomas B et al (2010) Exaggerated inflammation, impaired host defense and neuropathology in progranulin-deficient mice. J Exp Med 207:117–128
3. Martens LH, Zhang J, Barmada SJ et al (2012) Progranulin deficiency promotes neuroinflammation and neuron loss following toxin-induced injury. J Clin Investig 122:3955–3959
4. Inestrosa NC, Arenas E (2009) Emerging roles of Wnts in the adult nervous system. Nat Rev Neurosci 11:77–86
5. Ahmed Z, MacKenzie IR, Hutton M et al (2007) Progranulin in frontotemporal lobar degeneration and neuroinflammation. J Neuroinflammation 4:7
6. Yin F, Dumont M, Banerjee R et al (2010) Behavioral deficits and progressive neuropathology in progranulin-deficient mice: a mouse model of frontotemporal dementia. FASEB J 24:4639–4647
7. Van Damme P, Van Hoecke A, Lambrechts D et al (2008) Progranulin functions as a neurotrophic factor to regulate neurite outgrowth and enhance neuronal survival. J Cell Biol 181:37–41
8. Côté M, Poireir A, Aubé B et al (2015) Partial depletion of the proinflammatory monocyte population is neuroprotective in the myenteric plexus but not in the basal ganglia in a MPTP mouse model of Parkinson's disease. Brain Behav Immun 46:154–167
9. Petkau TL, Neal SJ, Milnerwood A et al (2012) Synaptic dysfunction in progranulin-deficient mice. Neurobiol Dis 45:711–722
10. Tanaka Y, Chambers JK, Matsuwaki T et al (2014) Possible involvement of lysosomal dysfunction in pathological changes of the brain in aged progranulin-deficient mice. Acta Neuropathol Commun 2:78
11. Tanaka Y, Matsuwaki T, Yamanouchi K et al (2013) Exacerbated inflammatory responses related to activated microglia after traumatic

brain injury in progranulin-deficient mice. Neuroscience 231:49–60

12. Tasker RA, Adams-Marriott AL, Shaw CA (2010) New animal models of progressive neurodegeneration: tools for identifying targets in predictive diagnostics and presymptomatic treatment. EPMA J 1:217–227
13. Xu J, Xilouri M, Bruban J et al (2011) Extracellular progranulin protects cortical neurons from toxic insults by activating survival signaling. Neurobiol Aging 32:2326.e5–2326.16
14. Ryan CL, Baranowski DC, Chitramuthu BP et al (2009) Progranulin is expressed within motor neurons and promotes neuronal cell survival. BMC Neurosci 10:130
15. Kanazawa M, Kawamura K, Takahashi T et al (2015) Multiple therapeutic effects of progranulin on experimental acute ischaemic stroke. Brain 138:1932–1948
16. Van Kampen JM, Baranowski DB, Kay DG (2014) Progranulin gene delivery protects dopaminergic neurons in a mouse model of Parkinson's disease. PLoS One 9:e97032
17. Kantor B, Bailey R, Wimberly K et al (2014) Methods for gene transfer to the central nervous system. Adv Genet 87:125–197
18. Kordower JH, Emborg ME, Bloch J et al (2000) Neurodegeneration prevented by lentiviral vector delivery of GDNF in primate models of Parkinson's disease. Science 290:767–773
19. Manfredsson FP (2016) Introduction to viral vectors and other delivery methods for gene therapy of the nervous system. Methods Mol Biol 1382:3–18
20. Naldini L, Blomer U, Gage FH et al (1996) Efficient transfer, integration, and sustained long-term expression of the transgene in adult rat brains injected with a lentiviral vector. Proc Natl Acad Sci U S A 93:11382–11388
21. Kordower JH, Bloch J, Ma SY et al (1999) Lentiviral gene transfer to the nonhuman primate brain. Exp Neurol 160:1–16
22. Jakobsson J, Ericson C, Jansson M et al (2003) Targeted trans-gene expression in rat brain using lentiviral vectors. J Neurosci Res 73:876–885
23. Franklin K, Paxinos G (2008) The mouse brain in stereotaxic coordinates, 3rd edn. Academic Press

Chapter 18

The Interaction Between Progranulin with Sortilin and the Lysosome

Xiaolai Zhou, Peter M. Sullivan, Daniel H. Paushter, and Fenghua Hu

Abstract

Accumulating evidence suggests that progranulin is essential for proper lysosomal function. Progranulin is a lysosomal resident protein and sortilin has been demonstrated to be the lysosomal trafficking receptor for progranulin. Here we describe the methods used to study the interaction between progranulin and sortilin, as well as the critical role of sortilin in mediating the lysosomal delivery of progranulin.

Key words Progranulin, Sortilin, Lysosome, Trafficking, Frontotemporal lobar degeneration, Neuronal ceroid lipofuscinosis

1 Introduction

Recent evidence points to a surprising but critical role of progranulin (PGRN) in regulating lysosomal function. Firstly, total loss of PGRN leads to neuronal ceroid lipofuscinosis (NCL), a severe lysosomal storage disease, in humans [1]. The accumulation of autofluorescent storage material (lipofuscin), which is the hallmark of NCL, is also seen in PGRN-deficient mice [2]. Secondly, bioinformatic studies have demonstrated that PGRN is transcriptionally co-regulated with a number of essential lysosomal genes and is a transcriptional target of transcription factor EB (TFEB), the master regulator of lysosomal biogenesis [3].

However, how PGRN regulates lysosomal function remains unclear. By searching for PGRN binding partners, we uncovered the interaction between PGRN and sortilin, a trafficking receptor of the VPS10 family [4]. We determined that the C-terminus of PGRN binds to sortilin [5], that sortilin mediates lysosomal delivery of PGRN, and that PGRN is a lysosomal resident protein [4]. Here we describe the detailed biochemical and cell biological techniques used to study the interactions between PGRN, sortilin, and the lysosome.

Andrew Bateman et al. (eds.), *Progranulin: Methods and Protocols*, Methods in Molecular Biology, vol. 1806, https://doi.org/10.1007/978-1-4939-8559-3_18,

2 Materials

2.1 PGRN/Sortilin Interaction Assay

2.1.1 COS-7 Cell Surface Binding Assay

Alkaline Phosphatase (AP) Fusion Protein Generation

1. Constructs: pAP5-tag vector (GenHunter, Nashville, TN), Human PGRN in the pCMV-Sport6 vector (Open Biosystems, GE Life Sciences). AP-PGRN is generated by subcloning the coding sequence for the human PGRN mature peptide (aa 18-593) into the pAP5-tag vector. AP-PGRNΔ3aa (lacking the sortilin-binding site) [5] is generated by subcloning the coding sequence for PGRN (aa 18-590) into the pAP5-tag vector.
2. Complete medium: DMEM (Dulbecco's Modified Eagle's Medium; Corning Cellgro) supplemented with 10% FBS (fetal bovine serum; Sigma) and 1% penicillin/streptomycin (Corning Cellgro).
3. 0.25% trypsin/EDTA (Corning Cellgro).
4. PEI (polyethylenimine): a 25 kDa linear polymer from Polysciences, Inc. (23966-2). Dissolve PEI in water at 1 mg/mL, neutralize with HCl to pH 7.2, and filter sterilize using a 0.22 μm filter. Aliquot and store at −80 °C. A 1 mL aliquot can be kept at 4 °C for up to 4 months.
5. HEK293T cells: cells are maintained in complete medium at 37 °C with 5% CO2 and passaged 1:3–1:10 upon reaching 80% confluency.
6. Tissue culture dish and plate.

AP Fusion Protein Concentration Measurement

1. pNPP (*p*-nitrophenyl phosphate) solution: dissolve 1 tablet of SIGMAFAST™ pNPP (Sigma, N2770) in 20 mL deionized water.
2. 96-well tissue culture plate.
3. Plate reader.

AP-Binding Assay in Sortilin-Expressing COS-7 Cells

1. COS-7 cells: cells are maintained in complete medium at 37 °C with 5% CO_2 and passaged at 1:3–1:10 upon reaching 80% confluency.
2. Human sortilin cDNA in the pCMV-Sport6 vector (Open Biosystems, GE Life Sciences).
3. HBSS (Hank's Balanced Salt Solution): 1 L 10× HBSS stock – dissolve 1.4 g calcium chloride ($CaCl_2$), 4.0 g potassium chloride (KCl), 0.6 g potassium phosphate monobasic (KH_2PO_4), 1.0 g magnesium chloride ($MgCl_2 \cdot 6H_2O$), 1.0 g magnesium sulfate ($MgSO_4 \cdot 7H_2O$), 80 g sodium chloride (NaCl), 3.5 g sodium bicarbonate ($NaHCO_3$), and 0.48 g sodium phosphate dibasic (Na_2HPO_4) in a final volume of 1 L with distilled water.
4. HBH: 100 mL 10× HBSS, 20 mL 1 M HEPES pH 7.3–7.4, 1 g BSA, and 880 mL water to make 1 L HBH. Filter with a 0.22 μm filter.

5. 3.7% formaldehyde: dilute 37% formaldehyde (Sigma) 1:10 in phosphate-buffered saline (PBS) (Sigma) [please check if PBS is included in the list of materials].
6. AP buffer: 100 mM Tris–HCl, 100 mM sodium chloride (NaCl), and 5 mM magnesium chloride ($MgCl_2$); adjust the pH to 9.5 with HCl.
7. Nitro blue tetrazolium (NBT)/5-bromo-4-chloro-3-indolyl phosphate (BCIP): dissolve 1 g NBT (Sigma, N5514) in 20 mL of 70% dimethylformamide (DMF). Dissolve 1 g BCIP (Sigma, B0274) in 20 mL of 100% DMF. Add 33 μL of BCIP and 66 μL of NBT to 10 mL of AP buffer just before adding to cells.

AP-Binding KD Calculation

1. Inverted microscope with 4× or 10× lens and transmitter light.
2. NIH ImageJ software.

2.1.2 Co-immunoprecipitation Assay

1. HEK293T cells.
2. NSC34, N2a, or other cells that have high sortilin expression levels.
3. Constructs: FLAG-tagged PGRN can be generated by adding the FLAG peptide coding sequence to the PCR primer to amplify the PGRN coding region (mature peptide, aa 18-593 or PGRNΔ3aa, aa 18-590) and cloned into pSectag2 vector (Invitrogen). Human sortilin cDNA in the pCMV-Sport6 vector (Open Biosystems, GE Life Sciences).
4. Lysis buffer: 50 mM Tris pH 8.0, 150 mM NaCl, 1% Triton X-100, 0.1% sodium deoxycholate, and 1× protease inhibitors (Roche).
5. Immobilon-FL polyvinylidene fluoride (PVDF) membrane (Millipore).
6. Protein electrophoresis and blotting system from Bio-Rad.
7. 2× SDS (sodium dodecyl sulfate) sample buffer, 12% SDS polyacrylamide gel, SDS running buffer, transfer buffer, and TBST from Bio-Rad [are all listed items from Bio-Rad YES If not please describe compositions of each item. TBS-T = transfer buffer? NO].
8. Odyssey blocking buffer (LI-COR Biosciences).
9. Mouse anti-FLAG antibody-conjugated beads (Sigma) and mouse anti-FLAG M2 antibodies (Sigma, F1804).
10. Goat anti-sortilin antibodies (R&D Systems, AF2934).
11. Alexa Fluor 680-conjugated donkey anti-goat IgG (Invitrogen).
12. IRDye 800-conjugated donkey anti-mouse IgG (LI-COR Biosciences).
13. Odyssey infrared imaging system (LI-COR Biosciences).

2.2 Analyze Sortilin-Mediated Lysosomal Delivery of PGRN In Vitro

2.2.1 Determine the Role of Sortilin in Regulating PGRN Trafficking by HEK293T Co-transfection

Sample Preparation

1. HEK293T cells.
2. Constructs: human PGRN and human sortilin in the pCMV-Sport6 vector (Open Biosystems, GE Life Sciences).
3. Lysis buffer: 50 mM Tris pH 8.0, 150 mM NaCl, 1% Triton X-100, and 1× protease inhibitors (Roche).
4. Trichloroacetic acid (TCA): 100% (w/v) TCA is prepared by dissolving 500 g TCA (Sigma) into 350 mL distilled water. Store at 4 °C.
5. 100% acetone (Mallinckrodt Pharmaceuticals).
6. 1× and 2× SDS sample buffer (2X: Tris-HCl (100 mM, pH 6.8), SDS (electrophoresis grade) (4%), Bromophenol blue (0.2%), glycerol (20%)). [please indicate compositions].
7. 1 M Tris pH 8.0.

Western Blot

1. 12% SDS polyacrylamide gel (W × L × thickness, 8.6 × 6.7 × 0.1 cm, Bio-Rad) [please indicate dimensions of the gel].
2. Western blot reagents (running buffer, PVDF membranes, methanol, transfer buffer, TBST buffer, Odyssey blocking solution) are the same as Subheading 2.1.2.
3. Goat anti-human PGRN antibodies (R&D Systems, AF2420).
4. Mouse anti-β-actin antibodies (Abcam, ab8226).
5. Chicken anti-human transferrin antibodies (Immunology Consultants Laboratory, CTX-80A).
6. Alexa Fluor 680-conjugated donkey anti-goat IgG (Invitrogen).
7. IRDye 800-conjugated donkey anti-mouse and chicken IgG (LI-COR Biosciences).
8. Odyssey infrared imaging system (LI-COR Biosciences).

2.2.2 Immuno-fluorescence Staining to Visualize Sortilin-Mediated PGRN Endocytosis

1. COS-7 cells.
2. HEK293T cells.
3. mCherry-PGRN construct: mCherry-tagged PGRN is generated in the pAP5 vector (GenHunter) by replacing the AP sequence with the mCherry sequence followed by inserting the mature human PGRN (aa 18–593) coding sequence.
4. Microscope cover glass (Fisherbrand, 12-545-81) and microscope slides (Fisherbrand, 12-544-2).
5. Complete medium for culturing HEK293T and COS-7 cells.
6. mCherry-PGRN conditioned medium is generated by transfecting the HEK293T cells with the mCherry-PGRN construct using PEI.
7. Blocking solution: 0.05% saponin (Alfa Aesar) in Odyssey blocking solution.

8. Rat anti-mouse LAMP1 antibodies (BD pharmingen, 553,792).
9. Donkey anti-rat 488 antibodies (Thermo fisher Scientific).
10. Fluoromount-G mounting solution (Southern Biotech, 0100-01).

2.3 Determine the Role of Sortilin in Lysosomal PGRN Trafficking In Vivo

2.3.1 Sample Preparation

1. Age-matched wild type and sortilin knockout adult mice (gift from Dr. Anders Nykjaer).
2. Isoflurane (Henry Schein Animal Health).
3. 15 mL syringe.
4. 25 gauge needles.
5. Perfusion pump.
6. Anatomic scissors.
7. Tweezers.
8. Forceps.
9. PBS.
10. 4% paraformaldehyde (PFA) (w/v) in PBS pH 7.4: dissolve 40 g PFA powder (Sigma) in 1000 mL milliQ H_2O with heating (65 °C), adjust pH to 7.4 with 10 N NaOH to help solubilize PFA, and filter through a 0.45 μm filter. Because PFA is toxic, this should be prepared in a fume hood.

2.3.2 Serum PGRN Level Measurement

1. Mouse PGRN ELISA kit (R&D Systems, DY2557).
2. Nunc 96-well ELISA microplates (Thermo Fisher Scientific).
3. 50 mL reagent reservoir (Corning Inc., 4870).
4. Multichannel micropipette (50–200 μL).
5. Plate reader (450 nm and 560 nm).

2.3.3 PGRN Levels in Brain Lysate

1. RIPA buffer: 50 mM Tris–HCl pH 8.0, 150 mM NaCl, 1% Triton X-100, 0.5% sodium deoxycholate, and 0.1% SDS plus protease inhibitors (Roche).
2. Western blot reagents are the same as in Subheading 2.2.1.
3. Odyssey blocking buffer (LI-COR Biosciences).
4. Sheep anti-mouse PGRN antibodies (R&D, AF2557).
5. Mouse anti-β-actin antibodies (Abcam, ab8226).
6. Alexa Fluor 680-conjugated donkey anti-sheep IgG (Invitrogen).
7. IRDye 800-conjugated donkey anti-mouse IgG (LI-COR Biosciences).
8. Odyssey infrared imaging system (LI-COR Biosciences).

2.3.4 Lysosomal PGRN Localization in Neurons

1. 4% PFA in PBS.
2. OCT (optimal cutting temperature) compound (Electron Microscopy Sciences, 62550-01).

3. Cryotome.
4. Superfrost Plus Microscope Slides (Fisherbrand, 12-550-15).
5. Dako pen (Dako, S2002).
6. Micro cover glass (VWR, 48393241).
7. Blocking solution: 0.05% saponin (Alfa Aesar) in Odyssey blocking solution.
8. Sheep anti-mouse PGRN antibodies (R&D, AF2557).
9. Rat anti-mouse LAMP1 antibodies (BD pharmingen, 553792).
10. Alexa Fluor 488-conjugated donkey anti-rat IgG (Invitrogen).
11. Alexa Fluor 594-conjugated donkey anti-sheep IgG (Invitrogen).
12. Fluoromount-G mounting solution (Southern Biotech, 0100-01).
13. Laser-scanning confocal microscope.

3 Methods

3.1 PGRN/Sortilin Interaction Assay

3.1.1 COS-7 Cell Surface Binding Assay

Alkaline phosphatase (AP)-based detection of ligand–receptor interactions on the cell surface (AP cell surface binding assay) was first developed by Dr. Flanagan's group [6, 7] and has been used by many labs to identify receptors for orphan ligands during the last few decades. In this method, receptor expression constructs will be transfected into COS-7 cells to allow the expression of the receptor (such as sortilin) on the cell surface. Transfected cells will then be incubated with AP-tagged ligand (such as AP-PGRN), and positive interactions will be detected using AP substrates (Fig. 1). Here we provide detailed protocols to analyze the PGRN–sortilin interaction using this assay.

AP Fusion Protein Generation

Day 1 - Plate HEK293T cells. Plate HEK293T cells at a density of 1×10^5 per mL into either a 6-well plate or 10 cm dish depending on the amount of AP fusion protein needed for the binding assay (*see* **Note 1**). The cells should be 70–80% confluent the next day.

Day 2 - Transfect HEK293T cells with AP fusion constructs. Use PEI according to the ratio below (Table 1) (*see* **Note 2**):

1. Warm DMEM to room temperature, and aliquot the medium into sterile Eppendorf tubes.
2. Dilute PEI in warm DMEM (*see* **Note 3**).
3. Add plasmid DNA (pAP5, pAP5-PGRN, or pAP5-PGRN Δ3aa) to PEI/DMEM, mix gently with a pipette, and spin briefly to collect droplets if necessary.
4. Incubate the mixture for 15–30 min at room temperature.

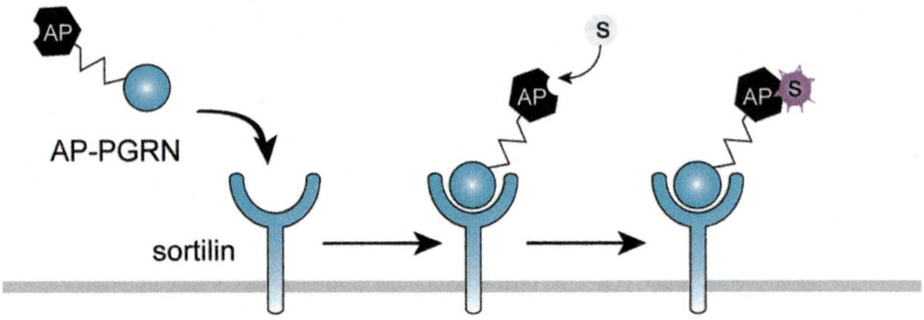

Fig. 1 Illustration of AP cell surface binding assay to visualize ligand–receptor interactions. An AP-fused ligand (such as AP-PGRN) binds to a receptor (such as sortilin) expressed on the surface of COS-7 cells. This interaction can be detected by the AP-dependent conversion of the substrates NBT/BCIP into insoluble purple precipitates on the cell surface

Table 1
Conditions for HEK293T transfections

Plate/dish	Cells	DMEM	DNA	PEI (1 mg/mL)
10 cm dish	1.0×10^6	1000 μL	10 μg	50 μL
6-well plate	2.0×10^5	200 μL	2 μg	10 μL

5. Add the transfection mixture drop by drop to the cells.

   Day 3 - Change medium.

6. Replace the medium with warmed complete medium 12–24 h after transfection.

   Day 8 or 10 - Collect conditioned medium.

7. Carefully collect the medium without disturbing the cells. Spin in a tabletop centrifuge at 10,000 × *g* for 5 min, and collect the supernatant.

AP Fusion Protein Concentration Measurement

The AP fusion protein concentration can be measured by the AP activity. AP catalyzes the hydrolysis of phosphate esters in an alkaline environment, resulting in the formation of an organic radical and inorganic phosphate. *p*-Nitrophenyl phosphate (pNPP) is hydrolyzed by AP to generate *p*-nitrophenol, which is deprotonated under alkaline conditions to produce *p*-nitrophenolate that has maximal absorbance at 405 nm.

1. Add 100 μL of pNPP solution to the 96-well plate.
2. Add 1–5 μL of conditioned medium (collected from above) into the wells containing the pNPP solution and mix well by pipetting.

3. Immediately read the plate at 405 nm, and repeat the reading every minute for 5 min.
4. Calculate the concentration of the AP fusion protein as follows:

   Protein concentration (nM) = slope of OD 405 nm changes vs time (min) × 3000/volume (μL).
5. Aliquot 1.0 mL of AP fusion protein conditioned medium in 1.5 mL Eppendorf tubes, and store at −20 °C up to 6 months or −80 °C for 1–2 years.

AP-Binding Assay in Sortilin-Expressing COS-7 Cells

1. The day before the transfection, dissociate COS-7 cells with 0.25% trypsin/EDTA, and count the cells. Plate the cells in complete growth medium as listed below in Table 2 (*see* **Note 4**).
2. Transfect the cells with the human sortilin construct using the amount of DNA listed in the table above following the procedure as described in the AP fusion protein generation section.
3. Allow the transfected cells to grow in complete medium for 24–48 h.
4. Wash cells with cold HBH two times to remove any cell debris.
5. Dilute the AP fusion protein conditioned medium with HBH to the desired concentration. For the AP-PGRN/sortilin interaction, 100 nM, 50 nM, 25 nM, 12.5 nM, 6.3 nM, and 3.125 nM can be used. AP conditioned medium of similar concentration can be used as a background control. Adjust the pH by adding 20 mM HEPES pH 7.4 (*see* **Note 5**).
6. Remove the HBH buffer from the wells, add the prepared AP conditioned medium (500 μL/well in a 24-well plate or 250 μL/well in a 48-well plate) to the cells, and incubate at room temperature for 2 h.
7. Remove the medium and wash the cells quickly with HBH three times.
8. After aspirating the HBH from the last wash, add 3.7% formaldehyde in PBS and incubate for 20 min at room temperature to fix the cells.
9. Wash the cells with HBH one-two times.

Table 2
Conditions for COS-7 transfections

Plate/dish	Cells	DMEM	DNA	PEI (1 mg/mL)
6-well	2.0×10^5	80 μL	0.8 μg	4 μL
24-well	0.5×10^5	20 μL	0.2 μg	1 μL
48-well	2.5×10^4	10 μL	0.1 μg	0.5 μL

10. Remove the HBH, fill the wells fully with fresh HBH, and then incubate the plate at 65 °C for 5 h or overnight to inactivate endogenous AP activity (*see* **Note 6**).
11. Wash the wells with AP buffer.
12. Remove the AP buffer from the wells, and add the freshly prepared NBT/BCIP solution (500 μL/well in a 24-well plate or 250 μL/well in a 48-well plate).
13. Incubate the cells with the AP reaction buffer in the dark until the positive cells turn purple, and stop the reaction by washing with water several times (*see* **Note 7**). An example of the AP-binding assay for PGRN and sortilin is shown (Fig. 2). Deletion of the last three residues of PGRN completely abolished the interaction (Fig. 2).

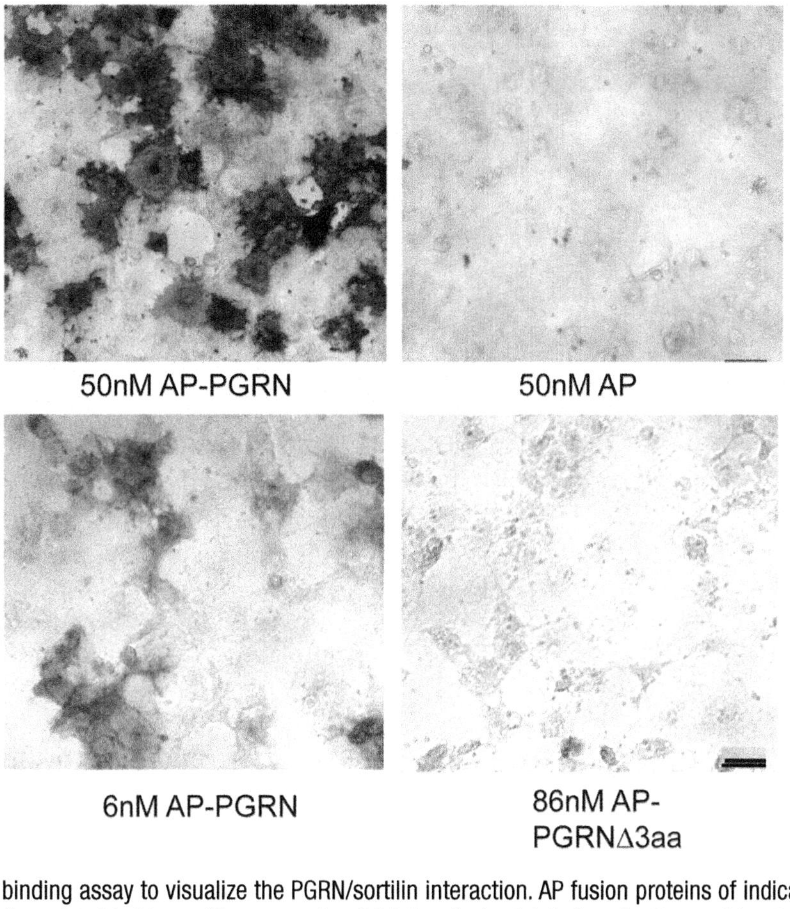

Fig. 2 COS-7 binding assay to visualize the PGRN/sortilin interaction. AP fusion proteins of indicated concentration were allowed to bind to sortilin-expressing COS-7 cells for 2 h before fixation and incubation with AP substrates. Deletion of the last three residues of PGRN (AP-PGRNΔ3aa) completely abolished the interaction. Scale bar = 100 μm

AP-Binding KD Calculation

1. Apply different concentrations of AP-PGRN (200, 150, 100, 75, 50 25, 12.5, and 6.25 nM) to sortilin- and vector (as control)-transfected COS-7 cells.
2. After finishing the binding assay, take four to six pictures from four fixed regions of each well under bright field.
3. Measure the graphic gray value of the picture by using the NIH Image J software according to its instructions. Subtract the signal from AP control as background.
4. Use AP-PGRN concentration (nM) as x and gray value (V, units) as y to generate the saturation binding curves and Scatchard plots. Calculate the binding Kd by using GraphPad Prism 5.0 software or Microsoft Excel.

3.1.2 Co-immunoprecipitation Assay

1. Generate conditioned medium containing FLAG-PGRN, FLAG-PGRNΔ3aa, or control (vector transfected) as in Subheading 3.1.1. Conditioned medium can be aliquoted and stored at −20 °C.
2. Centrifuge 500 μL conditioned medium containing FLAG-PGRN, FLAG-PGRNΔ3aa, or control medium at 10,000 × *g* for 10 min. Add 5 μL slurry of mouse anti-FLAG-antibody-conjugated beads to the supernatants, and incubate at 4 °C for 2–4 h with gentle shaking. Spin the samples at 4,000 × *g* for 30 s, and aspirate the lysate carefully without disturbing the beads. Add 1,000 μL PBS, and rinse the beads by gently inverting the tube several times, and then centrifuge 4,000 × *g* for 30 s. Repeat the washing three times. After the final wash, carefully remove the PBS with a 200 μL pipette without disturbing the beads.
3. To prepare NSC34 lysate (*see* **Note 8**), aspirate the medium, and wash the cells once with ice-cold PBS on ice.
4. Lyse the cells with cold lysis buffer (350 μL/well in 6-well plate, one well for one IP (immunoprecipitation)) on ice. Spin the cell lysate in a micro-centrifuge at 10,000 × *g* for 20 min at 4 °C. Collect the supernatant into a new 1.5 mL Eppendorf tube. Aliquot 15 μL into a new 1.5 mL Eppendorf tube, and mix it with 15 μL 2× SDS sample buffer; save it at −20 °C as input.
5. Transfer 300 μL of the collected cell lysate to the Eppendorf tubes containing FLAG-PGRN or FLAG-PGRNΔ3aa bound to anti-FLAG beads or control beads.
6. Incubate the cell lysate with the beads at 4 °C for 2 h with gentle shaking.
7. Wash the beads with 1 mL cold lysis buffer for 3 times as in **step 2**. After the final wash, carefully remove any wash buffer with a 200 μL pipette without disturbing the beads.

8. Add 30 μL of 2× SDS sample buffer, vortex, and spin.
9. Heat the immunoprecipitated (IP), and input samples at 95 °C for 5 min, and then load 15 μL of each onto a 12% SDS gel.
10. Run the sample at 90 V for the first 15 min, and then adjust to 120 V until the blue dye runs off the gel.
11. Activate an Immobilon-FL polyvinylidene fluoride (PVDF) membrane with pure methanol for 2 min. Equilibrate the membrane with transfer buffer for 1 min, then transfer proteins at 300 mA for 60 min at 4 °C.
12. Wash the membrane with PBS, and block with Odyssey blocking solution at room temperature for 1 h with mild shaking (*see* **Note 9**).
13. Incubate the membrane with mouse anti-FLAG (1:2,000) and goat anti-sortilin (1:2,000) antibodies in TBST buffer at 4 °C overnight with mild shaking [any recommended speed? No].
14. Wash membranes with TBST three times, 5 min each.
15. Add secondary antibodies: Alexa Fluor 680-conjugated donkey anti-goat (1:10,000) and IRDye 800-conjugated donkey anti-mouse (1:10,000) for 1 h at room temperature.
16. Wash the membranes three times with TBST, then scan the membranes and quantify the signals with an Odyssey infrared imaging system (LI-COR Biosciences) (*see* **Note 10**). An example of co-immunoprecipiation (co-IP) between PGRN and sortilin is shown in Fig. 3. Deletion of the last three residues of PGRN completely abolished the interaction (Fig. 3).

3.2 Analyze Sortilin-Mediated Lysosomal Delivery of PGRN In Vitro

The AP-binding assay and co-IP assay described above allow us to visualize and quantify the binding between PGRN and sortilin. In this section, we will introduce a method to study sortilin-mediated PGRN endocytosis and lysosomal delivery after binding to sortilin on the cell surface.

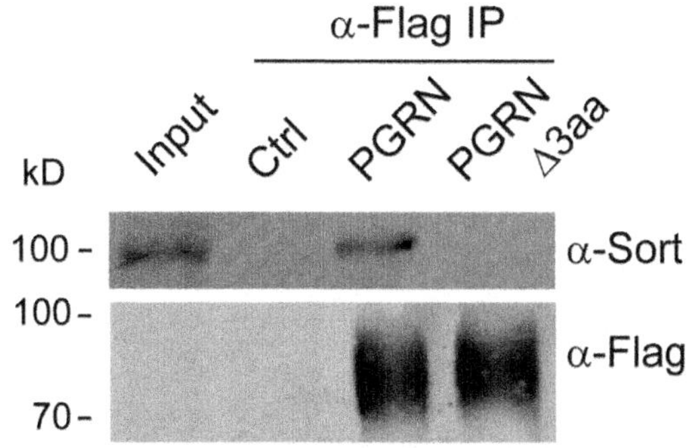

Fig. 3 Co-immunoprecipitation between PGRN and sortilin. FLAG-tagged PGRN but not PGRNΔ3aa or control beads pull down sortilin in the cell lysates of NSC-34, a motor neuron cell line

3.2.1 Determine the Role of Sortilin in PGRN Trafficking by HEK293T Cell Co-transfection

Sample Preparation

1. Plate HEK293T cells into a 6-well plate, and transfect the cells with the following groups of constructs (Table 3).
2. 40–48 h after the transfection, wash the cell 1X with PBS and change the medium to serum-free DMEM supplemented with 1% penicillin/streptomycin.
3. 24 h after changing the medium, collect the conditioned medium in 2 mL Eppendorf tubes and precipitate the protein from the medium following **steps 5–11**.
4. Wash the cells with PBS once, and then lyse the cells with 150 μL lysis buffer, and spin at 10,000 × *g* for 20 min at 4 °C as in Subheading 3.1.2. Mix 20 μL of the supernatant with 20 μL 2× SDS sample buffer. This input sample will be used to analyze PGRN levels in the lysate.
5. Spin down the conditioned medium at 10,000 × *g* for 5 min to remove cell debris. In a 1.7 mL Eppendorf tube, collect 1.5 mL of medium, and add 165 μL of 100% ice-cold TCA (to a final concentration of ~10%).
6. Mix and keep on ice for 45 min to 3 h, and then centrifuge for 30 min at 10,000 × *g* at 4 °C.
7. Remove the supernatant, and add 1 mL of cold 100% acetone (kept at −20 °C) to wash the pellet.
8. Vortex thoroughly, and then centrifuge as in **step 6**.
9. Remove the supernatant, and allow the pellet to air-dry for 10–20 min.
10. Resuspend the pellet in 30 μL 1× SDS sample buffer. Add 1 M Tris pH 8.0 until the color turns blue (~3 μL) (*see* **Note 11**). Vortex to mix.

Western Blotting

1. Heat the samples (both cell lysate and conditioned medium at 95 °C for 5 min), and spin briefly.
2. Load 15 μL of the cell lysate samples and 10 μL of the conditioned medium samples onto a 12% SDS gel.
3. Electrophoresis and Western blot should be performed as in Subheading 3.1.2.

Table 3
Conditions for HEK293T transfections

	Group 1	Group 2
Vector	1 μg	0 μg
hPGRN	1 μg	1 μg
hSort	0 μg	1 μg
Total DNA	2 μg	2 μg

4. Primary antibodies: blot whole cell lysate samples with goat anti-human PGRN antibodies (1:2,000) and mouse anti-β-actin antibodies (1:5,000) and conditioned medium samples with goat anti-human PGRN antibodies (1:2,000) and chicken anti-human transferrin antibodies (1:2,000).
5. Secondary antibodies: donkey anti-goat Alexa Fluor 680 or donkey anti-mouse or anti-chicken IRDye 800. An example of Western blot for PGRN levels in both cell lysate and conditioned medium is shown (Fig. 4).

3.2.2 Immunofluorescence Staining to Visualize Sortilin-Mediated PGRN Endocytosis

Days 1–2

1. Plate and transfect COS-7 cells in a 24-well plate containing coverslips at a density of 5.0×10^4 per well in 500 μL complete medium with the full-length human sortilin construct as described in Subheading 3.1.1 (*see* **Notes 12** and **13**).

Day 4

2. 24–48 h after transfection, wash the cells with ice-cold HBH one-two times on ice.
3. Add 500 μL mCherry-PGRN conditioned medium diluted 1:2 in HBH to the wells, and incubate the cells on ice in the dark for 4 h. Wash the cells quickly with HBH three times, and then add 500 μL warm complete medium to the wells (**Note 14**).
4. Transfer the plate to a 37 °C incubator with 5% CO_2.
5. At different time points (0, 15, 30, and 60 min), transfer one coverslip to a new 24-well plate, wash once with PBS, and fix the cells with 4% PFA for 15 min at room temperature.

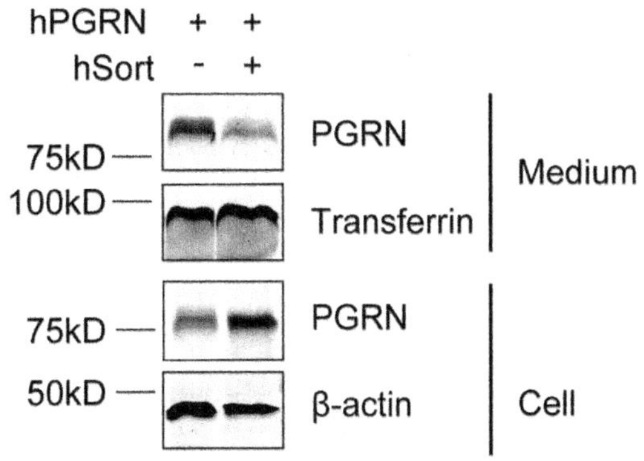

Fig. 4 Co-expression of sortilin with PGRN leads to decreased extracellular PGRN levels. HEK293T cells were transfected with PGRN with vector control (hPGRN+/hSort−) or sortilin expression ((hPGRN+/hSort+) constructs. Cells were changed to serum-free medium 40–48 h after transfection. Conditioned media was precipitated with TCA. PGRN levels in the medium and lysates (Cell) were analyzed by Western blot. Transferrin and β-actin were used as loading controls

6. Wash the cells with PBS three times after fixation, and keep the coverslips at 4 °C in the dark until all coverslips for each time point have been harvested and fixed.

Immunofluorescence Staining

The whole procedure should be carried out in a humidified chamber in the dark.

7. Permeabilization and blocking: incubate the cells with blocking solution at room temperature for 30 min to 1 h.
8. Primary antibody: dilute rat anti-mouse LAMP1 antibodies in blocking solution at 1:500, apply to the coverslips, and incubate at 4 °C overnight.
9. Wash the coverslips with PBS three times at 5-min intervals.
10. Secondary antibody: dilute donkey anti-rat 488 antibodies in blocking solution at 1:500, apply to the coverslips, and incubate for 2 h at room temperature in the dark (*see* **Note 15**).
11. Aspirate the secondary antibody and wash the coverslips with PBS three times at 5-min intervals.
12. Carefully dry the coverslips with Kimwipes, and mount them on a glass slide with Fluoromount-G solution.
13. Keep the microscope slides at room temperature until the mounting solution is dry. For long-term storage, keep the slides at −20 °C.
14. Take immunofluorescence images with a laser-scanning confocal microscope. An example image shows that the mCherry-PGRN (red) binds to the surface of sortilin-overexpressing cells at 4 °C and is trafficked into LAMP1-positive vesicles (lysosomes, green) upon 30-min incubation at 37 °C (Fig. 5).

3.3 Determine the Role of Sortilin in Lysosomal PGRN Trafficking In Vivo

Here, we introduce the following methods to determine the role of sortilin in lysosomal PGRN trafficking in vivo by comparing PGRN levels in wild type (WT) and sortilin-deficient mice: (1) Western blot to detect PGRN levels in mouse brain, (2) ELISA to measure PGRN levels in serum, and (3) immunofluorescence staining to visualize PGRN lysosomal localization.

3.3.1 Brain and Serum Collection

1. Anesthetize the mice with isoflurane.
2. Immobilize the mouse, and cut a 5–6 cm lateral incision through the integument and abdominal wall just beneath the rib cage.
3. Cut the right atrium, and collect the blood into 1.5 ml Eppendorf tubes by P1000 pipette for serum isolation (for details *see* Subheading 3.3.2).
4. Using a needle attached to a 15 mL syringe, penetrate the left ventricle and perfuse with 15 mL of cold PBS (*see* **Note 16**).

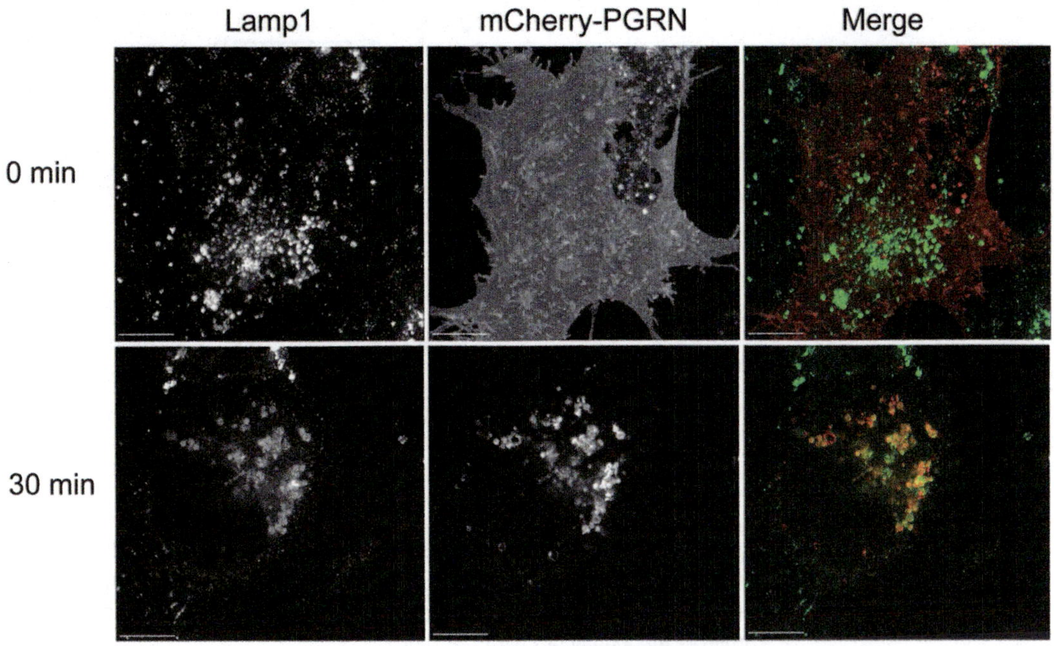

Fig. 5 PGRN is delivered to lysosomes after sortilin-mediated endocytosis. Sortilin-expressing COS-7 cells were incubated with conditioned medium containing mCherry-PGRN on ice for 4 h. After washing, cells were transferred to 37 °C and fixed at 0 and 30 min. Cells were stained with anti-LAMP1 antibodies to visualize lysosomes. The lower panels show that mCherry-PGRN (red) was trafficked into LAMP1-positive vesicles (lysosomes, green) upon 30-min incubation at 37 °C with sortilin-overexpressing cells, as seen by the yellow signal in the lower merged view panel. Scale bar = 10 μm

5. Dissect the brain directly to prepare brain lysate or perfuse the mouse with 15 mL 4% PFA in PBS for immunofluorescence staining.
6. For brain lysate preparation, brains will be immediately processed as in Subheading 3.3.3 or frozen in liquid nitrogen and stored at −80 °C.
7. For immunostaining, brains will be kept in 4% PFA.

3.3.2 PGRN Levels in Serum

1. Leave collected blood undisturbed for 1 h at room temperature or 15 min at room temperature plus overnight at 4 °C to allow it to clot.
2. Centrifuge the serum at 10,000 × *g* for 20 min at 4 °C.
3. Carefully collect the aqueous supernatant (serum) with a P200 pipette. If there are any clots in the collected supernatant, another centrifugation is needed to remove the clot.
4. Measure the PGRN levels in the serum by using the mouse PGRN ELISA kit according to the manufacturer's instructions (*see* **Note 17**). An example of ELISA results showing the serum PGRN levels in wild type, PGRN knockout, and sortilin knockout mice (Fig. 6).

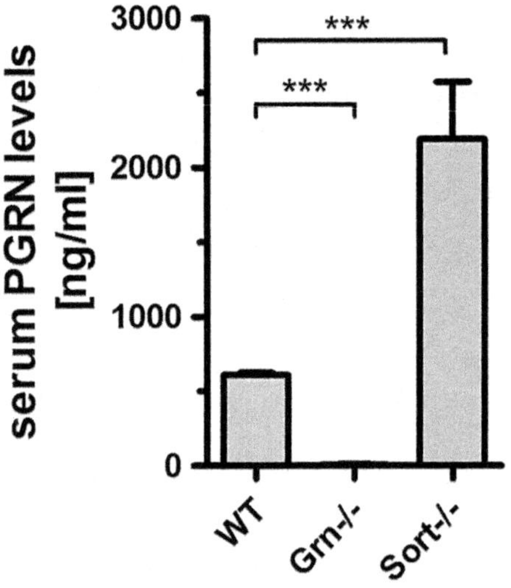

Fig. 6 Increased serum PGRN levels in sortilin-deficient mice. PGRN levels in serum samples collected from WT, $Grn^{-/-}$ and $sortilin^{-/-}$ mice were analyzed with ELISA. ***The changes observed between WT and $Grn^{-/-}$ and between WT and $sortilin^{-/-}$ are significant with $p < 0.001$

3.3.3 PGRN Levels in Brain Lysate

1. Weigh the brain tissue and homogenize in RIPA buffer at a 1:10 ratio (i.e., for 1 g of tissue, add 10 mL of RIPA buffer) (*see* **Note 18**).
2. Centrifuge the brain lysate in a microcentrifuge for 20 min at 14,000 × *g* and 4 °C.
3. Collect the supernatant, measure the protein concentration, and aliquot into 1.5 mL Eppendorf tubes.
4. Take a 50 μL aliquot, and add an equal volume of 2× SDS sample buffer. Save the remainder at −80 °C.
5. Heat the samples at 95 °C for 5 min, and then load 50 to 100 μg of protein onto a 10% SDS-PAGE gel.
6. Transfer and blot the samples with sheep anti-mouse PGRN and mouse anti-β-actin antibodies. An increase in PGRN levels, especially the highly glycosylated form, is expected in the sortilin knockout samples [4].

3.3.4 PGRN Lysosomal Localization in the Brain

1. After perfusion, fix the brain with 4% PFA for another 24 h at 4 °C with mild shaking.
2. Wash the brain with ice-cold PBS three times, and then gradually dehydrate the brain with 15% and 30% (w/v) sucrose for 8 h each sequentially at 4 °C with mild rocking.
3. Embed the brain with OCT compound, and keep the embedded blocks at −80 °C.

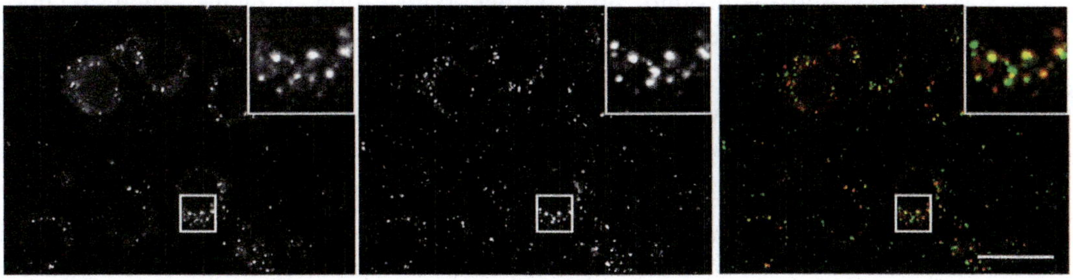

Fig. 7 Lysosomal localization of PGRN in mouse brain. Adult brain sections were stained with anti-PGRN, anti-LAMP1 antibodies. Scale bar = 20 μm

4. Section the brains with a microtome into 15 μm sections, and transfer them onto super frosted microscope slides. Keep the slides at −80 °C.
5. When ready for immunostaining, warm the slides to room temperature for 15 min.
6. Wash carefully with PBS four times, 5 min each.
7. Absorb the PBS carefully with Kimwipes tissue, and circle the brain tissue sections with a Dako sealing pen.
8. Carefully drop 100 μL of blocking solution on each brain section, and incubate 1 h at room temperature.
9. Add 100 μL diluted primary antibodies (sheep anti-mouse PGRN (1:200) and rat anti-mouse LAMP1 (1:200) in the blocking solution) to the section, and incubate at 4 °C overnight. Remove the primary antibodies, wash the sections with PBS three times, add 100 μL secondary antibodies (donkey anti-sheep 594 (1:250) and donkey anti-rat 488 (1:250) diluted in the blocking solution), and incubate for 2.5 h at room temperature in the dark (*see* **Note 19**).
10. Wash the sections with PBS four times, dry the slides with Kimwipes carefully, and mount the sections with micro cover glass (*see* **Note 20**).
11. Take images using a laser-scanning confocal microscope after the mounting solution is dry. An example image showing lysosomal PGRN in wild type mouse brain is shown (Fig. 7).

4 Notes

1. The HEK293T used for producing AP fusion proteins should be in very good condition. The cell quality directly affects the yield of AP fusion protein.

2. By the time of transfection, the cell confluency should be around 80%. Other transfection reagents, such as lipofectamine, can be used depending on availability.
3. Generally, PEI precipitates and sinks to the bottom of the tube. Therefore, it is important to mix well immediately prior to addition into transfection reactions.
4. Cells should be 50–80% confluent on the day of transfection.
5. After 5–7 days, the AP fusion protein-containing conditioned medium might appear yellow (acidic), which could negatively affect the viability of the COS-7 cells during the binding assay. Adding HEPES can help correct the pH of the medium.
6. The wells should be filled with HBH, especially for overnight heat inactivation, to prevent the cells from drying.
7. For the AP-PGRN and sortilin interaction, positive signals should appear within 10–20 min. However, for the low concentration of AP-PGRN used to measure the binding Kd, more time might be needed to see weak binding.
8. Other neuronal cell lines with high sortilin expression levels, such as N2a (available at ATCC), can also be used. Brain lysate can also be used (*see* Subheading 3.3 for preparation). Alternatively, HEK293T cells can be transfected with sortilin expression constructs to prepare cell lysates with high sortilin levels.
9. Use of Odyssey blocking solution will reduce the background for goat and sheep antibodies compared to the 5% nonfat milk in TBST.
10. If an Odyssey Li-COR infrared scanner is not available, Western blot can be carried out using HRP (horseradish peroxidase)-conjugated secondary antibodies and developed with ChemiLucent substrates and films (Pierce).
11. Residual acid has to be buffered with Tris; otherwise precipitated proteins cannot be solubilized.
12. Gently press the coverslips down with a sterile 200 μL pipette tip to prevent them from floating in the medium before adding the cells.
13. Adherent cells other than COS-7 (e.g., HeLa) can be used for the uptake assay.
14. This experiment can also be done with conditioned medium containing untagged PGRN, FLAG-tagged PGRN, or purified recombinant PGRN protein. Anti-PGRN or anti-FLAG antibodies will be used to immunostain PGRN together with LAMP1 in these cases. If using untagged PGRN, caution should be taken to distinguish endogenous PGRN expressed by the cells used in the assay (COS-7 in this case) and the

endocytosed PGRN. A control with no exogenous PGRN added should be used.

15. Most of the secondary antibodies contain precipitates which will bind to the cells or tissue nonspecifically. To avoid nonspecific binding, spin the diluted secondary antibody at 14,000 × *g* for 1–2 min to remove the precipitates.
16. Perfusion with PBS allows removal of blood from the brain. This step helps to analyze PGRN levels in the brain more accurately and clean up IgG in the Western blot. This will also remove residual blood which might clot and interfere with PFA perfusion, thus causing insufficient fixation.
17. The PGRN level in wild type mouse serum is around 550 ng/mL, while in sortilin knockout mouse serum, it is elevated ~5-fold. In order to make sure the measured PGRN concentration is within the range of the ELISA standard (0–8 ng/mL), the serum needs to be diluted 1:500 in the reagent diluent.
18. The ratio between the lysis buffer (RIPA) and the brain tissue is critical for the quality of the lysate. Too little lysis buffer results in insufficient lysis of the brain tissue, while too much lysis buffer will dilute the lysate.
19. Prior to adding the primary and secondary antibodies, complete removal of the blocking or washing buffer is necessary to avoid dilution of the antibodies. The unequal amount of residual buffer will lead to uneven dilution of antibodies, which will ultimately influence the intensities of immunofluorescence signals.
20. The micro cover glass should be slowly set down on the slide at an angle from one edge. If bubbles are present, gently press down on the surface of the cover glass with forceps to remove them.

Acknowledgments

We thank Dr. Anders Nykjaer for providing sortilin-deficient mice. This work is supported by funding to F.H. from the Weill Institute for Cell and Molecular Biology, the Alzheimer's Association, the Association of Frontotemporal Dementia (AFTD), the Muscular Dystrophy Association, and NINDS (R21 NS081357, R01NS088448, R01NS095954) and by funding to X. Z. from the Weill Institute Fleming Postdoctoral Fellowship.

References

1. Smith KR, Damiano J, Franceschetti S et al (2012) Strikingly different clinicopathological phenotypes determined by progranulin-mutation dosage. Am J Hum Genet 90:1102–1107
2. Ahmed Z, Sheng H, Xu YF et al (2010) Accelerated lipofuscinosis and ubiquitination in granulin knockout mice suggest a role for progranulin in successful aging. Am J Pathol 177:311–324
3. Belcastro V, Siciliano V, Gregoretti F et al (2011) Transcriptional gene network inference from a massive dataset elucidates transcriptome organization and gene function. Nucleic Acids Res 39:8677–8688
4. Hu F, Padukkavidana T, Vægter CB et al (2010) Sortilin-mediated endocytosis determines levels of the frontotemporal dementia protein, progranulin. Neuron 68:654–667
5. Zheng Y, Brady OA, Meng PS et al (2011) C-terminus of progranulin interacts with the beta-propeller region of sortilin to regulate progranulin trafficking. PLoS One 6:e21023
6. Cheng HJ, Flanagan JG (1994) Identification and cloning of ELF-1, a developmentally expressed ligand for the Mek4 and Sek receptor tyrosine kinases. Cell 79:157–168
7. Flanagan JG, Cheng HJ (2000) Alkaline phosphatase fusion proteins for molecular characterization and cloning of receptors and their ligands. Methods Enzymol 327:198–210

Chapter 19

Methods to Study the Role of Progranulin in Preimplantation Mouse Embryo Development

Laura Diaz-Cueto, Fabian Arechavaleta-Velasco, and George L. Gerton

Abstract

Progranulin is a 67–88 kDa glycoprotein, also known as acrogranin, PC-cell-derived growth factor, granulin-epithelin precursor, and proepithelin. This protein is present in a variety of mouse, rat, and human tissues. Progranulin, which is a growth factor, mediates cell cycle progression and cell migration in normal and pathological conditions. In several types of cancers, progranulin expression is upregulated, whereas function-interfering mutations in the granulin gene in humans have been linked to a subset of heritable cases of frontotemporal lobar degeneration. Also, progranulin has important effects on mouse preimplantation embryo development in vitro, including regulation of the appearance of the epithelium in the developing mouse blastocyst and growth of trophectoderm. Furthermore, progranulin promotes mouse blastocyst hatching, adhesion, and outgrowth in vitro. In this chapter, we describe some of the techniques that may be useful in the study of progranulin in embryo development.

Key words Progranulin, Embryo development

1 Introduction

Preimplantation mammalian embryo development includes a series of events or complex processes that allow a single cell fertilized egg to develop into a multicellular blastocyst that has the ability to implant in the uterus. Mammalian embryo development is a highly coordinated process in which cell division and activation of the embryonic genome at the appropriate stage of development ensure the formation of specific cell types such as the inner cell mass (ICM) and trophectoderm. Establishment of these cells allows the proper development of the embryo and the placenta, respectively [1]. Several growth factors participate in this process [2, 3], including progranulin that has positive effects on mouse embryo development in vitro [4, 5].

Different techniques have been used to study progranulin participation in development of preimplantation mouse embryo [4]. Semiquantitative RT-PCR was performed to follow progranulin

Andrew Bateman et al. (eds.), *Progranulin: Methods and Protocols*, Methods in Molecular Biology, vol. 1806, https://doi.org/10.1007/978-1-4939-8559-3_19, © Springer Science+Business Media, LLC, part of Springer Nature 2018

mRNA expression at different stages of development. Interestingly, progranulin transcripts are present in the oocytes and zygotes, but its level drops dramatically at the two-cell stage with a reactivation at four-cell stage. The highest point is reached in the blastocyst stage when the epithelium is first formed. Immunofluorescence and Western blot analyses were used to study progranulin protein. These techniques demonstrated that the progranulin is secreted into the surrounding medium by the trophectoderm of the blastocyst.

Biological function of progranulin in preimplantation mouse embryo was accessed by two different approaches. First, mouse embryos at eight-cell stage were cultured in the presence of progranulin. These experiments showed that progranulin increases the number of trophectoderm cells and blastocoel expansion and reduces the cavitation time. Thus, we concluded that progranulin stimulates blastocyst formation. The second approach included the inhibition of endogenous progranulin effect by a neutralizing anti-progranulin antibody. The data obtained with this strategy confirmed the previous results since this antibody inhibits 8-cell embryo progression to blastocyst stage in a dose-dependent manner. Interestingly, eight-cell embryo development can be restored when the antibody is removed from the medium. Additionally, blastocyst exposure to the neutralizing anti-progranulin antibody reduces the number of cells per embryo and trophectoderm proliferation as detected by nuclei staining with bisbenzimide and BrdU (5-Bromo-2′-Deoxyuridine) incorporation, respectively. These results suggest that progranulin has an autocrine effect that promotes mouse embryo growth and differentiation.

The data mentioned herein demonstrate that progranulin plays a significant role in embryo development; nonetheless, more information about the mechanisms by which this protein regulates this process is required. Therefore in this chapter, a description of some methods that may be useful in the study of progranulin in preimplantation embryos is presented.

2 Materials

2.1 Embryo Collection

1. CF-1 female mice 6 weeks of age (Harlan Industries).
2. B6D2F1/J male mice (Jackson Laboratory).
3. Pregnant mare serum gonadotropin (PMSG, Sigma).
4. Human chorionic gonadotropin (hCG, Sigma).
5. Insulin syringe (1 mL) with 28 G × 1/2″ permanently attached needle.

6. MEM medium: Bicarbonate-free minimal essential medium pH 7.2 (Earl's salts), 100 μg/mL sodium pyruvate, 10 μg/mL gentamicin, and 25 mM HEPES.
7. MEM/PVP: MEM medium containing 3 mg/mL polyvinylpyrrolidone (PVP).
8. MEM/PVP/H: MEM/PVP medium containing 0.5 mg/mL hyaluronidase type IV from bovine testes (Sigma).
9. KSOM medium: 95 mM NaCl, 0.25 mM KCl, 0.35 mM KH_2PO_4, 0.20 mM $MgSO_4$, 25 mM $NaHCO_3$, 1.71 mM $CaCl_2$, 10 mM sodium lactate, 0.20 mM sodium pyruvate, 0.01 mM EDTA, 0.20 mM glucose, 1.0 mM glutamine, 1.0 mg/mL BSA, 100 U/mL penicillin G, and 0.05 mg/mL streptomycin with or without 0.01 mg/mL phenol red.
10. KSOM/AA: KSOM medium containing amino acids. Add 0.5 mL nonessential amino acids (NEAA) and 1 mL of essential amino acids (EAA) to 100 mL of KSOM medium. Another option is KSOM/AA commercial medium (GlobalStem or Caisson).
11. Embryo-handling pipette: Consisting of mouth pipette assembly and pulled Pasteur pipette.
12. Bunsen or alcohol burner for pulling embryo-handling pipettes.
13. Forceps, small scissors, fine watchmaker forceps, and needle for fine dissection.
14. Plastic Petri dish (35 mm or 60 mm) and watch glass for embryo collection.
15. Mineral oil embryo-tested (Sigma).
16. Stereomicroscope.
17. Water-jacketed CO_2 incubator.

2.2 Progranulin RNA Quantification

2.2.1 RNA Extraction and Reverse Transcription

1. Embryos: The number depends on experiment design (100 embryos per group).
2. TRIzol reagent (Invitrogen).
3. DEPC water: Add 1 mL of diethyl pyrocarbonate (DEPC) in 1.0 L of water, shake, and incubate at 50 °C overnight. Autoclave for 20 min and bring to room temperature before use. DEPC is an eye, skin, and respiratory irritant. Avoid contact with the skin and eyes. Use safety glasses, gloves, and mask. Another option is nuclease-free water (Promega).
4. RQ1 RNAase-free DNase (Promega).
5. dNTPs.
6. Oligo-dT.
7. RNase out (Invitrogen).

8. SuperScript® II Reverse Transcriptase (Invitrogen).
9. Thermocycler.
10. UV-Vis spectrophotometer.

2.2.2 Quantitative PCR for Progranulin

1. cDNA.
2. SYBR Green PCR master mix (Applied Biosystems).
3. Progranulin primers (forward and reversed): 5′-GTCTCA GGGACTTCCAGTTGCT-3′ and 5′-CGGGCATTCGAAC TGACTTATC-3. Stock solution (100 μM) and working solution (10 μM) of primers are stored at −20 °C.
4. Actin primers (forward and reversed): 5′-TCTAGACT TCGAGCAGGAGATGG-3′ and 5′-TCGTTGCCAATAG TGATGACTTG-3. Stock solution (100 μM) and working solution (10 μM) of primers are stored at −20 °C.
5. 96-well optical reaction plate (Applied Biosystems).
6. Optical caps (Applied Biosystems).
7. Ice bucket (to keep all reagents and samples cold).
8. Sterile water.
9. ABI 7300 Real-Time PCR system (Applied Biosystems).
10. ImageJ software (National Institutes of Health, http://imagej.nih.gov/ij/).

2.3 Immunofluorescence and Laser-Scanning Confocal Microscopy

1. Embryos: The number depends on the experiment design (20 embryos per group).
2. Acid Tyrode's solution: 2.68 mM KCl, 0.137 M NaCl, 0.49 mM $MgCl_2$, 1.63 mM $CaCl_2$, 5.55 mM glucose, and 0.01 mM polyvinylpyrrolidone (PVP). This solution can be prepared at room temperature and adjusted to pH 2.5 with HCl. Filter-sterilize and store in aliquots at −20 °C.
3. PBS 1×: Dilute one part of 10× phosphate-buffered saline buffer (Invitrogen) to nine parts of deionized water.
4. PBS/PVP: 1× PBS containing 3 mg/mL polyvinylpyrrolidone (PVP).
5. 3.7% paraformaldehyde (PFA): Slowly dissolve 0.925 g of PFA in 25 mL of 1× PBS over low heat (use a stirring hot plate in a fume hood) until solution clears. Place it in cold water and add 5 μL of 1 N NaOH to neutralize. Adjust to pH 7.4 and filter. It is important to make this solution fresh every time. Avoid contact with the skin and eyes. Use safety glasses, gloves, and mask.
6. PBS/Triton: 1× PBS containing 0.1% Triton X-100.
7. Blocking solution: 1× PBS containing 0.1% bovine serum albumin (BSA) and 0.01% Tween-20.

8. Rabbit monoclonal antihuman progranulin antibody (Abcam).
9. Fluorescein (FITC), rhodamine (TRITC), or Texas Red-conjugated anti-rabbit IgG (Jackson ImmunoResearch Lab).
10. Humidified chamber.
11. Slides and coverslips.
12. Vectashield® Mounting Medium (Vector Labs).
13. Nail polish.
14. Laser-scanning confocal microscope.

2.4 Differential Staining of Inner Cell Mass and Trophectoderm Cells

1. Embryos: The number depends on the experiment design (20 embryos per group).
2. Acid Tyrode's solution: Prepare as outlined in Subheading 2.3.
3. M2 medium: 94.7 mM NaCl, 4.78 mM KCl, 1.19 mM $MgSO_4{\cdot}7H_2O$, 1.19 mM KH_2PO_4, 1.71 mM $CaCl_2{\cdot}2H_2O$, 5.56 mM glucose, 0.33 mM sodium pyruvate, 4.15 mM $NaHCO_3$, 30.85 mM HEPES, 4.0 g/L BSA, 0.06 g/L penicillin, 0.05 g/L streptomycin sulfate, and 0.01 g/L phenol red.
4. Medium M2/BSA 5%: M2 medium containing 5% BSA.
5. Medium M2/BSA 1%: M2 medium containing 1% BSA.
6. Anti-mouse T-cell antiserum (Thy-1) (Accurate Chemical and Scientific Corp).
7. Guinea pig complement (Sigma).
8. Bisbenzimide Hoechst 33258.
9. Propidium iodide.

2.5 Determination and Quantification of Progranulin in the Embryo and Embryo Culture-Conditioned Media

1. Embryos: The number depends on the experiment design, 500 embryos per group.
2. Lysis buffer: 50 mM HEPES pH 7.5, 150 mM NaCl, 1.5 mM $MgCl_2$, 1 mM EGTA, 10% glycerol, and 1% Triton X-100. Other option is RIPA buffer 10× (Cell Signaling).
3. Protease inhibitor cocktail tablets (Complete Mini, Roche): Add one tablet to 10 mL of lysis buffer.
4. Loading buffer: 4% SDS, 125 mM Tris–HCl pH 6.8, 20% glycerol, 0.02% bromophenol blue, and 80 mM dithiothreitol. Filter and store at room temperature.
5. 10% sodium dodecyl sulfate-polyacrylamide gel.
6. Hybond-P PVDF membranes (GE Healthcare Life Sciences).
7. Electro-transfer equipment.
8. Blocking buffer: 20 mM Tris–HCl, 500 mM NaCl pH 7.5, and 5% powdered skim milk.

9. Rabbit monoclonal antihuman progranulin antibody (Abcam).
10. Horseradish peroxidase-conjugated goat anti-Rabbit IgG (Jackson ImmunoResearch Lab).
11. Chemiluminescent kit West Pico (Pierce).
12. ImageJ software (National Institutes of Health, http://imagej.nih.gov/ij/).
13. Progranulin (h) ELISA kit (AdipoGen).

2.6 Bromo-deoxyuridine Incorporation Assay for Cell Proliferation in Embryo

1. Embryos: The number depends on the experiment design (20 embryos per group).
2. 5-Bromo-2′-deoxyuridine (BrdU).
3. KSOM/AA medium: Prepare as outlined in Subheading 2.1.
4. Mineral oil embryo-tested (Sigma).
5. PBS/PVP: 1× PBS containing 3 mg/mL polyvinylpyrrolidone (PVP).
6. 3.7% paraformaldehyde: Prepare PFA as outlined in Subheading 2.3.
7. 2 N HCl.
8. 0.1 M sodium borate buffer (pH 8.5).
9. Triton /PBS: 1× PBS containing 0.1% Triton X-100.
10. Blocking solution: 1× PBS, 0.1% BSA, and 0.01% Tween-20.
11. Rat monoclonal anti-BrdU antibody (Abcam).
12. Fluorescein (FICT), rhodamine (TRITC), or Texas Red-conjugated goat anti-rat IgG (Jackson ImmunoResearch Lab).
15. Humidified chamber.
16. Slides and coverslips.
13. Vectashield® Mounting Medium (Vector Labs).
14. Nail polish.
15. Laser-scanning confocal microscope.

2.7 Blastocyst Hatching, Adhesion, and Outgrowth Assay

1. Early blastocyst: The number depends on the experiment design (ten embryos per group).
2. MEM/PVP: Prepare as outlined in Subheading 2.1.
3. Flat-bottom 96-well plates.
4. Six-well plates pre-coated with 50 μg/mL of fibronectin (Roche) (*see* **Note 1**).
5. Recombinant progranulin.
6. Inverted phase-contrast microscope.
7. Photo editing software.
8. ImageJ software (National Institutes of Health, http://imagej.nih.gov/ij/).

3 Methods

3.1 Embryo Collection

1. Inject female mice with PMSG (5 IU/mouse) intraperitoneally using an insulin syringe.
2. Forty-six to forty-eight hours later, inject the mice with hCG (5 IU/mouse) in the same way to induce superovulation (it is recommended to inject the animals between 2 and 4 pm).
3. Mate single females with single male mice of proven fertility.
4. Next morning at approximately 9:00 am, check each female mouse for a vaginal plug, which is suggestive that mating has occurred and the female is pregnant.
5. At appropriate times post-hCG injection, collect embryos from the oviducts and uteri as follows: One-cell, 16–18 h; two-cell, 44 h; four-cell, 60 h; eight-cell, 67 h; morula, 78 h; and blastocyst, 88 h.
6. For best results, it is recommended to sacrifice one mouse at a time. The number of embryos harvested per mouse will depend on superovulation efficiency.
7. Two hours before embryo collection, prepare culture plates by placing some microdrops (20–50 μL) of KSOM/AA medium under mineral oil in a 35-mm or 60-mm Petri dish. Equilibrate and prewarm media in a water-jacketed CO_2 incubator at 37 °C and 5% CO_2 (*see* **Note 2**).
8. Euthanize mice by CO_2. Wipe the abdomen with 70% ethanol, open the abdominal cavity, and localize the ovaries, oviducts, and uterine horns. Anatomical localization of each stage of development of the embryo is shown in Fig. 1.
9. One-cell stage embryo collection: Remove the oviduct, and place it in MEM/PVP/H medium (it is better to have two to six oviducts in each drop of medium). Under stereomicroscope, hold the ampulla with the forceps and puncture with the needle so that all the embryos and surrounding cumulus cell come out. Remove the oviduct and handle the embryos with a mouth pipet. Wash the embryos through five drops of MEM/PVP medium to remove the remaining hyaluronidase and cumulus cells. Select embryos that have pronuclei, and transfer them to the prewarmed and equilibrated culture medium KSOM/AA (*see* **step 7** of Subheading 3.1). Incubate them in a water-jacketed CO_2 incubator at 37 °C and 5% CO_2 (*see* **Note 3**).
10. Two-cell to eight-cell stage embryo collection: Remove the oviducts and flush them using 1 mL syringe with MEM/PVP in a 35-mm plastic Petri dish. Collect and transfer the embryos to the prewarmed and equilibrated KSOM/AA culture

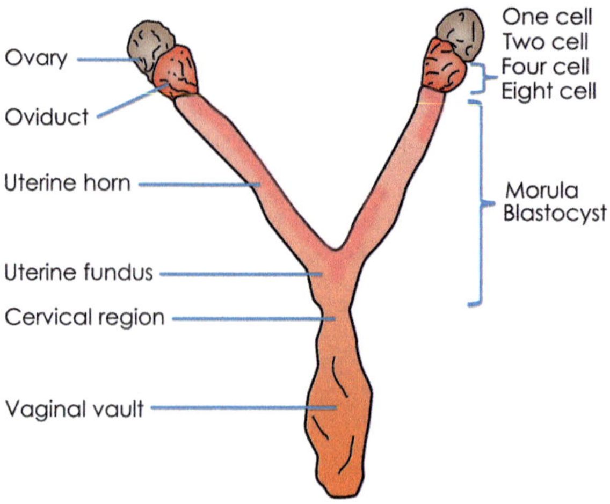

Fig. 1 Schematic representation of the anatomical sites used to collect preimplantation mouse embryos. One-cell, two-cell, four-cell, and eight-cell embryos are collected from the oviduct; meanwhile morula stage and blastocyst are collected from the uterine horn and fundus

medium (*see* **step** 7 of Subheading 3.1), and place them in a water-jacketed CO_2 incubator at 37 °C and 5% CO_2.

11. Morula and blastocyst stage collection: Remove the uterus and place in MEM/PVP, and flush them with the same medium to obtain the embryos. Place 20–30 embryos in a drop of KSOM/AA medium under mineral oil previously equilibrated in the water-jacketed CO_2 incubator at 37 °C and 5% CO_2.
12. All experiments performed with the embryos can be done using KSOM/AA medium.

3.2 Progranulin RNA Quantification

3.2.1 RNA Extraction

1. Collect embryos as outlined in Subheading 3.1.
2. Transfer the embryos into RNase-free micro-centrifuge tubes containing TRIzol reagent (0.5 mL), and incubate at room temperature for 10 min.
3. Add 0.1 mL of chloroform and centrifuge at 12,000 × *g* for 15 min.
4. After centrifuge, remove and save the colorless upper aqueous phase containing total RNA.
5. Precipitate with 0.25 mL of isopropanol and centrifuge at 12,000 × *g* for 10 min.
6. Wash with 75% ethanol and centrifuge at 7500 × *g* for 5 min.
7. At this point, the RNA pellet should be visible. Resuspend it with DEPC-treated H_2O (12–20 μL) according to the pellet size. Incubate at 55–60 °C for 10 min and quantify by UV spectrophotometry.

3.2.2 Reverse Transcription (Using SuperScript® II Reverse Transcriptase)

1. In a sterile micro-centrifuge tube, add one microgram of total RNA, 1 μL of RQ1 RNase-free DNase 10× buffer, 1 μL of RQ1 RNase-free DNase, and DEPC-treated water to a final volume of 10 μL.
2. Incubate at 37 °C for 30 min, and add 1 μL of RQ1 stop solution. Heat at 65 °C for 10 min and spin briefly.
3. Add 1 μL of oligo(dT) and 1 μL of dNTPs. Incubate at 65 °C for 5 min and place promptly on ice.
4. Add 4 μL of 5× first strand buffer, 2 μL of 0.1 M DTT, and 1 μL of RNase out. Mix gently and incubate at 42 °C for 2 min.
5. Add 1 μL of SuperScript TM® Reverse Transcriptase (200 U/μL), mix gently, and incubate at 42 °C for 50 min. Heat reaction mixture at 70 °C for 15 min.
6. Dilute cDNA synthesis reaction with 179 μL of DEPC-treated H_2O to obtain a final concentration of ≈5 ng/μL.
7. Store cDNA at −20 °C. It can be used for quantitative PCR using the forward and reverse primers specified in Subheading 2.2.2.

3.2.3 Quantitative Progranulin PCR

1. Calculate how many reactions will be done, and add one or two extra reactions to account for pipetting error. Include positive control, water as non-template control, and standards for each set of primers (progranulin and actin).
2. Mix the following for each PCR reaction (keep all your reagents on ice): 12.5 μL SYBR Green PCR master mix, 1 μL forward primer, and 1 μL reverse primer (*see* **Note 4**).
3. Add 14.5 μL of the reaction mixture to each well in a 96-well optical reaction plate.
4. Add 1 μL of each cDNA sample to separate wells in the plate.
5. Add 9.5 μL of sterile water to each well to have a final volume of 25 μL.
6. Securely close each well with optical caps, and centrifuge reaction plate briefly to collect all liquid at the bottom.
7. Place plate in the ABI 7300 Real-Time PCR System.
8. PCR parameters consist of an initial denaturation step at 95 °C for 10 min follow by 40 cycles at 95 °C for 15 s and 60 °C for 60 s.
9. Determine the number of copies of progranulin and actin in each sample by extrapolation from the standard curve generated for each set of primers (*see* **Note 5**).
10. Normalize sample data by dividing the number of copies of progranulin by the number of copies of actin.

3.3 Immunofluorescence and Laser-Scanning Confocal Microscopy

1. Collect embryos as outlined in Subheading 3.1.
2. Remove zona pellucida by incubating embryos in groups of five or less in a drop of acid Tyrode's solution. Observe the embryos under a stereomicroscope during incubation time (1–3 min), and remove them immediately after the first sign of zona pellucida degradation. Note that embryos without zona pellucida are sticky and can adhere to the inside of the transfer pipette during handling.
3. Wash embryos twice with 2 mL of PBS/PVP in a watch glass.
4. Fix embryos by placing them in 2 mL of 3.7% paraformaldehyde solution in a watch glass at 4 °C overnight.

All subsequent steps should be done in a humidified chamber at room temperature.

5. Wash embryos with one drop (100 μL) of PBS/PVP twice, and permeabilize in one drop (100 μL) of 0.1% Triton X-100/PBS for 15 min.
6. Wash embryos with one drop (100 μL) of PBS/PVP twice and place in 100 μL of blocking solution for 15 min.
7. Incubate in 100 μL of progranulin antibody, previously diluted 1:100 in blocking solution, for 60 min.
8. Wash four times with one drop (100 μL) of blocking solution for 15 min.
9. Incubate in 100 μL of the secondary antibody with the specific dye, previously diluted 1:500–1:1000 in blocking solution, for 60 min in the dark.
10. Wash embryos five times for 15 min with one drop (100 μL) of blocking solution.
11. Mount embryos on microscope slides with ~50 μL of Vectashield® Mounting Medium, and seal coverslip with nail polish.
12. Observe embryos under a laser-scanning confocal microscope. Progranulin localization by immunofluorescence in embryos is shown in Fig. 2.

3.4 Differential Labeling of the Inner Cell Mass and Trophectoderm

Inner cell mass and trophectoderm nuclei of mouse blastocyst can be stained using a modification of the selective antibody-mediated complement lysis immunodissection procedure of Handyside and Hunter [6]. Using this method, ICM cells remain viable and their nuclei stain blue with bisbenzimide, while trophectoderm nuclei are stained pink with both bisbenzimide and propidium iodide.

1. Collect embryos as outlined in Subheading 3.1.
2. Remove zonae pellucidae by incubating embryos in groups of five or less in a drop of acid Tyrode's solution. Observe the embryos under a stereomicroscope during the incubation time

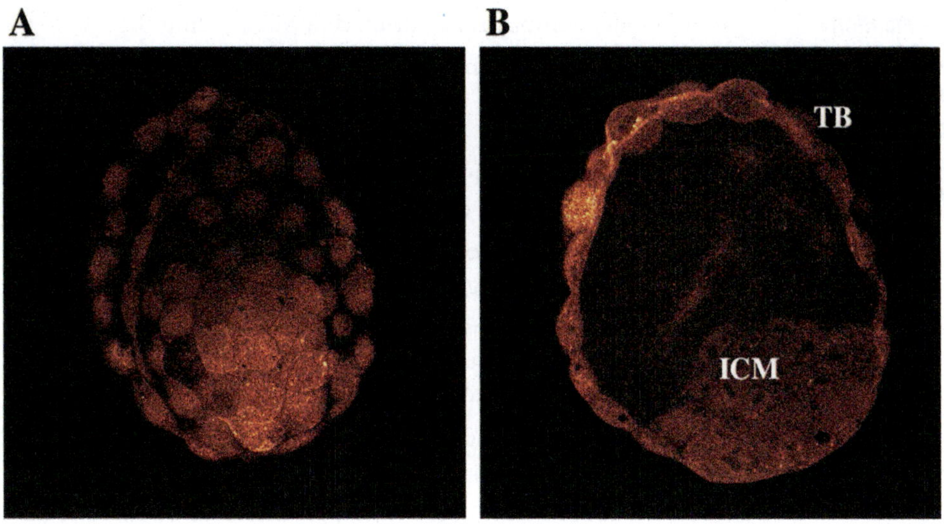

Fig. 2 Immunofluorescence localization of progranulin in morula (**a**) and early blastocyst (**b**). *ICM* inner cell mass, *TB* trophoblast

(1–3 min), and remove them immediately after the first sign of zona pellucida degradation. Note that embryos without zona pellucida are sticky and can adhere to the inside of the transfer pipette during handling.

3. Wash embryos twice with 2 mL of medium M2/BSA 5% in a watch glass.
4. Incubate embryos in one drop (100 μL) of 10% anti-mouse T-cell antiserum (Thy-1) for 10 min at 37 °C.
5. Wash embryos six times with one drop (100 μL) of medium M2/BSA 5%.
6. Incubate them in 100 μL of guinea pig complement previously diluted 1:10 in M2/BSA 5% medium containing 40 μg/mL bisbenzimide and 20 μg/mL propidium iodide for 30 min at 37 °C.
7. Wash several times with one drop (100 μL) of medium M2/BSA 1%.
8. Mount embryos on microscope slides with ~50 μL of Vectashield® Mounting Medium, gently compress the coverslip and seal with nail polish (*see* **Note 6**).
9. Observe embryos under an immunofluorescence microscope.

3.5 Detection of Progranulin in the Embryo or Secreted Progranulin in Embryo-Conditioned Medium

3.5.1 In the Embryo

1. Collect embryos as outlined in Subheading 3.1.
2. Place embryos in round-bottom micro-centrifuged tubes, and add 50 μL of lysis buffer containing protease inhibitor cocktail.
3. Determine protein concentration by a reliable method such as Bradford, BCA, or Lowry.
4. Load 25 μg of total proteins on a 10% sodium dodecyl sulfate-polyacrylamide gel, and run it for 2 h at 25–30 mA.
5. Transfer proteins to Hybond-P PVDF membrane. Time and current may require some optimization. We recommend overnight at 4 °C and 25 mA.
6. Block membrane for 1 h at room temperature using 5 mL of blocking buffer.
7. Incubate membrane with appropriate dilution of progranulin antibody (optimum working antibody concentration should be determined in each lab) overnight at 4 °C in 5 mL of blocking buffer.
8. Wash membrane three times for 15 min with 15 mL of TBS.
9. Incubate membrane with the recommended dilution of the secondary antibody (usually at 1:5,000–1:10,000) at room temperature for 60 min in blocking buffer.
10. Rinse the membrane five times for 15 min with 15 mL of TBS.
11. For immunodetection, apply the chemiluminescence substrate to the membrane as recommended by the manufacturer.
12. Acquire chemiluminescent Western blot image with an imaging system.
13. Quantify and analyze the band intensities of progranulin and housekeeping protein using the ImageJ software.

3.5.2 In Embryo-Conditioned Medium

Quantification of progranulin in the embryo conditioned-medium can be performed by a commercial enzyme-linked immunosorbent assay (ELISA).

1. Collect embryos as outlined in Subheading 3.1.
2. Incubate embryos in 50 μL of KSOM/AA medium in a humidified atmosphere of 5% CO_2 in air for at least 12 h.
3. Collect the medium and store at −70 °C until use.
4. Quantify progranulin concentration in embryo-conditioned medium by ELISA kit according to the manufacturer's instructions.

3.6 Embryos and Bromodeoxyuridine Incorporation Assay

1. Collect embryos as outlined in Subheading 3.1.
2. Culture embryos in the presence of different concentrations of anti-progranulin antibody for the desired time (usually 14 h is recommended; however, optimum time should be determined in each lab).
3. Label them with 2 mL of 10 μM bromodeoxyuridine (BrdU) for 45 min in a watch glass.
4. Wash embryos three times for 15 min with 2 mL of PBS/PVP in a watch glass.
5. Remove zona pellucida as outlined in Subheading 3.3.
6. Fix embryos in 2 mL of 3.7% paraformaldehyde at 37 °C in a watch glass for 30 min.
7. Wash embryos three times with 2 mL of PBS/PVP in a watch glass.
8. For DNA denaturation, incubate embryos in 2 mL of 2 N HCL at 37 °C for 1 h in a watch glass.
9. Add 10 μL of borate buffer to neutralize the acid and incubate for 20 min at 37 °C.

All subsequent steps should be done in a humidified chamber at room temperature.

10. Wash embryos with one drop (100 μL) of PBS/PVP twice, and permeabilize in one drop (100 μL) of 0.1% Triton X-100/PBS for 15 min.
11. Wash embryos twice with one drop (100 μL) of PBS/PVP, and place in one drop (100 μL) of blocking solution for 15 min.
12. Incubate with 100 μL of BrdU antibody previously diluted 1:50 in blocking solution for 60 min.
13. Wash four times for 15 min with one drop (100 μL) of blocking solution.
14. Incubate in 100 μL of anti-mouse IgG antibody conjugated with Texas Red dye previously diluted 1:500 in blocking solution for 45 min.
15. Wash the embryos five times for 15 min with one drop (100 μL) of blocking solution.
16. Mount embryos on microscope slides with ~50 μL drop of Vectashield® Mounting Medium, gently compress the coverslip, and seal with nail polish.
17. Observe embryos under a laser-scanning confocal microscope. Figure 3 shows BrdU incorporation into preimplantation mouse embryos after treated with anti-progranulin antibody.

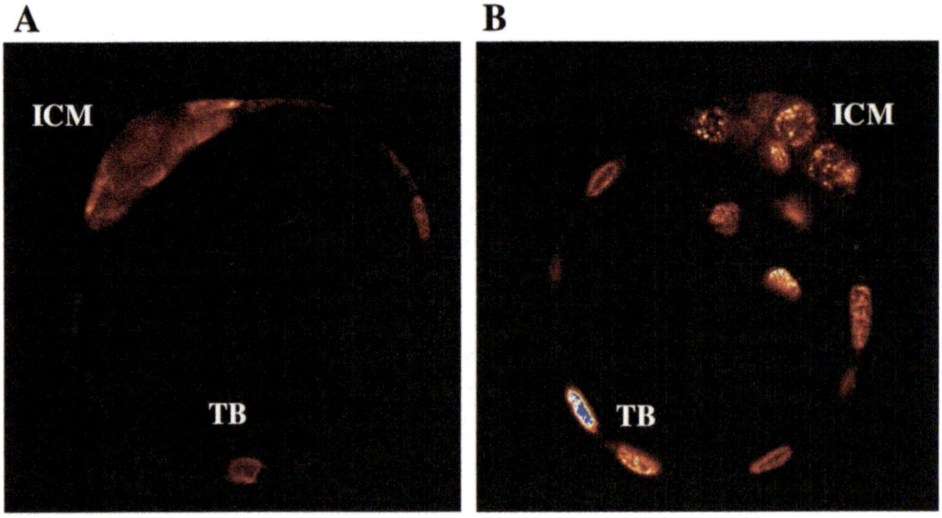

Fig. 3 BrdU incorporation into preimplantation mouse embryos incubated in the presence of anti-progranulin antibody (**a**) and after removal of the antibody and subsequent culture (**b**). DNA synthesis was not affected in the inner cell mass (ICM) by the anti-progranulin antibody treatment. Trophoblast (TB) proliferation was restored after the antibody was removed

3.7 Blastocyst Hatching, Adhesion, and Outgrowth Assay

1. Collect blastocyst as outlined in Subheading 3.1.
2. Divide the blastocysts into as many groups as required.
3. Culture these blastocysts in 50 μL of prewarmed and CO_2-equilibrated KSOM/AA medium containing 3% of fetal bovine serum and different concentrations of progranulin for 60 h at 37 °C and 5% CO_2 (*see* **Note 7**).
4. For hatching assay: Place ten embryos per well in a flat-bottom 96-well plate. Assess hatching at 24, 36, 48, and 60 h after initiation of in vitro culture using an inverted phase-contrast microscope. Embryos completely free from the zona pellucida are classified as hatched blastocysts. Record the numbers of blastocysts undergoing hatching in each condition and calculate the ratios of hatched blastocysts relative to the total number of embryos.
5. For adhesion assay: Culture ten blastocyst per well in a 96-well plate. Examine adhesion at 48 h after the initiation of the incubation period. Blastocyst adhesion is examined once, since gentle pipetting of the medium is required to determine whether the embryo would detach from the bottom of the well. Under an inverted phase-contrast microscope, count the embryos that are not floating in each condition and calculate the ratios of attached blastocysts relative to the total number of embryos.
6. For outgrowth assay: Place one blastocyst per well in a 6-well plate pre-coated with fibronectin. Evaluate outgrowth every

24 h until 5 days after initiation of the incubation period using an inverted phase-contrast microscope. Acquire digital images of the outgrowth area each day. Improve image quality with a photo editing software, and measure the area of the blastocyst outgrowth using the ImageJ software. Calculate the mean area of at least ten embryos.

3.8 Regulation of Progranulin Expression in Embryos by Proteins and Other Factors

1. Before embryo collection, equilibrate several micro-drops (20–50 μL) of KSOM/AA medium containing compound(s) or factor(s) that are to be tested, in a water-jacketed CO_2 incubator at 37 °C and 5% CO_2 for 30–60 min. Micro-drops should be placed under mineral oil in a 35-mm Petri dish.
2. Collect embryos as outlined in Subheading 3.1.
3. Culture embryos for the desired time (usually 24–60 h depending on the effect to be analyzed).
4. Harvest embryos and medium at the end of the incubation time.
5. Perform the appropriate technique to evaluate progranulin expression.

4 Notes

1. Dissolve fibronectin according to the manufacturer's instructions. Dilute and appropriate aliquot with sterile Hank's balanced salt solution (HBSS) (Gibco) to a final concentration of 50 μg/mL. Pipette 600 μL per well of a 6-well plate, and leave overnight at 37 °C. Remove solution carefully, and wash the coated surface three times with HBSS containing 4 mg/mL of bovine serum albumin. Plates may be used immediately or may be stored at 2–8 °C, damp or air dried, if sterility is maintained.
2. The size of the micro-drop depends on the number of embryos that will be cultured in it. Embryos in culture secrete growth factors that reduce apoptosis and stimulate cell proliferation [4, 7]. Thus, it is recommended to culture a high density (one embryo/μL) to induce adequate development.
3. Time between female mouse sacrifice and embryo collection and culture should not be more than 20–30 min. Extended exposure of embryos to hyaluronidase and prolonged incubation in the washing medium prior to transfer into the culture medium compromises their subsequent development.
4. You can design your own set of primers for progranulin and housekeeping gene. The primers herein *work* fine for quantitative and semiquantitative PCR.

5. Standard curves can be created by PCR amplification of progranulin and actin cDNA with the same set of primers as the samples. The resulting amplicon must be purified and quantified by UV spectrophotometry. The number of copies is calculated by using the molecular weight of each gene amplicon. Serial dilutions of the amplified transcripts at known concentrations are used as templates for the standard curve. Alternatively, the template can be a plasmid containing the amplicon generated by RT-PCR with the same set of primers.
6. After photographing and recording the staining pattern, the embryos could be carefully squashed and disaggregated by gently pressing the coverslip with a pencil eraser, in order to facilitate embryonic cell counting.
7. In our experience, a concentration of 0.2 μg/mL of recombinant progranulin is optimal for blastocyst hatching, adhesion, and/or outgrowth.

Acknowledgment

The authors thank Manuel Mendez Montes for his assistance with figure preparation.

References

1. Fleming TP, Johnson MH (1988) From egg to epithelium. Annu Rev Cell Biol 4:459–485
2. Diaz-Cueto L, Gerton GL (2001) The influence of growth factors on the development of preimplantation mammalian embryos. Arch Med Res 32:619–626
3. Kaye PL (1997) Preimplantation growth factor physiology. Rev Reprod 2:121–127
4. Diaz-Cueto L, Stein P, Jacobs A et al (2000) Modulation of mouse preimplantation embryo development by acrogranin (epithelin/granulin precursor). Dev Biol 217:406–418
5. Qin J, Diaz-Cueto L, Schwarze JE et al (2005) Effects of progranulin on blastocyst hatching and subsequent adhesion and outgrowth in the mouse. Biol Reprod 73:434–442
6. Handyside AH, Hunter S (1984) A rapid procedure for visualising the inner cell mass and trophectoderm nuclei of mouse blastocysts in situ using polynucleotide-specific fluorochromes. J Exp Zool 231:429–434
7. Brison DR, Schultz RM (1997) Apoptosis during mouse blastocyst formation: evidence for a role for survival factors including transforming growth factor alpha. Biol Reprod 56:1088–1096

Chapter 20

Establishment of a Modified Collagen-Induced Arthritis Mouse Model to Investigate the Anti-inflammatory Activity of Progranulin in Inflammatory Arthritis

Jian-Lu Wei and Chuan-Ju Liu

Abstract

Progranulin (PGRN) was found to play an anti-inflammatory and protective role in both inflammatory and degenerative arthritis (Tang et al., Science 332:478–484, 2011; Zhao et al., Ann Rheum Dis 74:2244–2253, 2015). We recently published a visualized protocol to demonstrate a surgically-induced mouse model for examining the protective role of PGRN in degenerative osteoarthritis (Zhao et al., J Vis Exp:e50924, 2014). Herein we describe a modified collagen-induced arthritis (CIA) mouse model to investigate the anti-inflammatory activity of PGRN in inflammatory arthritis. CIA model is the most commonly used autoimmune model of inflammatory arthritis which shares both immunological and pathological features with human rheumatoid arthritis. Autoimmune inflammatory arthritis is induced by immunization with an emulsion of complete Freund's adjuvant and chicken type II collagen (CII) using a modified procedure in PGRN deficient mice and control littermates. Using the protocol described here, the investigator should be able to reproducibly induce a high incidence of CIA in PGRN deficient mice and also learn how to critically evaluate the severity and incidence of this disease model.

Key words Progranulin, TNFα, Collagen-induced arthritis, Autoimmune disease, Animal model

1 Introduction

Progranulin (PGRN), also known as granulin-epithelin precursor (GEP), PC cell-derived growth factor (PCDGF), proepithelin, and acrogranin, plays an important role in a variety of physiological and pathological processes, including early embryogenesis, wound healing, inflammation, host defense, and neuronal and skeletal degenerative disorders [1–12]. It is known that PGRN has anti-inflammatory activity in various conditions [6, 13–19]. PGRN's anti-inflammatory activity is through, at least in part, its ability to directly bind to tumor necrosis factor receptors (TNFRs) and to disturb the TNFα-TNFR interaction [5, 20–26]. In addition, PGRN's anti-inflammatory activities in various conditions have been demonstrated by several laboratories with corresponding animal

Andrew Bateman et al. (eds.), *Progranulin: Methods and Protocols*, Methods in Molecular Biology, vol. 1806,
https://doi.org/10.1007/978-1-4939-8559-3_20,

models [18, 27–29]. Our lab also reported the anti-inflammatory action of PGRN in inflammatory arthritis models, including collagen-induced arthritis (CIA), collagen antibody induced arthritis (CAIA), and spontaneous TNF transgenic mice model [5]. Among them, CIA model has been used extensively to identify potential pathogenic mechanisms of autoimmunity and inflammation, including the role of individual cell types in disease onset and progression [30–32]. In addition, CIA model has been instrumentally used in the testing and development of the new biologically based therapeutics, such as those that target TNFα [33–35]. In general, CIA mouse model is widely induced in DBA/1 mouse strain because of its high incidence and featured symptoms [36]. Nowadays, CIA is also widely elicited in genetically modified susceptible mice by immunization with chick type II collagen (CII) emulsified in complete Freund's adjuvant (CFA) in C57BL/6 mice strain [30]. We have established a modified CIA model using PGRN-deficient mice [5, 37]. Briefly, we used the magnetic stirring system to emulsify chick type II collagen with complete Freund's adjuvant in an ice-cold container. After single injection of the emulsion, we could observe the high incidence and severe phenotype in PGRN null mice despite the C57BL/6 genetic background. PGRN's anti-inflammatory activity in this model was further manifested by the incidence, clinical score, and histological analysis.

2 Materials

1. Chick type II collagen (Catalog No. 20012, Chondrex, LLC, Seattle, WA).
2. Complete Freund adjuvant (Catalog No. 7001, Chondrex, LLC, Seattle, WA).
3. Magnetic stirring system (Catalog No. 90-691-16, Fisher Scientific, Pittsburgh, PA).
4. Ketamine and xylazine (Catalog No. 23076-35-9, AnaSed, LLOYD, Shenandoah, IA).
5. Mice: PGRN knockout (PGRN−/−) C57BL/6 mice and wild-type C57BL/6 (PGRN+/+) mice, 10 weeks old, male [5].
6. 1 mL sterilized syringe with 27 G needle (REF 309623, BD eBioscience, San Jose, CA).

3 Methods

3.1 Preparation of Emulsion

The whole procedure should be performed on ice or 4 °C in dark to avoid the degradation of CII (*see* **Note 1**).

1. Determine the emulsion volume required. Calculate the volumes of Chick CII and CFA required for making the emulsion,

keeping their ratios at 1:1. While 100 μL of emulsion is required for each mouse, extra volumes must be made owing to both losses during processing and the dead volume in the syringes (*see* **Notes 2** and **3**).

2. Select an appropriate container. We recommend to use a 5 mL glass container with a small magnetic rotor to mix the Chick CII and CFA.
3. Make the mixture. Place the container in the center of an ice-cold beaker during emulsification to keep it at the low temperature. Turn on the magnetic system to stir the magnetic rotor, keeping the speed at 100 rpm. Drop the required volume of CFA first and add CII by dripping slowly to CFA (*see* **Note 4**).
4. Time 30 min to get a good mixture of CII and CFA. Always keep the emulsion in dark and at low temperature (*see* **Note 5**).
5. Check the quality of the emulsion. Stop the magnetic system, and check it visually to ensure that the emulsion appears to be thoroughly mixed. A good mixture has the viscosity of dense whipped cream in appearance. If the degree of emulsification is not satisfactory, further mixing by the magnetic system should be taken place to improve the quality of the emulsion (*see* **Note 6**).

3.2 Immunization of Mouse

1. Preparation of the anesthetic drugs. A dose of ketamine 150 mg/kg and xylazine 10 mg/kg of body weight can be given intraperitoneally. Use 0.75 mg ketamine and 0.5 mg xylazine diluted in 5 mL sterilized saline to get 0.15 mg/mL ketamine and 0.1 mg/mL xylazine (*see* **Note 7**).
2. Anesthetize the mouse. Although mice do not require to be anesthetized for immunization, it is much easier to inject the emulsion in the proper tissue layer while the mice are under anesthesia. The anesthetic drugs should be administered by intraperitoneal Injection (*see* **Note 8**).
3. Transfer the emulsion to a 27 G syringe. Depress the syringe plunger slowly to transfer the emulsion to the immunization syringe. This step should be performed slowly and carefully, and no air is allowed to be included in the syringe (*see* **Notes 9** and **10**).
4. Induction of collagen-induced arthritis mouse model: It is recommended to choose an injection site about 1.5 cm distal from the base of the tail, and be careful to choose a tissue site instead of a blood vessel. It is essential to avoid injecting into blood vessels in tail; if the anesthetic agent is injected into blood vessels, the mice will die immediately. Slowly inject a 100 μL emulsion intradermally (i.d.) into the tail (Fig. 1). There should be noticeable tissue resistance to the injection. A rapid, easy injection can be indicative of a subcutaneous or

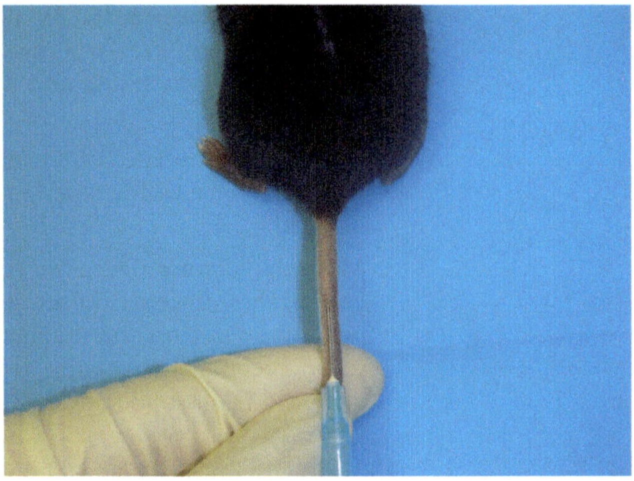

Fig. 1 Proper technique for immunization of mice for the induction of CIA. Emulsified CII in CFA is injected intradermally in the skin of the tail at approximately 1.5 cm distal to the base of the tail

deep tissue or even blood vessel injection which can result in a poor incidence of disease or even death (*see* **Notes 11–15**).

3.3 Evaluation of the Incidence and Clinical Score of Arthritis

Mice should be evaluated every other day for arthritis incidence and clinical score (Fig. 2). Each paw should be evaluated and scored individually according to the criteria as shown in Table 1 on a scale of 0–4, with 4 indicating the most severe inflammation and totally the highest score would be 16. In addition, CIA is considered to have been induced when the score is higher than 1 in more than two limbs or higher than 2 in more than one limb (*see* **Notes 16–18**).

3.4 Troubleshooting

Following our protocol, investigators can successfully establish the CIA mice model in this genetically modified mice strain. We list some problems investigators may meet in the process of CIA induction, and we give the solutions which may help accordingly (*see* Table 2).

4 Notes

1. The whole procedure of preparation of emulsion must be taken place on ice and in dark. Put the regents at 4 °C to melt.
2. Plan on at least a 400–500 μL loss of emulsion during preparation, and transfer to the immunization syringe.
3. Vortex CFA vigorously before use to thoroughly suspend the *M. tuberculosis* in the Freund's adjuvant.
4. Add CII to CFA drop by drop, and keep them in the magnetic stirring system and then wait until the two agents mix evenly.

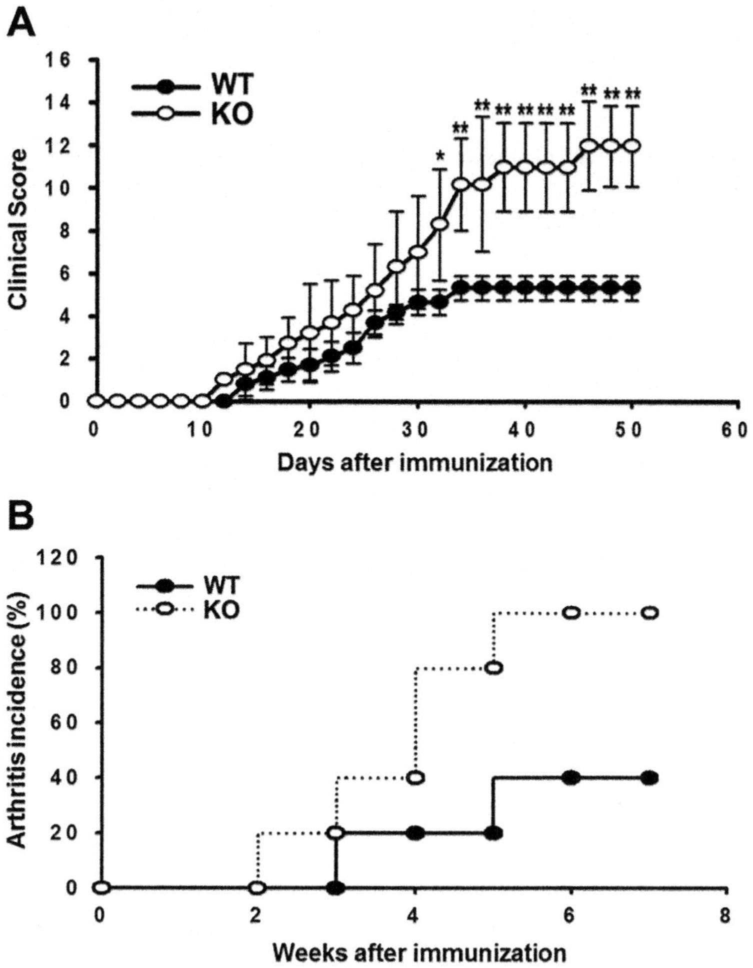

Fig. 2 Clinical score and arthritis incidence of wild-type mice and PGRN knockout mice. After induction of CIA, PGRN knockout mice (KO) exhibit significant higher clinical score and arthritis incidence compared to wild-type (WT) control mice

Table 1
Criteria of clinical score evaluation

Score	Degree of inflammation
0	No evidence of erythema and swelling
1	Erythema and mild swelling confined to the tarsals or ankle joint
2	Erythema and mild swelling extending from the ankle to the tarsals
3	Erythema and moderate swelling extending from the ankle to metatarsal joints
4	Erythema and severe swelling encompass the ankle, foot and digits, or ankylosis of the limb

Table 2
Troubleshooting

Problem	Possible cause	Suggested solutions
Low incidence	Denatured CII	Keep the collagen cold, and do not store at 4 °C for more than 2 months. Consider storing the collagen in small aliquots so that experimental quantities can be thawed when needed. Collagen should be a clear solution. Cloudy collagen suggests that it has crosslinked and precipitated or that it may be contaminated and should be discarded. In addition, please keep in mind for the C57BL/6 genetic background mice strain, only chick type II collagen can be used to induce the inflammatory arthritis model
	Subcutaneous injection	Precise injection site is required. Make sure to keep the injection shallow to maintain intradermal site. Avoid injections that go deep into the tail tissue
Unstable emulsion	Insufficient mixing	Increase the speed of the magnetic stirring system and always keep the reagents cold and in dark during emulsification. Add CII into CFA slowly and dropwise while mixing
Ulceration	Injection sequelae	Emulsion should stay in the intradermal region in the tail, and do not move the mouse during injection. In addition, draw the syringe quickly to avoid the emulsion flowing out
Mouse die	Injection or anesthesia sequelae	Avoid injecting emulsion into the tail vein, as this can be fatal. Monitor anesthesia carefully during procedure and do not overdose the anesthetic drugs

5. Keep the magnetic system nonstop and in dark at required low temperature during the preparation of the emulsion.
6. Use a sterilized tip to draw 10 μL emulsion and drop it into water. The emulsion should have the viscosity of a dense droplet rather than a diffused drop in water.
7. Dose the anesthetic drug at 0.1 mL/10 g of body weight.
8. Perform carefully and never damage the organs. Wait 3–5 min to make sure the mouse is well under anesthesia. In addition, never over-inject the anesthetic drugs; otherwise the mouse will die.
9. Make use of a 27 G syringe to draw the emulsion.
10. Because the dense of the emulsion is very high, draw the emulsion carefully and slowly so that no air is allowed to be included in the immunization syringe.
11. Intradermal injection in tail and the position of injection should be 1.5 cm distal from the base of the tail. Too close to the base of the tail will cause some complications such as orchitis.
12. There should be noticeable resistance when injection happens; otherwise, the injection should be stopped immediately.

13. Avoid injecting the emulsion into the blood vessels in the tail; otherwise, the mouse will die immediately.
14. After injection, be quick to withdraw the syringe; otherwise, the agent will flow out.
15. It is optional to put some antibiotics such as erythromycin on the injection site to prevent further infection.
16. Carefully observe the paws of induced mice every other day after induction, and evaluate incidence and clinical score of arthritis according to criteria. Determine the severity of arthritis according to the clinical score: mild arthritis (score 4–6), moderate arthritis (score 7–9), and severe arthritis (score 10 or even higher). In addition, here, the scores mean the total score of four paws. Collect all the scores, and do the statistical analysis for each time point.
17. CIA incidence is determined by observation of mice limbs, according to the number of arthritic limbs and the arthritis scores (scoring criteria is in Table 1). CIA is considered to have been induced when the score is higher than 1 in more than two limbs or higher than 2 in more than one limb. Incidence = number of CIA mice/total number of mice × 100 [38].
18. If the operation is performed appropriately, no booster or secondary immunization is needed.

Acknowledgment

This work is supported partly by NIH research grants R01AR062207, R01AR061484, and R56AI100901, a Disease Targeted Research Grant from the American College of Rheumatology Research and Education Foundation.

References

1. Feng JQ, Guo FJ, Jiang BC et al (2010) Granulin epithelin precursor: a bone morphogenic protein 2-inducible growth factor that activates Erk1/2 signaling and JunB transcription factor in chondrogenesis. FASEB J 24:1879–1892
2. Hrabal R, Chen Z, James S et al (1996) The hairpin stack fold, a novel protein architecture for a new family of protein growth factors. Nat Struct Biol 3:747–752
3. Toh H, Chitramuthu BP, Bennett HPJ et al (2011) Structure, function, and mechanism of progranulin; the brain and beyond. J Mol Neurosci 45:538–548
4. Zhao YP, Liu B, Tian QY et al (2015) Progranulin protects against osteoarthritis through interacting with TNF-alpha and beta-Catenin signalling. Ann Rheum Dis 74:2244–2253
5. Tang W, Lu Y, Tian QY et al (2011) The growth factor progranulin binds to TNF receptors and is therapeutic against inflammatory arthritis in mice. Science 332:478–484
6. Zhao YP, Tian QY, Frenkel S et al (2013) The promotion of bone healing by progranulin, a downstream molecule of BMP-2, through interacting with TNF/TNFR signaling. Biomaterials 34:6412–6421
7. Zhu J, Nathan C, Jin W et al (2002) Conversion of proepithelin to epithelins: roles of SLPI and elastase in host defense and wound repair. Cell 111:867–878

8. Guo F, Lai Y, Tian Q et al (2010) Granulin-epithelin precursor binds directly to ADAMTS-7 and ADAMTS-12 and inhibits their degradation of cartilage oligomeric matrix protein. Arthritis Rheum 62:2023–2036
9. Ryan CL, Baranowski DC, Chitramuthu BP et al (2009) Progranulin is expressed within motor neurons and promotes neuronal cell survival. BMC Neurosci 10:130
10. Bateman A, Bennett HPJ (2009) The granulin gene family: from cancer to dementia. Bioessays 31:1245–1254
11. Kessenbrock K, Frohlich L, Sixt M et al (2008) Proteinase 3 and neutrophil elastase enhance inflammation in mice by inactivating antiinflammatory progranulin. J Clin Invest 118:2438–2447
12. Zhao Y, Liu B, Liu CJ (2014) Establishment of a surgically-induced model in mice to investigate the protective role of progranulin in osteoarthritis. J Vis Exp:e50924. https://doi.org/10.3791/50924
13. Guo Z, Li Q, Han Y et al (2012) Prevention of LPS-induced acute lung injury in mice by progranulin. Mediat Inflamm 2012:540794
14. Thurner L, Zaks M, Preuss KD et al (2013) Progranulin antibodies entertain a proinflammatory environment in a subgroup of patients with psoriatic arthritis. Arthritis Res Ther 15:R211
15. Zhao YP, Tian QY, Liu CJ (2013) Progranulin deficiency exaggerates, whereas progranulin-derived Atsttrin attenuates, severity of dermatitis in mice. FEBS Lett 587:1805–1810
16. Egashira Y, Suzuki Y, Azuma Y et al (2013) The growth factor progranulin attenuates neuronal injury induced by cerebral ischemia-reperfusion through the suppression of neutrophil recruitment. J Neuroinflammation 10:105
17. Kawase R, Ohama T, Matsuyama A et al (2013) Deletion of progranulin exacerbates atherosclerosis in ApoE knockout mice. Cardiovasc Res 100:125–133
18. Zhou M, Tang W, Fu Y et al (2015) Progranulin protects against renal ischemia/reperfusion injury in mice. Kidney Int 87(5):918–929
19. Yamamoto Y, Takemura M, Serrero G et al (2014) Increased serum GP88 (Progranulin) concentrations in rheumatoid arthritis. Inflammation 37:1806–1813
20. Tian QY, Zhao S, Liu CJ (2014) A solid-phase assay for studying direct binding of progranulin to TNFR and progranulin antagonism of TNF/TNFR interactions. Methods Mol Biol 1155:163–172
21. Uddin SMZ, Mundra JJ, Jian JL et al (2014) Progranulin inhibition of TNF alpha. Immunol Cell Biol 92:299–300
22. Jian JL, Zhao S, Tian QY et al (2013) Progranulin directly binds to the CRD2 and CRD3 of TNFR extracellular domains. FEBS Lett 587:3428–3436
23. Liu C (2012) Progranulin meets Tnf in inflammatory arthritis. Ann Rheum Dis 71:35–35
24. Wei FH, Zhang YY, Jian JL et al (2014) PGRN protects against colitis progression in mice in an IL-10 and TNFR2 dependent manner. Sci Rep 4:7023
25. Wei F, Zhang Y, Zhao W et al (2014) Progranulin facilitates conversion and function of regulatory T cells under inflammatory conditions. PLoS One 9:e112110
26. Li ML, Liu YN, Xia F et al (2014) Progranulin is required for proper ER stress response and inhibits ER stress-mediated apoptosis through TNFR2. Cell Signal 26:1539–1548
27. Hwang HJ, Jung TW, Hong HC et al (2013) Progranulin protects vascular endothelium against atherosclerotic inflammatory reaction via Akt/eNOS and nuclear factor-kappaB pathways. PLoS One 8:e76679
28. Vezina A, Vaillancourt-Jean E, Albarao S et al (2014) Mesenchymal stromal cell ciliogenesis is abrogated in response to tumor necrosis factor-alpha and requires NF-kappaB signaling. Cancer Lett 345:100–105
29. Huang K, Chen A, Zhang X et al (2015) Progranulin is preferentially expressed in patients with psoriasis vulgaris and protects mice from psoriasis-like skin inflammation. Immunology 145:279–287
30. Brand DD, Latham KA, Rosloniec EF (2007) Collagen-induced arthritis. Nat Protoc 2:1269–1275
31. Azizi G, Jadidi-Niaragh F, Mirshafiey A (2013) Th17 cells in immunopathogenesis and treatment of rheumatoid arthritis. Int J Rheum Dis 16:243–253
32. Park KS, Park MJ, Cho ML et al (2009) Type II collagen oral tolerance; mechanism and role in collagen-induced arthritis and rheumatoid arthritis. Mod Rheumatol 19:581–589
33. Mukai T, Gallant R, Ishida S et al (2015) Loss of SH3 domain-binding protein 2 function suppresses bone destruction in in tumor necrosis factor-driven and collagen-induced arthritis in mice. Arthritis Rheumatol 67:656–667
34. Price S (2011) Inflammation: new approach to TNF inhibition. Nat Rev Rheumatol 7:253
35. Monaco C, Nanchahal J, Taylor P et al (2015) Anti-TNF therapy: past, present and future. Int Immunol 27:55–62
36. Asquith DL, Miller AM, McInnes IB et al (2009) Animal models of rheumatoid arthritis. Eur J Immunol 39:2040–2044

37. Tian Q, Zhao Y, Mundra JJ et al (2014) Three TNFR-binding domains of PGRN act independently in inhibition of TNF-alpha binding and activity. Front Biosci (Landmark Ed) 19:1176–1185

38. Yim YK, Lee H, Hong KE et al (2007) Electroacupuncture at acupoint ST36 reduces inflammation and regulates immune activity in collagen-induced arthritic mice. Evid Based Complement Alternat Med 4:51–57

Chapter 21

Methods for Studying the Function of Progranulin in Atherosclerosis Using Both Knockout Mice Models and In Vitro Studies

Tohru Ohama, Ryota Kawase, Yasushi Sakata, and Shizuya Yamashita

Abstract

Progranulin (PGRN) has been reported to be associated with cell proliferation, cell growth, wound healing, and inflammation. PGRN mutations are known to be related to dementia. However, the association between PGRN and atherosclerosis remains to be elucidated. Therefore, we generated $PGRN^{-/-}ApoE^{-/-}$ mice to analyze the effect of PGRN on the development of atherosclerosis.

Key words Progranulin, Atherosclerosis, Inflammation

1 Introduction

A growing body of evidence indicates that cardiovascular diseases are the main contributor to mortality and morbidity. The pathophysiological mechanism of cardiovascular diseases is the development of atherosclerosis, which is characterized by chronic inflammation and excessive accumulation of modified lipids and lipoproteins. We previously reported that progranulin (PGRN) derived from human macrophages binds to apolipoprotein A1 and suppresses its conversion to proinflammatory granulin [1]. PGRN has multiple protein functions including cell proliferation, cell growth, wound healing, and inflammation [2–6]. Mutations in PGRN are also well-known to be associated with frontotemporal dementia [7, 8]. Moreover, PGRN, which is known to have anti-inflammatory action, can be cleaved into granulin, which is also known to be a pro-inflammatory molecule by a variety of proteases

Tohru Ohama and Ryota Kawase contributed equally to this work.

Andrew Bateman et al. (eds.), *Progranulin: Methods and Protocols*, Methods in Molecular Biology, vol. 1806, https://doi.org/10.1007/978-1-4939-8559-3_21,

[9, 10]. However, there is little information on the association between PGRN and atherosclerosis. Therefore, we generated PGRN$^{-/-}$ApoE$^{-/-}$ mice (double knockout, DKO mice) to evaluate the effect of PGRN on the progression of atherosclerosis. Deletion of PGRN exacerbates the progression of atherosclerosis partly through enhancing systemic inflammation. In addition, macrophages in the DKO mice are prone to becoming foam cells as a result of altered intracellular cholesterol homeostasis. Here, we describe methods to use in vitro assays and progranulin knockout mice to analyze the effect of progranulin on atherosclerosis.

2 Materials

2.1 *In Vitro Assay in HUVECs*

1. Human umbilical vein endothelial cells (HUVECs) (Lonza®).
2. Endothelial Growth Medium-2 Bullet Kit (EGM™-2 Medium) (Lonza®).
3. M-199 phenol red-free medium (GIBCO®).
4. Bovine serum albumin (BSA).
5. Tissue culture dish (100 × 20 mm).
6. 6-well plates.
7. Human recombinant PGRN (Otsuka®).
8. RNeasy Mini Kit (Qiagen®).

2.2 *Quantitative Real-Time PCR*

1. cDNA (made with Transcriptor First Strand cDNA Synthesis Kit) (Roche®).
2. Primers: Primer sequences are shown in Table 1.
3. Distilled water (dH_2O).
4. DyNAmo HS SYBR Green qPCR Kits (Thermo Fisher Scientific®).
5. Opticon 2 system (Bio-Rad Laboratories®).
6. 96-well plates.

2.3 *Cholesterol Efflux Assay*

1. Peritoneal macrophages from wild-type mice (C57BLK6/J).
2. 24-well plates.
3. PBS.
4. Dulbecco's modified Eagle medium (DMEM) containing 0.1% BSA.
5. [^{3}H]-cholesterol.
6. Polyethylene glycol (PEG)-HDL (Determiner HDL®: Kyowa Medex®) from ApoE KO mice and DKO mice.
7. 0.1 N NaOH.

Table 1
Primers for quantitative RT-PCR analysis of mouse genes and human genes

Gene name	Accession number	5′–3′ primer sequence (forward) / 5′–3′ primer sequence (reverse)	Amplicon size
		Primers for quantitative RT-PCR analysis of mouse genes	
TNF-α	NM_013693	TGTGCTCAGAGCTTTCAACAAC	101
		GCCCATTTGAGTCCTTGATG	
IL-6	M20571.2	ACAACCACGGCCTTCCCTACTT	129
		CACGATTTCCCAGAGAACATGTG	
IL-1β	M15131.1	TGAAGTTGACGGACCCCAAA	101
		TGATGTGCTGCTGCGAGATT	
MCP-1	NM_011333.3	CCACTCACCTGCTGCTACTCAT	76
		TGGTGATCCTCTTGTAGCTCTCC	
IL-10	NM_010548.2	TTTGAATTCCCTGGGTGAGAA	98
		GCTCCACTGCCTTGCTCTTATT	
ICAM-1	BC008626.1	TTCACACTGAATGCCAGCTC	182
		GTCTGCTGAGACCCCTCTTG	
VCAM-1	BC029823.1	GGAGCCTGTCAGTTTTGAGAATG	106
		TTGGGGAAAGAGTAGATGTCCAC	
SRA	AF203781.1	AGTGTAGGCGGATCAACCC	245
		TCACTTCATTCAGCCATATTGG	
CD36	L23108.1	GAGCAACTGGTGGATGGTTT	204
		GCAGAATCAAGGGAGAGCAC	
ABCA1	AF287263.1	AGGTTTGGAGATGGTTATACAATAGTTG	82
		CTTTTAGGACACTTCCCGGAAA	
ABCG1	AF323659.1	TTCCCCTGGAGATGAGTGTC	124
		CAGTAGGCCACAGGGAACAT	
GAPDH	GU214026.1	ACTCCACTCACGGCAAATTC	171
		TCTCCATGGTGGTGAAGACA	
MTP	NM_001163457.1	CATGTCAGCCATCCTGTTTG	247
		CTCTCGATACCACAGACTGA	
SREBP2	NM_033218	CACCTGTGGAGCAGTCTCAA	155
		TGGTAGGTCTCACCCAGGAG	

(continued)

Table 1 (continued)

Gene name	Accession number	5′–3′ primer sequence (forward) / 5′–3′ primer sequence (reverse)	Amplicon size
ACAT2	NM_009338	TGTCACAGAACAGGGCAGAG	189
		TGACAGTTCCTGTCCCATCA	
HMG-CoA-r	NM_008255	CCTGGGCCCCACATTCA	66
		GACATGGTGCCAACTCCAATC	
LDL-r	NM_010700	TTGGGTTGATTCCAAACTCCAC	58
		CCGATTGCCCCCATTGA	
SREBP1c	NM_010701	TTGGGTTGATTCCAAACTCCAC	59
		CCGATTGCCCCCATTGA	
		Primers for quantitative RT-PCR of human genes	
ICAM-1	NM_000201.2	GGCTGGAGCTGTTTGAGAAC	202
		ACTGTGGGGTTCAACCTCTG	
VCAM-1	NM_001078.3	CAGACAGGAAGTCCCTGGAA	212
		TTCTTGCAGCTTTGTGGATG	
GAPDH	NM_002046.4	GAGTCAACGGATTTGGTCGT	238
		TTGATTTTGGAGGGATCTCG	

8. Liquid scintillation cocktail.
9. Liquid scintillation counter.

2.4 Western Blotting

1. 8–16% precast polyacrylamide gel (SDS-PAGE mini, TEFCO®).
2. Electrophoresis chamber (TEFCO®).
3. Blotting chamber (TEFCO®).
4. PVDF membrane (Immobilon®-P Transfer Membrane).
5. Power supply unit.
6. Running buffer: Dissolve 90.7 g Tris base, 432.6 g glycine, and 30 g SDS in 3 L H_2O to make 10× running buffer. Dilute 1:10 with distilled water to make 1× running buffer.
7. Transfer buffer: Dissolve 87 g Tris base and 435 g glycine in 3 L H_2O to make 10× transfer buffer. Dilute 1:10 with distilled water to make 1× transfer buffer.
8. TBS-T buffer: Dissolve 363 g Tris base and 522 g NaCl in 3 L H_2O, and adjust pH to 7.6 with 1 N HCl to make 20×

TBS. Then mix 500 mL 20× TBS, 10 mL Tween20 and 9490 mL dH_2O to make TBS-T.

9. Methanol.
10. Blocking One (Nacalai Tesque®).
11. Primary antibodies: eNOS (1:1000, #9586; Cell Signaling Technology); phospho-eNOS (Ser1177) (1:1000, #9571; Cell Signaling Technology); β-Actin (13E5) (1:2000; Cell Signaling Technology); Mouse PGRN (1:3000, AF2557; R&D Systems); ApoA-I (1:800, FL267; Santa Cruz Biotechnology, Inc.); platelet-activating factor acetylhydrolase (1:1000,PAF-AH; Santa Cruz Biotechnology, Inc.); mouse serum amyloid A1 (1:1000, SAA1; R&D Systems); monoclonal anti-mouse MTP (1:2500, #612022; BD Transduction Laboratories).
12. Secondary antibodies: ECL anti-rabbit IgG, horseradish peroxidase-linked species-specific whole antibody (from donkey) (NA934V, GE Healthcare), ECL anti-mouse IgG, horseradish peroxidase-linked species-specific whole antibody (from sheep) (NA931V, GE Healthcare), sheep IgG horseradish peroxidase-conjugated antibody (HAF016, R&D Systems), donkey anti-goat IgG-HRP (Sc2020, Santa Cruz Biotechnology, Inc.).
13. Shaker.
14. Filter paper.
15. Chemiluminescence detection system (ECL Select, GE Healthcare®).
16. High-performance chemiluminescence film (GE Healthcare®).

2.5 Quantification of Atherosclerotic Area

2.5.1 Enface Analysis

1. Whole mouse aorta from ApoE KO mice and DKO mice.
2. 4% paraformaldehyde phosphate buffer solution (4% PFA in PBS).
3. Isopropanol.
4. 6-well plates.
5. Oil Red O (Nacalai Tesque®): Prepare a stock solution by dissolving 0.5 g Oil Red O in 100 mL isopropanol. Make Oil Red O working solution by mixing 60 mL Oil Red O stock solution with 40 mL distilled water.
6. Distilled water.
7. Filter paper.
8. 14 mL polypropylene round-bottom tube.
9. Insect pins (100 WHITE No.00: Shiga®).
10. Rubber board.

11. Micro tube rotator.
12. Stereo microscope.
13. Image analysis software: WCIF ImageJ®.

2.5.2 Aortic Root Analysis

1. Mouse aortic root from ApoE KO mice and DKO mice.
2. O.C.T. compound.
3. Cryomold® Intermediate (Tissue-Tek®).
4. Liquid nitrogen.
5. Silanized slides.
6. Research cryostat.
7. Isopropanol.
8. Oil Red O (Nacalai Tesque®).
9. Distilled water.
10. 10% formaldehyde neutral buffer solution (Nacalai Tesque®).
11. Microscope (AX80, Olympus®) with a CCD camera (DP70, Olympus®).
12. CC/Mount (Diagnostic Biosystems®).
13. Mayer's hematoxylin solution (Wako®).

2.6 Uptake of Oxidized LDL in Macrophages

1. Macrophages derived from bone marrow of ApoE KO mice and DKO mice.
2. PBS.
3. Dulbecco's Modified Eagle medium (DMEM).
4. FBS.
5. 6-well plates.
6. Human oxidized LDL.
7. Recombinant Mouse M-CSF (R&D systems®).
8. All-in-one fluorescence microscope (Keyence®).
9. Isopropanol.
10. Oil Red O working solution (*see* Subheading 2.5.1).
11. Red blood cell lysis buffer (Pharm Lyse, BD Biosciences®).
12. 10% formaldehyde neutral buffer solution.
13. Mayer's hematoxylin solution (Wako®).

2.7 Double Immunostaining

1. Frozen sections of sample on glass slides.
2. PBS (0.01 M).
3. Acetone.
4. Kimwipes.
5. Primary antibodies: sheep anti-mouse PGRN antibody, rat anti-mouse macrophage-specific antibody (F4/80).

6. Secondary antibodies: Alexa Fluor® 488-conjugated donkey anti-sheep IgG antibody (Life Technologies), Alexa Fluor® 594-conjugated goat anti-rat IgG antibody (Life Technologies).
7. BSA (bovine serum albumin).
8. Serum for blocking:

   Normal donkey serum (Jackson ImmunoResearch®).

   Normal goat serum (DakoCytomation).
9. DAPI solution (Dojindo Molecular Technologies, Inc.®).
10. Aqueous mounting medium (CC/Mount).
11. Cover glass (Micro cover glass NEO: Matsunami®).
12. Dako Pen (DAKO®).

3 Methods

3.1 In Vitro Assay in HUVEC

1. 2 × 10^5 HUVECs were transplanted to a 10-cm dish and cultured in EGM™-2 medium (*see* **Note 1**).
2. Cells were subcultured into 6-well plates before they reached 80% confluence (2 × 10^5 cells/well).
3. On the next day, confluent HUVECs were starved in M-199 phenol red-free medium containing 1% BSA for 4 h before stimulation.
4. Cells were stimulated with the various doses (0, 10, 50, 100 ng/mL) of human recombinant PGRN (*see* **Note 2**) for 30 min before TNF-α (5 ng/mL) challenge for 4 h.
5. Collect total RNA from HUVECs stimulated with various does of PGRN using RNeasy Mini Kit for quantitative real-time PCR analysis.

3.2 Quantitative Real-Time PCR

3.2.1 Preparation of cDNA

One microgram of total RNA isolated from tissues, macrophages, and HUVEC was primed with 50 pmol of oligo (dT)-20 and reverse-transcribed with SuperScript III (Invitrogen®) for first-strand cDNA synthesis, according to the protocol supplied by the manufacturer.

1. To 1000 ng RNA in 12 μL (diluted with dH_2O), add 1 μL anchored oligo $(dT)_{18}$ primer.
2. Mix and incubate 65 °C 10 min.
3. Place on ice.
4. Add 4 μL Transcriptor RT Reaction Buffer (5×), 2 μL Deoxynucleotide Mix, 0.5 μL Protector RNase Inhibitor, and 0.5 μL Transcriptor Reverse Transcriptase.
5. Incubate at 55 °C for 30 min.

6. Incubate at 85 °C for 5 min.
7. Place on ice.

3.2.2 q-RT PCR

Quantitative real-time PCR was performed using an Opticon 2 system (Bio-Rad Laboratories, Inc.) according to the protocol provided along with the DyNamo HS SYBR Green quantitative PCR kit (Finnzymes Oy). q-RT PCR was performed to analyze the mRNA expression levels of genes related to intracellular cholesterol contents of the liver and inflammation-related genes of the aorta and macrophages from ApoE KO mice and DKO mice.

In each assay (total 20 μL/well):

5 μL cDNA from Subheading 3.2.1 diluted by 50-fold.

1 μL forward primer (5 μM).

1 μL reverse primer (5 μM).

3 μL dH_2O.

10 μL DyNAmo HS SYBR Green.

PCR program:

Step1: Denaturation – incubate at 95 °C for 2 min on initial cycle and for the rest 10 s.

Step 2: Annealing – incubate at 60 °C for 30 s.

Step 3: Extension – incubate at 72 °C for 30 s.

Step 4: Number of cycles (**steps 1–3**) – 40.

Step 5: 4 °C ∞ (forever).

3.2.3 Data Analysis

The relative gene expression was quantified using glyceraldehyde-3-phosphate dehydrogenase (GAPDH) as an internal control.

3.3 Cholesterol Efflux Assay

[Cholesterol efflux assay was performed to investigate the effect of progranulin deficiency on the HDL function.]

1. Isolate the resident peritoneal macrophages from wild-type mice (*see* **Note 3**).
2. Plate 2×10^6 macrophages in a 24-well plate, and culture overnight in DMEM containing 0.1% BSA and 1.0 μCi/mL [^{3}H]-cholesterol.
3. Wash with PBS three times.
4. Incubate in DMEM containing 0.1% BSA with/without HDL (50 μg/mL) (*see* **Note 4**) for 9 h.
5. Collect the medium and determine radioactivity (radioactivity of [^{3}H]-cholesterol in the efflux medium):
   (a) Collect the medium and place into an Eppendorf tube.
   (b) Centrifuge at 15K rpm for 2 min.

(c) Place 250 μL of the supernatant into 5 mL of liquid scintillation cocktail.

(d) Determine radioactivity using a liquid scintillation counter.

6. Collect the macrophages and determine radioactivity (total cell-labeled radioactivity):

   (a) Wash with 1000 μL PBS three times.

   (b) Add 500 μL of 0.1 N NaOH and collect the macrophages.

   (c) Keep at room temperature for 30 min.

   (d) Pipet well and place 250 μL of the supernatant into 5 mL of liquid scintillation cocktail (*see* **Note 5**). Determine radioactivity using a liquid scintillation counter.

7. Calculation

   (a) Cellular cholesterol efflux was expressed as the percentage of radioactivity of [^{3}H]-cholesterol in the efflux medium relative to the total cell-labeled radioactivity.

   (b) HDL-mediated cholesterol efflux was calculated by subtracting the background from the total cellular cholesterol efflux. The background values are the radioactivity in the medium without HDL.

3.4 Western Blotting

We performed Western blotting to analyze the effect of progranulin deficiency on the expressions of eNOS and phosphorylated form of eNOS in the aorta as well as HDL composition. We also checked MTP protein expression levels to clarify the mechanisms relating to the significant lower level of plasma VLDL cholesterol in the DKO mice compared with that in ApoE KO mice.

1. Sample preparation:

   (a) Extract total protein lysates isolated from the aortas or the HDL fractions in lysis buffer. Vortex and centrifuge at 15K rpm for 10 min. Measure protein concentration in the supernatant.

   (b) Adjust the sample protein amount to 18 μL by adding H_2O, followed by the addition of 6× SDS (3.6 μL).

   (c) Incubate each sample at 95 °C for 5 min.

2. SDS-PAGE:

   (a) Apply samples aorta, HDL and liver of ApoE KO mice and DKO mice on 8–16% precast polyacrylamide gel.

   (b) Electrophorese the samples in running buffer at approximately 100 V.

3. Transfer:

   After immersing the membrane in methanol for 5 min and subsequently in water for 5 min, transfer the proteins to a nitrocellulose membrane.

4. Blocking:

   Block the membrane with 10 mL Blocking One at 4 °C overnight.

5. Hybridization with primary antibodies:
   (a) Wash the membrane(s) in about 100 mL TBS-T buffer on shaker three times for 5 min.
   (b) Incubate with each primary antibody in TBS-T Buffer at 4 °C overnight.

6. Hybridization with second antibodies:
   (a) Wash in 100 mL TBS-T buffer on shaker three times for 5 min.
   (b) Incubate with each second antibody in TBS-T buffer for 30 min at room temperature.

7. Chemiluminescence detection:
   (a) Wash in 100 mL TBS-T buffer three times for 5 min.
   (b) Develop the membrane with ECL according to the procedure provided in the ECL detection kit.

3.5 Quantification of Atherosclerotic Plaque Area

3.5.1 Enface Analysis

1. Extract whole aortas from PGRN$^{-/-}$ApoE$^{-/-}$ (DKO) and ApoE KO mice. Place the aortas on filter paper.
2. Soak the samples in 4% PFA in a polypropylene round-bottom tube.
3. Rotate the samples for 24 h at 4 °C with a micro tube rotator.
4. Peel off gently as much fat around aortas as possible under a stereo microscope (*see* **Note 6**).
5. Cut the aorta longitudinally with a pair of small scissors (*see* **Note 7**) and transfer the cut sections to 6-well plates.
6. Wash samples in 8 mL distilled water for 30 s.
7. Soak samples in 8 mL 60% isopropanol for 1 min.
8. Incubate with 8 mL Oil Red O working solution in 6-well plates at 4 °C on a shaker for 24 h.
9. Wash samples in 8 mL 60% isopropanol for 2 min.
10. Wash samples in 8 mL distilled water for 30 s (*see* **Note 8**).
11. Attach aortas to rubber board with insect pins under stereo microscope.
12. Take pictures of the plaques in the aortas under stereo microscope.
13. After scanning the pictures, determine plaque area relative to the total aortic area for each sample using WCIF ImageJ® software.

3.5.2 Analysis for Aortic Root

1. Extract the upper half of the heart including the aortic root from both the PGRN$^{-/-}$ApoE$^{-/-}$ (DKO) and ApoE KO mice.

2. Embed the samples in O.C.T. compound and Cryomold® as soon as possible.
3. Freeze the samples in liquid nitrogen.
4. Prepare 10-μm-thick frozen cross sections of the aortic sinus using a research cryostat.
5. Place the sections on silanized slides (*see* **Note 9**).
6. Air-dry the slides for 30 min.
7. Fix with 10% formaldehyde neutral buffer solution for 10 min on ice.
8. Wash with dH_2O twice for 5 min.
9. Stain with Oil Red O working solution for 10 min.
10. Wash with 60% isopropanol for 5 min.
11. Wash in running tap water for 10 min.
12. Counterstain with Mayer's hematoxylin solution for 3 min.
13. Wash in running tap water for 15 min.
14. Mount with aqueous mounting media (CC/Mount).
15. Take pictures under a microscope with a CCD camera.
16. Calculate the plaque area in the aortic wall per section using WCIF ImageJ® software.

3.6 Uptake of Oxidized LDL in Macrophages

1. Isolation of bone marrow-derived cells from both the PGRN$^{-/-}$ApoE$^{-/-}$ (DKO) and ApoE KO mice:
   (a) Suspend bone-marrow-derived cells isolated from the femur and tibia in 10 mL DMEM containing 10% FBS.
   (b) After treatment with 10 mL DMEM containing 300 μL red blood cell lysis buffer, culture 10^6 cells with 3 mL DMEM containing 10% FBS and 5 ng/mL M-CSF in a 6-well plate for 10 days.
2. Incubate macrophages collected from DKO and ApoE KO mice with human oxidized LDL (50 μg/mL) at 37 °C for 20 h.
3. Aspirate medium and wash twice with 4 mL PBS.
4. Fix with 10% formaldehyde neutral buffer solution for 10 min.
5. Wash with PBS for 1 min.
6. Wash with 60% isopropanol for 15 s.
7. Stain with Oil Red O for 1 min at 37 °C under cover by aluminum foil.
8. Wash with 60% isopropanol for 15 s.
9. Wash with PBS three times for 3 min.
10. Stain with hematoxylin solution for 3 min.
11. Wash with PBS three times for 3 min.

12. Take pictures using an all-in-one fluorescence microscope (Keyence®).
13. Count the number of Oil Red O-positive foam cells in the DKO and ApoE KO samples.

3.7 Double Immunostaining

The following are the methods for double staining of mouse PGRN antibody and F4/80:

ApoE KO mice [please indicate the mice used for the experiment] were perfused with 10 mL PBS containing heparin via the left ventricle under anesthesia. After cutting the heart, the aortic sinus was embedded and frozen in OCT compound.

1. Air-dry 10-μm thickness Cryostat sections for 30 min.
2. Fix samples in cold acetone for 10 min.
3. Wash with PBS two times for 5 min.
4. After removing excess water, draw a circle around the samples with a Dako Pen.
5. Blocking: Block nonspecific staining using 10% non-immune serum from the host species of a secondary antibody. Add 50 μL of donkey serum (diluted 1:50 by PBS) for 30 min in a moist chamber.
6. After removing the blocking solution, add 100 μL sheep anti-mouse PGRN antibody (diluted 1:100 by PBS containing 1% BSA), and incubate for 60 min in a moist chamber.
7. Wash with PBS three times for 5 min.

All the following procedures must be performed under cover by aluminum foil:

8. After removing excess water, dilute fluorescence-labeled secondary antibodies, Alexa Fluor® 488-conjugated donkey anti-sheep IgG antibody 1:200 by PBS containing 1% BSA, and add 50 μL to sample slide(s).
9. Wash with PBS three times for 5 min.
10. Incubate with 50 μL blocking goat serum solution (diluted by 50-fold with PBS) for 30 min in a moist chamber.
11. After removing the blocking solution, incubate with rat anti-mouse macrophage-specific antibody (F4/80) (diluted 1:100 by PBS containing 1% BSA) for 60 min in a moist chamber.
12. Wash with PBS three times for 5 min.
13. Incubate with 50 μL Alexa Fluor® 594-conjugated goat anti-rat IgG antibody (diluted 1:200 by PBS containing 1% BSA) for 30 min in a moist chamber.
14. Incubate with DAPI (diluted 5000-fold) for 2 min.
15. Wash with PBS three times for 5 min.

16. After adding two drops of CC/Mount, cover with cover glass.
17. Photograph the slide with a fluorescence microscope equipped with CCD camera.

4 Notes

1. HUVECs were used within the first six passages.
2. The PGRN concentration to be added was determined according to the physiological concentration, which ranged from 33.0 to 80.8 ng/mL in the data sheet from the Human Progranulin Quantikine ELISA Kit (DPGRN0, R&D).
3. It might be better to use macrophages that have not been stimulated with thioglycollate. Treatment with thioglycollate might interfere with the qPCR results in genes related to inflammation in macrophages.
4. PGRN is easily detached from HDL by ultracentrifugation; therefore, HDL was used by PEG precipitation. We performed PEG precipitation to make HDL according to the procedures supplied by the kits. Briefly, the same amount of reagents as serum was added and mixed with a vortex. After centrifugation at 1500 × *g* for 10 min, the supernatant was collected as an HDL fraction.
5. The recovery is sticky; therefore, it is important to pipet carefully and not stir vigorously.
6. After soaking the sample in PFA overnight, fat around the aorta can be easily detached. This step is very important to evaluate the development of atherosclerosis.
7. The aorta should be longitudinally cut with sharp scissors under the stereo microscope.
8. Oil Red O outside the aorta should be removed as much as possible.
9. Before staining with Oil Red O, check that the cross section contains aortic valve.

References

1. Okura H, Yamashita S, Ohama T et al (2010) HDL/apolipoprotein A-I binds to macrophage-derived progranulin and suppresses its conversion into proinflammatory granulins. J Atheroscler Thromb 17:568–577
2. He Z, Ong CH, Halper J et al (2003) Progranulin is a mediator of the wound response. Nat Med 9:225–229
3. He Z, Bateman A (2003) Progranulin (granulin–epithelin precursor, PC-cell-derived growth factor, acrogranin) mediates tissue repair and tumorigenesis. J Mol Med 81:600–612
4. Eriksen JL, Mackenzie IR (2008) Progranulin: normal function and role in neurodegeneration. J Neurochem 104:287–297

5. Toh H, Chitramuthu BP, Bennett HP et al (2011) Structure, function, and mechanism of progranulin: the brain and beyond. J Mol Neurosci 45:538–548
6. De Muynck L, Van Damme P (2011) Cellular effects of progranulin in health and disease. J Mol Neurosci 45:549–560
7. Baker M, Mackenzie IR, Pickering-Brown SM et al (2006) Mutations in progranulin cause tau-negative frontotemporal dementia linked to chromosome 17. Nature 442:916–919
8. Cruts M, Gijselinck I, van der Zee J et al (2006) Null mutations in progranulin cause ubiquitin-positive frontotemporal dementia linked to chromosome 17q21. Nature 442:920–924
9. Zhu J, Nathan C, Jin W et al (2002) Conversion of proepithelin to epithelins: roles of SLPI and elastase in host defense and wound repair. Cell 111:867–878
10. Kessenbrock K, Fröhlich L, Sixt M et al (2008) Proteinase 3 and neutrophil elastase enhance inflammation in mice by inactivating anti-inflammatory progranulin. J Clin Invest 118:2438–2447

Chapter 22

Methods to Investigate the Roles of Progranulin in Angiogenesis Using In Vitro Strategies and Transgenic Mouse Models

Huishi Toh, Eugene Daniels, and Andrew Bateman

Abstract

Endothelial cells activate the expression of progranulin during angiogenesis. Here we discuss methods to investigate progranulin activity on endothelial cells in vitro and on aortic explants. We then discuss methods to generate transgenic mice in which progranulin expression is targeted to endothelial cells. These mice can be used to study the influence of progranulin on angiogenesis during development in vivo. The transgenic strategy summarized here could be readily adapted to investigate the roles of progranulin in other cell types and tissues by use of appropriate targeting vectors to drive the expression of progranulin in the cell type of choice.

Key words Endothelial cells, Migration, Tube formation, Matrigel, Aortic rings, Tie2 promoter, Immunohistochemistry

1 Introduction

Angiogenesis is the outgrowth of new blood vessels from existing vessels [1]. In the adult, it is intimately associated with tissue remodeling, for example, in wound healing [2, 3] in the remodeling of the mammary gland [4], placenta [5], and endometrium [6] that takes place during pregnancy and menstruation and in complications of diabetes such as microvascular disease, retinopathy, and renal failure [7]. Angiogenesis is required by solid tumors to provide a blood supply to the cancer cells [1].

Angiogenesis is regulated by highly specialized vascular growth factors such as the vascular endothelial growth factors (VEGFs) and the angiopoietins [1], but it is also regulated by less specialized regulatory proteins that act on a broad cross section of cell types in addition to vascular cells. These include soluble factors such as platelet-derived growth factor [8] and fibroblast growth factor-2 (FGF2) [9] as well as matrix adhesion proteins including certain

Andrew Bateman et al. (eds.), *Progranulin: Methods and Protocols*, Methods in Molecular Biology, vol. 1806, https://doi.org/10.1007/978-1-4939-8559-3_22, © Springer Science+Business Media, LLC, part of Springer Nature 2018

integrins [10, 11]. There is increasing evidence that progranulin modulates angiogenesis and other aspects of vascular biology. Quiescent endothelial cells express little or no progranulin [12], but endothelial cells in tissues that are undergoing angiogenesis often display elevated progranulin expression. This is evident, for example, in wounds [13], the embryo [14], or the developing placenta [15]. Similarly, tumor blood vessels sometimes possess strongly progranulin-positive endothelial cells [16–21]. This may be clinically relevant. For example, in patients with glioblastoma, survival is poorer than otherwise if progranulin levels are high in the tumor vasculature [20]. Progranulin influences angiogenesis in at least two ways. It increases the secretion of VEGF from cancer cells [22], but it also acts directly on isolated endothelial cells to, for example, enhance their proliferation, migration, and tube formation in vitro [13]. Expression of a progranulin transgene (*Tie2-Pgrn*) by murine endothelial cells significantly influences angiogenesis during development [23]. The vasculature of progranulin knockout mice is grossly normal, but they develop more severe atherosclerosis than their wild-type siblings [24]. Detailed analysis of the cerebral vascular microstructure from progranulin knockout mice revealed that the blood-brain barrier is compromised and when challenged by cerebral ischemia and reperfusion, the resulting stroke-related edema caused by vascular leaking is far more pronounced than in wild-type mice [25].

The vascular biology of progranulin is an emerging area of research. Here we will first outline methods that have been used to investigate the activity of progranulin on the angiogenic behavior of endothelial cells first using cell culture methods in vitro, then by using explant assays, and finally by transgenic expression of progranulin in mouse endothelial cells in vivo.

1.1 Cell Culture Methods

There are many endothelial cell sources available according to the investigators specific experimental needs. These can be categorized as (1) commercial human primary cultures that will be exemplified here by HMVEC cells. The commercial cells are high quality and are human in origin, but they are expensive especially if larger numbers are required. (2) Cell lines represented here by the EJG cell line that can be propagated to give large numbers of cells. (3) Primary endothelial cell cultures that can be prepared in the investigator's laboratory. The method we discuss for making primary endothelial cell cultures does not require specialized manipulations and should work in any laboratory with experience in cell culture. It is better if possible to use at least two independent cell sources to perform the cell culture experiments as this gives confidence that the results obtained will be representative of the behavior of endothelial cells and not specific to a given cell line. We will discuss two assays, first the migration of endothelial cells through collagen matrices since angiogenesis is associated with the movement of endothelial cell through biological matrices. We will then discuss the formation of tube-like structures by endothelial cells that are growing on a Matrigel matrix. Matrigel is a commercial preparation of basement membrane proteins. The formation of tube-like structures by endothelial cells confirms that the test factor has endothelial morphogenic activity consistent with angiogenesis [26]. Some cancer cells also exhibit tube-forming morphogenic responses in Matrigel [27] which may be related to the phenomenon of vasculogenic mimicry [28].

1.2 Explant Methods

We will discuss a method that employs mouse aortic ring explant assays to investigate the ability of progranulin to support angiogenic sprouting. In this ex vivo assay, the structure of the vascular tissue with its endothelial lining and vessel wall is retained while permitting the controlled administration of test substances such as progranulin to the assay system. The aortic ring assay method was first developed for use with rat aortic rings which will undergo angiogenic sprouting in serum-free medium [29]. Mouse aortic rings sprout less readily unless the culture medium is supplemented with autologous serum, although methods for serum-free sprouting of mouse aortic rings have been developed. The method we discuss does not use autologous mouse serum medium supplements. It shows low spontaneous sprouting but is responsive to progranulin.

1.3 A Transgenic Strategy to Investigate the Role of Progranulin on Vessel Development In Vivo

We will discuss a strategy based on the transgenic expression of progranulin in endothelial cells that uses the endothelial-specific promoters (*Tie2*) to target progranulin expression to murine endothelia [23]. The transgenic strategy probes how endothelial progranulin impacts the formation and growth of the vasculature in an otherwise unperturbed animal and how, if at all, progranulin modifies interactions between endothelial cells and other vessel wall components such as vascular smooth muscle cells.

The expression of the progranulin transgenes in murine endothelial cells was found to influence blood vessel development and structure [23]. In *Tie2-Pgrn* embryos from 10-day-post conception onward, and in neonates, blood vessels are often enlarged, the basement membrane is poorly formed, and there is a reduction in the coverage of vessels by pericytes and vascular smooth muscle cells [23]. Many *Tie2-Pgrn* mice die from vascular leakage. Mortality is the highest near or during birth [23]. The *Tie2-Pgrn* vessels are reminiscent of the tortuous, thin-walled capillaries seen during tumor angiogenesis [23]. In wild-type mice, progranulin expression by endothelial cells is normally off and is induced only during active angiogenesis. In the *Tie2-Pgrn* mice, progranulin expression is always on, and the blood vessels exhibit a phenotype consistent with the persistent activation of the early stages of angiogenesis [23]. The *Tie2-Pgrn* mice reveal the importance of the on/off regulation of progranulin expression in endothelial cells.

It should be noted that the transgenic approach summarized here can be generalized to study any aspect of progranulin biology by the use of other targeting vectors to drive the expression of progranulin in any cell type of choice.

2 Materials

2.1 Animals

All experiments involving animals were subject to institutional regulatory approval. The transgenic mice are maintained in a specific pathogen-free (SPF) barrier facility.

1. B6C3F1 mice (The Jackson Laboratory) for microinjection use, female.
2. Use C57/BL6, male and female (The Jackson Laboratory), for aortic ring assays and backcrossing of *Tie2-Pgrn* founder lines.
3. CD-1 rats, male or female, 12 weeks (The Jackson Laboratory) to isolate RE cells.

2.2 In Vitro Cell Culture Assays

2.2.1 General Laboratory Equipment for Cell Culture and Explant Assays

1. A level II laminar flow biological safety cabinet.
2. Humidified temperature-controlled incubator with a constant atmosphere of 5% CO_2.
3. An inverted microscope preferably with camera attachment.
4. A standard microscope for cell counting.
5. Hemocytometer for cell counting.
6. Individually wrapped sterile disposable plastic pipettes, 10 and 2 ml volume.
7. Battery-operated or plug-in pipette aid.
8. T75 tissue culture flasks surface treated for adherent cells, with angled necks and filter caps.
9. 50 ml sterile polypropylene tubes.
10. A vacuum aspirator, either with a mechanical pump or a water vacuum.
11. Variable volume pipettors and sterile pipette tips.

2.2.2 Culturing and Maintaining Endothelial Cells

1. Human dermal microvascular endothelial cells (HMVEC-D, Lonza).
2. Bovine adrenal medullary endothelial cells (EJG cells, ATCC® CRL-8659).
3. Rat cardiac microvascular endothelial cells (RE) that are prepared in the laboratory (methods for their preparation are given below).
4. 70% ethanol.

5. 1 and 10 ml graduated sterile disposable pipettes, individually wrapped, with cotton wool plugs at top.
6. Portable Pipet-Aid (Drummond) to fill the pipettes.
7. T75 75 cm^2 tissue culture flasks, vented caps (Corning Life Sciences).
8. Trypin-EDTA 0.25% with phenol red (Fisher Scientific).
9. Eagle's minimum essential medium (MEM).
10. Penicillin-streptomycin solution (5000 U/ml penicillin, 5 mg/ml streptomycin).
11. L-Glutamine solution liquid, 200 mM.
12. Heat inactivated fetal bovine serum (FBS).
13. Trypan blue (0.4%) (Sigma-Aldrich).

2.2.3 Preparation of Rat Cardiac Endothelial RE Cells

This requires access to a fluorescence microscope.

1. Young CD1 rats, either sex.
2. 50 ml glass beaker with magnetic stirrer, sterilized by autoclaving.
3. Glass or plastic funnel (60 mm diameter, short wide stem, 10 mm diameter), lined with cotton gauze at least four single layers thick. Wrap in aluminum foil and autoclave.
4. Sterile 50 ml polypropylene centrifuge tubes with conical bottoms.
5. 70% ethanol.
6. Collagenase (Worthington, type 2).
7. Eagle's minimum essential medium (MEM).
8. Heat inactivated fetal bovine serum (FBS).
9. Rabbit anti-factor VIII antibody (Sigma-Aldrich).
10. Fluorescein-labeled goat anti-rabbit IgG antibody (Sigma-Aldrich).
11. Sterile phosphate-buffered saline (PBS) pH 7.4.

2.2.4 Reagents for the Transwell Cell Migration Assay

A standard microscope is required.

1. 24-well tissue culture plates (Nunc).
2. 24-well Transwell inserts (Corning Costar).
3. Human type I collagen (Life Technology, Rockville, MD).
4. Progranulin (R&D Systems).
5. Paraformaldehyde powder (Fisher).
6. Giemsa stain (Fisher).

2.2.5 Matrigel Tube Formation Assay

This assay requires an inverted tissue culture microscope equipped with a camera.

1. Reduced growth factor Matrigel (BD Bioscience).
2. HMVEC cells (*see* above).
3. 24-well culture plates.
4. Endothelial growth medium (EGM, Clonetics, Lonza).
5. Heat inactivated fetal bovine serum (FBS).
6. Progranulin (*see* above).
7. Dimethyl sulfoxide, DMSO (optional).
8. Mitogen-activated protein kinase inhibitor PD98059, store as a 100 mM stock solution in DMSO (optional).
9. AKT/protein kinase B inhibitor wortmannin (optional).
10. P38 mitogen-activated protein kinase inhibitor SB203580, store as a 2.6 mM stock solution in DMSO (optional).

2.3 Aortic Ring Assay

All manipulations involving cells are performed in a sterile laminar flow biological safety cabinet.

1. Mice, C57Bl/6 at around 12 weeks, either sex.
2. A dissecting microscope.
3. An inverted tissue culture microscope equipped with a camera.
4. Fine dissecting forceps and scissors.
5. Collagen R (Serva 47254). 2 mg/ml rat collagen solution in 0.1% acetic acid.
6. $NaHCO_3$ and NaOH.
7. Endothelial basal medium EBM2 (Clonetics, Lonza).
8. Penicillin-streptomycin solution, 5000 U/ml (optional).
9. Supplements for endothelial basal medium, EGM-2 (Clonetics, Lonza) comprising SingleQuots (10 ml FBS, 0.2 ml FGF2, 0.5 ml VEGF, 0.5 ml R3-IGF-1, 0.5 ml ascorbic acid, 0.5 ml hEGF, 0.5 ml GA-1000 (gentamycin amphotericin B), 0.5 ml heparin).
10. Sterile 4-well culture plates.

2.4 Generation and Analysis of Tie2-Pgrn Mice that Express a Progranulin Transgene Targeted to Endothelial Cells

2.4.1 General Laboratory Reagents Required

1. Tris (tris(hydroxymethyl)aminomethane).
2. EDTA (ethylenediaminetetraacetic acid).
3. Glacial acetic acid.
4. HCl.
5. Phenol.
6. Chloroform.

2.4.2 Plasmids Used to Generate the Tie2-Pgrn Targeting Vector

1. The *Tie2-vector.* The 6.8 kb *Tie2-promoter/enhancer* vector, *pSPTg.T2FpAXK* is as reported in [30] (obtained from Montreal Clinical Research Institute, originally from Dr. T. Sato, Texas Southwestern University at Dallas). The parental vector is an ampicillin-resistant pSPTg.F (vector size is about 2.4 kb) and contains as DNA inserts, the *Tie2* promoter (2 kb), polyA signal (0.3 kb) and *Tie2* enhancer (1.6 kb *XhoI-KpnI* fragment). The unique cloning sites are *HindIII* and *NotI.*
2. The murine *pgrn* (progranulin) insert was derived from RIKEN clone 21900053G23 which contains the full-length mouse *pgrn* cDNA (1.769 kb) in a modified bluescript2 vector backbone. It is ampicillin resistant, and the cloning sites are *SalI/XhoI* and *BamHI.* It is available from Source Bioscience. The insert was sequenced before use using the primers; PgrnFwdSeq, mPgrnRvsSeq, and mPgrnInternalSeq (*see* below). A *XhoI/BamHI* double restriction digestion releases the mouse progranulin cDNA.

2.4.3 Oligonucleotide Primers Used

Several oligonucleotide primers are required at different stages of this project. The structure and key features of the primers are summarized here. Primers were obtained from AlphaDNA (Montreal, QC):

1. Sequencing primers for mouse *progranulin:*

   Forward primer: mPgrnFwdSeq: TCTGCTGCTCAGACAA CCAGCATT

   Reverse primer: mPgrnRvsSeq: GAACAGCAGATGGCATT GGGCATT

   Internal primer: mPgrnInternalSeq: TCCTGTGACCTGGT TCAC
2. Primer sequences for adding *HindIII* and *NotI* overhangs onto murine pgrn DNA insert for sticky-end ligation into the cloning site of the *Tie2-promoter/enhancer* vector, *pSPTg.T2FpAXK:*

   HindIIIFwd: ACCCAAGCTT**ATG**TGGGTCCTGATGA GCTG

   The *HindIII* site is underlined; the progranulin start codon is in bold font.

   NotIRvs: ATAAGAATGCGGCCGC**TTA**CAGTAGCGGT CTTGGGA

   The *NotI* site is underlined; the progranulin stop codon is in bold font.
3. Primer sequences for qRT-PCR estimation of *Tie2-Pgrn* transgene copy number:

   Forward primer: *Tie2-Pgrn* Fwd, TCTTCTGGACACATG GCCTA. (The forward primer spans the exon 2–3 junction of progranulin).

Reverse primer: *Tie2-Pgrn* Rvs: ATCTGACATCTGGAA GCAGGAT. (The reverse primer spans the exon 4–5 junction of progranulin).

The mouse *β-actin* gene primers used to normalize the reactions.

mActin Fwd: TGTGATGGTGGGAATGGGTCAGAA

mActin Rvs: TGTGGTGCCAGATCTTCTCCATGT

4. Primers used for genotyping:

   For genotyping transgene-specific primers are used:

   Genotyping forward primer: AGCGGGAAGTCGCAAAGT TGTGAGTTGT

   Genotyping reverse primer 1: CTTGCCAAGAAGGTTGAG AAGCCATAC

   Genotyping reverse primer 2: AGAAATGCCCTTCTCCAC ACTCCA

   Both the reverse primers are paired with the same forward primer. The forward primer anneals within the *Tie2* promoter region of the *Tie2-Pgrn* construct, whereas the reverse primers are located within the full-length mouse *Pgrn* cDNA. Forward primer with reverse primer 1 yields a PCR product of 2.1 kb, and forward primer with reverse primer 2 yields a PCR product of 1.6 kb.

   Mouse β-actin is used as the housekeeping control using primers described above.

5. Primers for PCR of *Tie2-Pgrn* mRNA and endogenous *Pgrn* mRNA:

   To selectively detect *Tie2-Pgrn* mRNA expression and exclude detection of endogenous *Pgrn* expression, we use the following primers:

   Tie2-Pgrn-forward primer: TTGCGAAAGAAGATTCCTCG CTGG (located close to 3′-UTR region, exon 13).

   Tie2-Pgrn-reverse primer: ACCTCTACAGATGTGATATGG CTG (from the polyA region on the transgene construct).

   To specifically detect endogenous progranulin mRNA, we use a primer pair close to 3′-UTR in the end of exon 13:

   Pgrn-forward primer: TCTTCCGGTTTCTGTGGACCTTGT

   Pgrn-reverse primer: TTATTGGAGCAACACACGCACACG

   Mouse β-actin is used as the housekeeping control using primers described above.

2.4.4 PCR Amplification of Tie2-Pgrn Transgene and Preparation for Ligation

1. 0.2 ml PCR tubes, free from human DNA, RNases, and DNases (Thermo Scientific, Fisher).
2. Diethyl pyrocarbonate (DEPC, Sigma): Add 1 ml of DEPC in 1 L of water, incubate at 50 °C overnight, autoclave for

20 min, and cool to room temperature. DEPC is toxic. Avoid contact with the skin and eyes. Use safety glasses, gloves, and mask.

3. HindIIIFwd and NotIRvs PCR primers (*see* above).
4. 10× PCR buffer, 25 mM $MgCl_2$, deoxynucleoside triphosphates (NTP) (Invitrogen).
5. Platinum Taq DNA polymerase high fidelity (Invitrogen).
6. UltraPure™Agarose (Fisher).
7. Ethidium bromide (Fisher, toxic, handle with care, dispose as toxic chemical waste).
8. Loading buffer, bromophenol blue, 0.25% xylene cyanol FF, 30% glycerol (Fisher).
9. 1 kb DNA ladder (Fisher)
10. Restriction enzymes *HindIII* and *NotI* (New England Biolabs).
11. Zymoclean gel DNA recovery kit (Zymo Research).
12. PCR thermal cycler.

2.4.5 Ligation of HindIII-Progranulin-NotI PCR Amplification Product into the Tie2 Vector

1. The purified *HindIII-progranulin-NotI* insert.
2. The *Tie2-promoter/enhancer* vector, *pSPTg.T2FpAXK*.
3. Restriction enzymes *HindIII*, *NotI*, *SalI*, and *BamHI*.
4. Bovine serum albumin, BSA, 10 μg/μl (NEB).
5. 10× NEBuffer2 (50 mM NaCl, 10 mM Tris–HCl, 10 mM $MgCl_2$, 1 mM dithiothreitol, pH 7.9 (New England Biolabs).
6. Shrimp alkaline phosphatase (Roche).
7. Dephosphorylation buffer (supplied with shrimp alkaline phosphatase).
8. 10× ligase buffer (New England Biolabs).
9. 1 μl of T4 DNA ligase (New England Biolabs).
10. *E. coli* XTRACells-A (Bio S&T, Montreal).
11. LB Agar.
12. Ampicillin.
13. LB broth.
14. QIAprep Spin Miniprep Kit (Qiagen).

2.4.6 Production of Tie2-Pgrn Founder Mice

The founder *Tie2-Pgrn* transgenic mice were generated by microinjection of pseudopregnant eggs in a dedicated mouse genetics facility. The reagents listed here are only those used to isolate the *Tie2-Pgrn* insert prior to microinjection. Consult with your mouse genetic facility to confirm the purification steps and analysis they require.

1. The *Tie2-Pgrn* vector.
2. Restriction enzyme *SalI* NEB.
3. Agarose, ethidium bromide as in Subheading 2.4.4.
4. Zymoclean gel DNA recovery kit (Zymo Research).
5. Elutip column (Schleicher & Schuell).
6. B6C3F1 mice (*see* Subheading 2.1 above).

2.4.7 Genotyping

1. Roche LightCycler for quantitative real-time PCR.
2. LightCycler® 480 SYBR Green I Master.
3. Tail snips or embryo yolk sacs (can be frozen in liquid N_2 or used fresh).
4. 1.5 ml autoclaved Eppendorf microcentrifuge tubes.
5. Tail lysis buffer: 10 mM Tris, pH 8.0, 100 mM NaCl, 10 mM EDTA, 0.5% SDS.
6. Proteinase K (Sigma-Aldrich). Make 50× stock solution, 20 mg/ml in water.
7. Phenol chloroform (Fisher).
8. Tris-EDTA (TE) buffer, pH 8.0. To make a 10 mM Tris–Cl, 1 mM EDTA, pH 8.0 buffer, mix 10 ml 1 M Tris–Cl pH 7.5 per L and 2 ml 0.5 M EDTA pH 8.0. Check the pH and adjust as necessary. Use DEPC-treated water. Regents from Fisher.
9. For PCR use the following primers (*see* above for details): genotyping forward primer, genotyping reverse primer 1, and genotyping reverse primer 2. Mouse *β-actin* is used as the housekeeper gene to control for recovery and gel loading and to ensure the PCR reaction works.
10. Reagents for PCR amplification from the genomic DNA are as given in Subheading 2.4.4.

2.4.8 Quantification of Tie2-Pgrn Expression in Tissues

1. RNA extraction kit such as RNeasy Mini Kit (Qiagen).
2. RNA extraction kit such as RNeasy Fibrous Tissue Mini Kit (Qiagen).
3. RT-PCR as Invitrogen's SuperScript III First-Strand Synthesis System for RT-PCR.
4. Reagents for qRT-PCR as above.

2.5 Materials and Reagents for Immunohistochemistry

1. Neutral buffered formalin.
2. Slides and coverslips.
3. Microwave oven.
4. Glass slide holder.
5. Metal slide tray.

6. Humidified chamber. For small numbers of slides, a plastic box with a sealable lid can be used, place absorbent paper in the bottom of the box, saturate with water, and stand the slides in a tray above the paper.
7. Platform rocker.
8. CitriSolv™ Hybrid Solvent and Clearing Agent (Fisher).
9. Ethanol.
10. Antibodies: Anti-mouse progranulin antibody from R&D systems, anti-laminin antibody purchased Abcam, anti-smooth muscle α-actin antibody, and anti-desmin antibody were purchased from Dako; the anti-PECAM1 antibody was purchased from Santa Cruz.
11. Hydrophobic pen 5 mm tip width (Dako).
12. Dual endogenous enzyme block (Dako).
13. Biotinylated link universal solution (Dako).
14. Streptavidin HRP (Dako).
15. DAB substrate (3,3′-diaminobenzidine). Dako liquid DAB+ substrate chromogen system.

3 Methods

3.1 Endothelial Cell Culture (See Note 1)

It is, in general, better to use at least two independent cell lines to perform the cell culture experiments as this gives more confidence that the results obtained will be representative of the behavior of endothelial cells and not specific to a given cell line. Here, we have chosen a commercial primary human cell line (HMVEC), primary rat cardiac endothelial cells (RE), and a bovine cell line (EJG). The rationale for selecting these cells is explained in the Introduction (Sub-section 1.1, Cell Culture Methods).

All experiments involving cells must be conducted in a laminar flow biological safety cabinet using sterile equipment and maintain aseptic technique at all times:

1. Carefully clean the work space of the laminar flow biosafety hood by wiping with 70% ethanol.
2. All liquid manipulations are performed using sterile disposable pipettes and a battery-operated Pipet-Aid to fill or empty the pipettes.
3. HMVEC-D cells are grown in endothelial growth medium (Clonetics EGM™, Endothelial Growth Medium, Lonza) with 5% fetal bovine serum (FBS). Optionally the medium can be supplemented with penicillin and streptomycin to a final concentration of 100 IU penicillin and 100 μg/ml of streptomy-

cin if there are concerns about risk of bacterial contamination of the cultures.

4. EJG and RE cells are grown in Eagle's MEM supplemented with 10% FBS. Add L-glutamine (1ml per 100 ml medium). Optionally the medium can be supplemented with penicillin and streptomycin as above.
5. Stock cultures are maintained in T-75 75 cm^2 plastic tissue culture flasks with vented caps in a humidified incubator with 5% CO_2 atmosphere. Change the medium twice a week. Split the cells by trypsinization before the cells become confluent.

3.1.1 Subculturing and Cell Preparation Prior to Experimentation

1. Remove the medium from the T75-flask using a 10 ml disposable pipette or vacuum aspirator; wash the cells in 2 ml serum-free Eagle's MEM (or sterile PBS). Remove wash solution, and add 1.5 ml 0.25% Trypsin-EDTA ensuring that the cells are fully covered.
2. Remove the Trypsin-EDTA after no more than 30 s. This is to inactivate residual trypsin inhibitors from the FBS that may not have been removed in the first wash.
3. Immediately add 2–3 ml 0.25% Trypsin-EDTA and incubate at 37 °C for 3 min. Check for cell detachment by visual inspection using an inverted microscope.
4. If cells are not detaching gently, knock the side of the culture flask two or three times with the palm of your (gloved) hand. Place in 37 °C incubator for 2 min, and inspect for detachment using an inverted microscope. Continue to incubate until all cells are detached. If necessary, gently disperse any remaining cell clumps by pipetting up and down in a 2 ml sterile pipette between 5 and 10 times.
5. Remove the Trypsin-EDTA cell suspension using a 1 ml pipette, and place in a sterile 15 ml tube, and spin at 200 g for 5 min (*see* **Note 2**). A cell pellet should be visible after the suspension has been centrifuged.
6. Remove Trypsin-EDTA from tube and discard. Replace with 1 ml EBM or Eagle's MEM supplemented with 10% serum. Resuspend the pelleted cells. To disperse cells into new T75 flasks, bring the volume up to 10 ml and aliquot 5 ml of the cell suspension into each of two T-75 flasks, and bring the volume to 20 ml with medium with 10% FBS.
7. For cells that are to be used directly in a bioassay after trypsinization, it is necessary to obtain a cell count. Keep the volume of the resuspended cells at 1–2 ml. Mix gently by up and down pipetting with a 1 ml sterile pipette. Remove a 10 μl aliquot into a microfuge tube or similar container, add 10 μl trypan Blue, and mix well by pipetting up and down. Apply 5 μl to a hemocytometer, gently remove excess with a tissue, and count

living cells in the hemocytometer gratings under a microscope. Living cells will exclude the trypan blue dye. Dead cells will take up the dye and appear dark. No more than 15% of the cells should be dead. Do not continue with the bioassay if many dead cells are present. Calculate total cell number accordingly, and dilute the cell in medium supplemented with FBA to provide the required number of cells.

3.1.2 Preparation of RE Cells

The RE cells are prepared as follows [31].

1. Sacrifice CD-1 rats. Surface sterilize the sacrificed animals by thoroughly wetting the fur in 70% ethanol. Open the ventral midline with scissors, and remove the hearts.
2. Perfuse the hearts with ice-cold serum-free Eagle's MEM to remove the blood. Keep the hearts in ice-cold serum-free Eagle's MEM.
3. Remove the left ventricular tissue and discard all other heart tissues.
4. Place the cleaned ventricle into 70% ethanol (Fisher) for 30 s. Wash thoroughly in serum-free Eagle's MEM. Dissect the outer quarter of the free wall of the ventricle, and remove the epicardial arteries and penetrating blood vessels. Work in a sterile cabinet.
5. Finely mince the retained heart tissue in 0.2% collagenase in PBS. Stir for 30 min at 37 °C in a sterile covered beaker. After 30 min add 0.025% trypsin and 0.01% EDTA in PBS, and shear ten times by pipetting the partially suspended cells up and down in a sterile 10 ml pipette rapidly expelling back into the beaker. Incubate for 30 min.
6. Filter through sterile gauze in a small funnel into a sterile 50 ml polycarbonate tube, and centrifuge at $450 \times g$ in a bench top centrifuge for 10 min.
7. Resuspend the pellet in Eagle's MEM with 20% FBS. Repeat the centrifugation and resuspension as above twice.
8. Grow the isolated cells in Eagle's MEM with 20% FBS for the first 24 h and then in Eagle's MEM with 10% FBS.
9. To confirm the endothelial identity of the RE cells, immunostain with factor VIII antibody and perform a tube-forming assay (*see* below). To immunostain for factor VIII, grow an aliquot of cells on a glass coverslip in six-well Petri dishes for 24 h or more. Fix in −20 °C methanol for 5 min, wash in PBS, and incubate with rabbit factor VIII antibody in PBS for 1 h at 37 °C. Wash three times in PBS, and then visualize for factor VIII staining by incubation with a fluorescein-labeled anti-rabbit IgG. Inspect for immunostaining with a fluorescence microscope.

10. To ensure that RE cells are used in early passage, a stock of early passage cells should be established for cryopreservation in liquid N_2.

3.2 Endothelial Migration Through Collagen (*See* Fig. 1)

If possible use at least two different endothelial cell types in this assay, for example, RE and HMVEC or EJG and HMVEC or all three cell types. This ensures that the results are generally applicable and not restricted to a specific cell line. For statistical purposes,

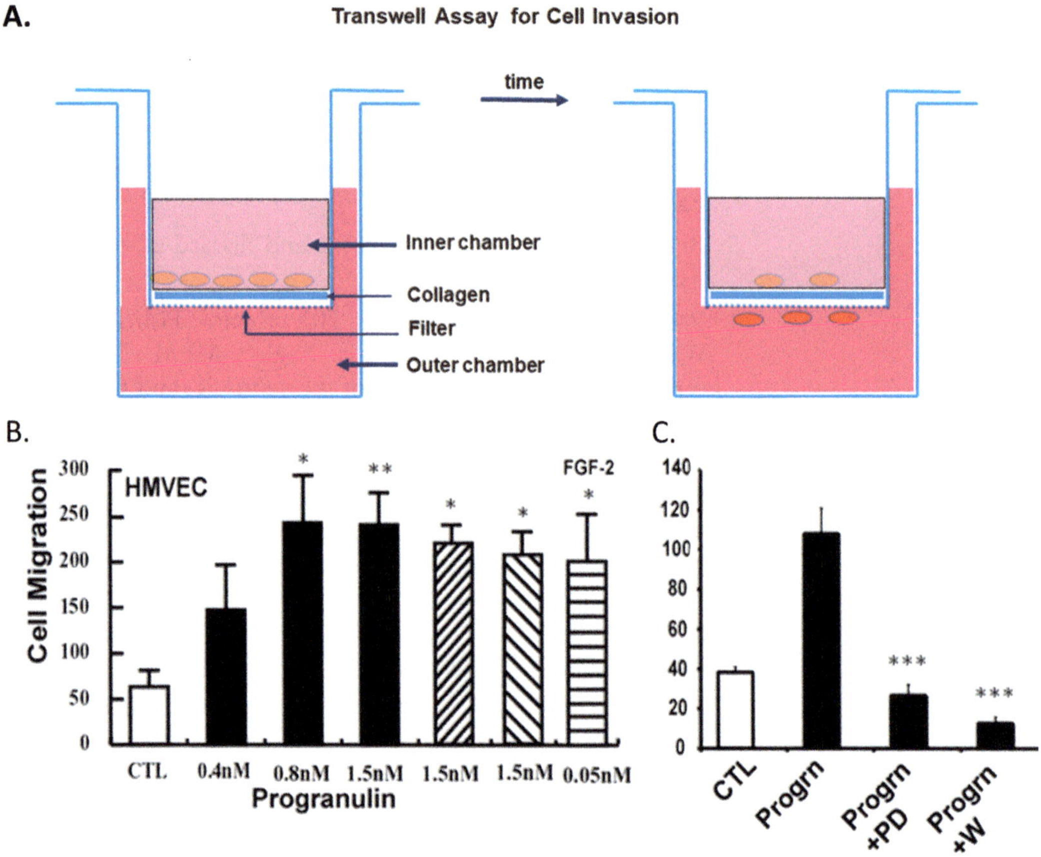

Fig. 1 An outline of the endothelial migration assay. (**a**) Cells are plated on a collagen-coated filter placed inside a tissue culture well insert. Test substances may be added to the inner chamber, the outer chamber, or both chambers. After a suitable period, the filter is removed, and the cells that have migrated through the collagen to the lower surface of the filter are counted. (**b**) HMVEC dermal microvascular cells respond to progranulin by migration through collagen. The black bars are cells stimulated with progranulin at the indicated concentrations in the upper and lower Transwell chambers simultaneously. The left-sloping hatched bar is cells stimulated with progranulin in the upper chamber only; the right-sloping hatched bar is cells stimulated with progranulin in the lower chamber only. FGF2 is the positive control. (**c**) Progranulin stimulated cell migration of HMVEC was assessed with progranulin in both chambers, progranulin with PD098905 (PD), or wortmannin (W); the controls were tested in the absence of progranulin. Results are the mean of at least four independent determinations performed in duplicate ± s.e.m.; $^{*}P < 0.05$, $^{**}P < 0.01$, *** and $P < 0.001$. Differences in the mean value of different groups were assessed by analysis of variance followed by alternative *t*-test. *See* [13] for further details

repeat each concentration of progranulin at least four independent times in duplicate using separate batches of cells. **Steps 11–14** test the sensitivity of progranulin action to signal transduction inhibitors and are optional:

1. Coat the upper surfaces of Transwell cell culture chamber filters (5.0 μm pore size) with 50 μg/ml of human type I collagen in 0.5 M NaCl, 0.05 M Tris–HCl, pH 7.2. The filter must be completely and evenly coated.
2. Set the collagen at 37 °C for at least 30 min.
3. Place 800 μl serum-free Eagle's MEM in the lower chamber, and place the Transwell filter into place in the plate well.
4. Trypsinize and count endothelial cells as described above.
5. Suspend 5×10^4 endothelial cells in 200 μl of Eagle's MEM, and gently add to the top chamber of the Transwell culture unit. (*See* above for how to propagate, trypsinize, and count the cells.)
6. Add progranulin (from 0.4 to 1.5 nM). To test for chemokinesis, progranulin is added to the upper and the lower chambers together at a final volume 200 and 800 μl, respectively, for each chamber. To test for chemotaxis, progranulin is added to the lower chamber only. To perform baseline controls, omit the progranulin from both wells. To perform positive controls, add 0.05 nM FGF2 in the Transwell chambers. FGF2 is a well-established angiogenic protein.
7. Incubate the Transwell chambers for 5 h at 37 °C in a humidified incubator in a 5% CO_2 atmosphere.
8. Remove the Transwell inserts, gently wash twice in pH 7.4 phosphate-buffered saline (PBS), and then fix using 4% paraformaldehyde.
9. To visualize the cells that migrate through the collagen matrix to the lower surface of the filter, stain the filters using 0.4% Giemsa stain for 2 min, and wash three times with pH 7.4 PBS.
10. Transfer the filter to a microscope slide. Count cells on the lower surface of the filter in five visual fields per filter at 400× total magnification.
11. To investigate relevant signaling pathways, set up Transwell assays as above **steps 1–5**.
12. Add selected pharmacological cell signaling inhibitors to both chambers of the Transwell culture apparatus with or without progranulin. Dose-response curves must be performed to determine the optimal concentration of each of the inhibitors. The values presented below worked well in our experiments.
13. Dilute a 100 mM stock solution of PD98059 to 10 mM in DMSO. To obtain a final test concentration of 10 μM, add

0.2 μl or 0.8 μl to 200 μl or 800 μl of serum-free Eagle's MEM in the top and bottom chambers as above. Dilute a 1 mM stock solution of wortmannin to 10 μM in DMSO. To obtain a final test concentration of 10 nM, add 0.2 μl or 0.8 μl to 200μl or 800 μl of serum-free Eagle's MEM in the top and bottom chambers as above. To obtain a 10 μM final test concentration of SB203580, add 0.8 μM or 3 μl of a 2.6 mM stock solution to 200 μl or 800 μl serum-free Eagle's MEM in the upper and lower chambers as above. Add equal volumes of DMSO without the inhibitor compounds to the control wells.

14. Perform the migration assay as above, **steps 7–10**.

3.3 Tube-Forming Assays

An inverted microscope with camera attachment is required. Perform all steps in a laminar flow biosafety cabinet. For statistical purposes, repeat each concentration of progranulin at least four independent times in duplicate using separate batches of cells. *See* Fig. 2 for representative results.

1. Thaw the Matrigel to 4 °C and keep on ice until ready to use. Matrigel is liquid at 4 °C and forms a gel at room temperature.
2. Pipette 250 μl of ice-cold Matrigel into 24-well tissue culture plates. Leave to set at room temperature overnight in a biological safety cabinet with the sterile air flow on. Suspend 5×10^4 HMVEC cells in endothelial growth medium (EGM) supplemented with 5% FBS with or without progranulin (up to 7.5 nM). Use 2.5 nM FGF2 for a positive control and zero progranulin as the baseline control. Layer the cells above the Matrigel. Incubate the cells for 12 h at 37 °C in an atmosphere of 5% CO_2. Repeat each data point at least three times in duplicate.
3. Quantify the tube-like structures after 12 h or any time after that (*see* **Note 3**) by photography in an inverted microscope at 16× total magnification. Measure the following parameters: (1) the length of individual branches, (2) the total length of each tube-like structure including all branches, and (3) the number of branches per tube-like structure.

3.4 Aortic Ring Assay

3.4.1 Prepare Endothelial Cell Culture Medium

1. Before beginning the experiment, prepare endothelial culture medium by adding EGM-2 SingleQuot supplements to 500 ml EBM2 medium.

3.4.2 Preparation of the Aortic Rings

1. Sacrifice the mice (C56Bl/6 strain) at about 12 weeks old. Surface sterilize the animals by thoroughly wetting the fur in 70% ethanol before dissection. Open the thorax and rib cage by cutting a line from the neck toward the abdomen. Remove the thoracic aorta. Place in a petri dish with cold serum-free EBM2 or Eagle's MEM.

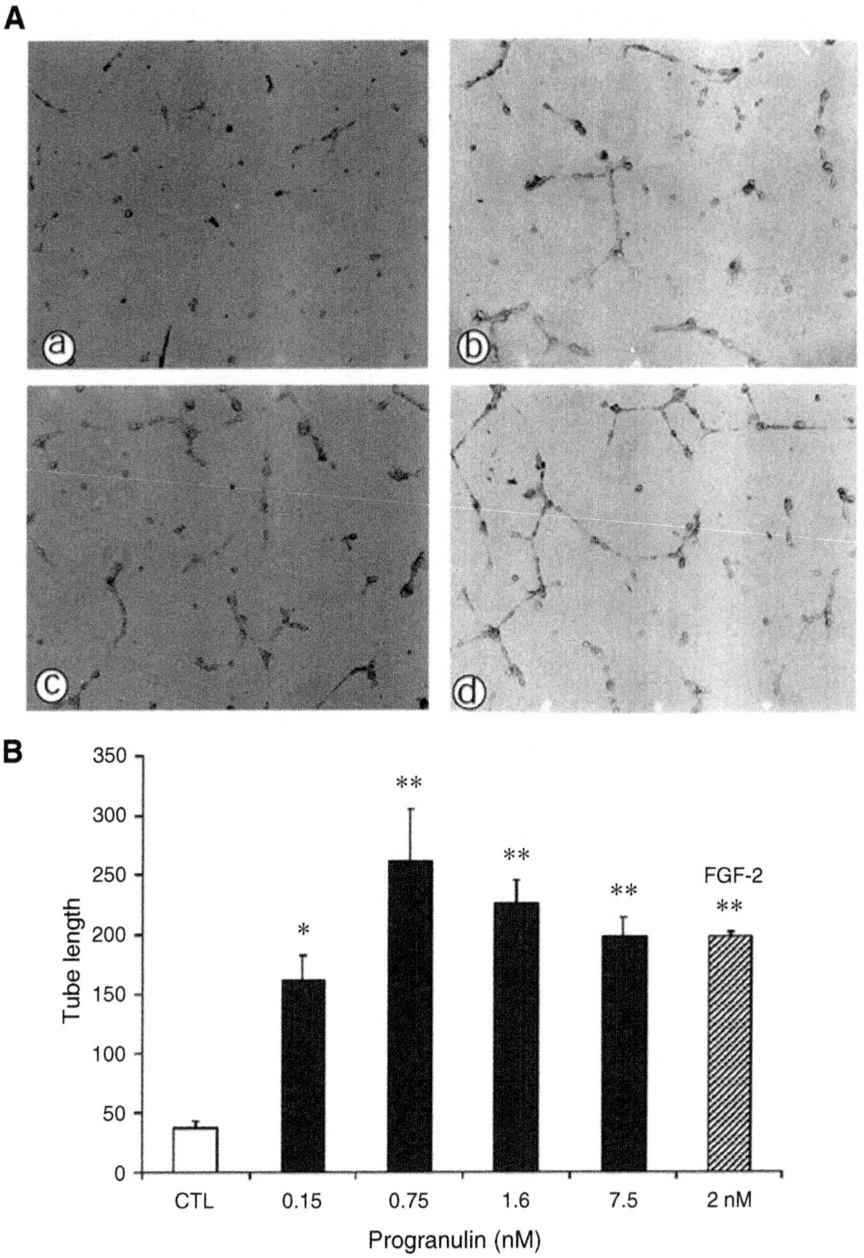

Fig. 2 (**A**) The appearance of HMVEC dermal microvascular cells after 12 h incubation on Matrigel, (**a**) unstimulated cells, (**b**, **c**) cells incubated with progranulin, and (**d**) cells incubated with FGF2 (**d**). (**B**) Formation of tubule-like structures by HMVEC on Matrigel in response to saline (baseline control), progranulin, or the positive control FGF2. Results are the mean of three independent determinations performed in duplicate s.e.m.; $^{*}P < 0.05$ and $^{**}P < 0.01$. Differences in the mean value of different groups were assessed by analysis of variance followed by alternative *t*-test (panel B is adapted from [13])

2. Under a dissecting microscope, carefully remove excess tissue to leave the aorta clean and undamaged.
3. Wash the aorta three times with 5 ml of EBM2 or Eagle's MEM to remove all blood.
4. Slice the aorta into rings each about 1 mm in width. Approximately 10–15 rings per aorta can be obtained. Do not allow aortic rings to dry at any stage.
5. Rinse the rings ten times with 1 ml of EBM2 or Eagle's MEM. Make sure any residual blood contamination is removed.
6. Embed the rings in collagen (*see* below).

3.4.3 Embedding the Rings into Collagen

1. On ice, mix 100 μl of 10× Eagle's MEM and 100 μl of 23.4 mg/ml $NaHCO_3$ in an autoclaved 1.5 ml Eppendorf tube. Add 2.3 μl of 1 N NaOH. The change in pH should cause the medium to become orange.
2. Add 800 μl collagen R (this is a commercially available preparation of 2 mg/ml collagen type I from rat trails solubilized in 0.1% acetic acid).
3. Drop 30 μl of the collagen mixture into the center of each well of a four round-well culture plate.
4. Gently pull each ring across the plastic surface to remove excess media.
5. Place individual aortic rings into the collagen droplets ensuring that the luminal axis lies parallel to surface of the wells.
6. With a pipet tip, gently spread collagen mixture into a uniform disk of approximately 8 mm in diameter.
7. Incubate culture plates at 37 °C for 10 min to induce collagen gelling.
8. Carefully drip 500 μl of endothelial basal medium-2 (EBM2), supplemented with EGM-2 SingleQuot reagents, progranulin, or other test compounds as required into the well to over the collagen.
9. Change the medium every second day.
10. To record sprouting, photograph under a 4× objective lens in an inverted microscope (*see* Fig. 3).

3.5 Preparation of Transgenic Mice with Endothelial Cell-Targeted PGRN Expression

3.5.1 Generation of the Tie2-Pgrn Vector

This methods employs the *Tie2-promoter/enhancer* vector, *pSPTg.T2FpAXK*, and RIKEN clone 21900053G23 which contains the full-length mouse *pgrn* cDNA. (*See* above for details of both plasmids.) The mouse *pgrn* cDNA is inserted in the multiple cloning site between the *Tie2* promoter and the polyA signal of the *Tie2-promoter/enhancer* vector (*see* Fig. 4). The mouse *pgrn* insert is released from the 21900053G23 vector by a *XhoI/BamHI* double restriction digestion. Other sources of *pgrn* insert can be used and

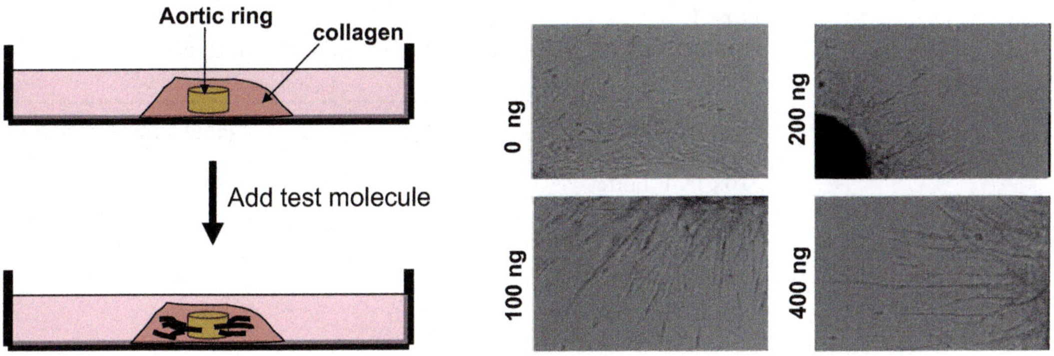

Fig. 3 (**a**) An outline of the aortic ring vessel sprouting assay. An aortic ring is embedded in collagen and incubated in medium which may be supplemented with test substances. After a suitable period, the ring is examined for vessel sprouts. (**b**) The appearance of aortic rings incubated for 9 days with 0–400 ng of progranulin in 500 μl of endothelial basal medium-2 and visualized in an inverted microscope using the 4× objective

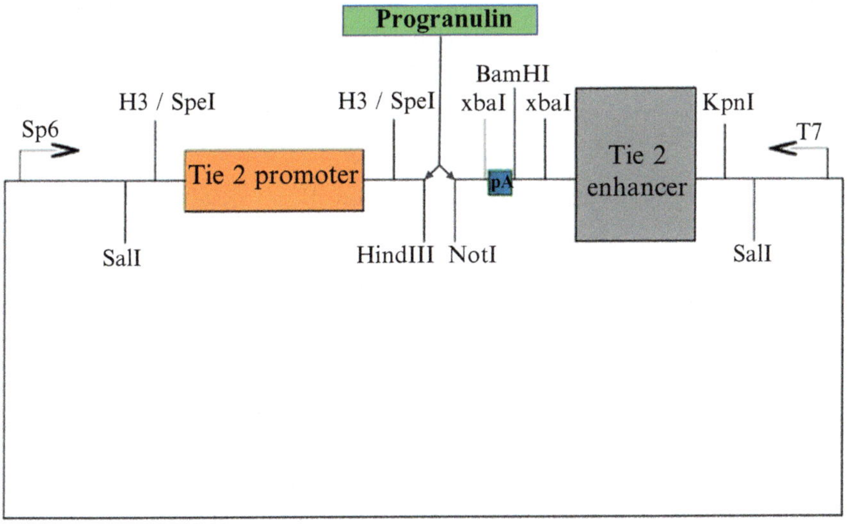

Fig. 4 Position of the progranulin insert in the 6.8 kb *Tie2-promoter/enhancer* vector, *pSPTg.T2FpAXK*

may require other restriction digestions to release the full-length *pgrn* insert. The *XhoI/BamHI* double restriction digestion used to release the mouse progranulin cDNA from RIKEN clone 21900053G23 is incompatible with the *HindIII/NotI* cloning sites in the *Tie2* vector. Therefore, *HindIII* and *NotI* restriction sites were added to mouse progranulin cDNA to enable cloning into the *Tie2* targeting vector by PCR-based modification. Details of the PCR reaction are as follows:

1. Primer sequences used are *Hin*dIIIFwd and *Not*IRvs (*see* above for details).
2. Reagents for the PCR reaction are prepared in a 0.2 ml PCR tube, free from human DNA, RNases, and DNases (Table 1).

Table 1
PCR reagents for the preparation of the progranulin insert

Components	Vol. (μl)	Final concentration
10× PCR buffer	5	1×
25 mM $MgCl_2$	2	2 mM
10 mM dNTP mix	2	2 mM
Primer FW (20 μM)	1	0.4 μM
Primer RV (20 μM)	1	0.4 μM
Taq DNA polymerase	0.3	0.4 μM
Template DNA (10 ng/μl)	2.5	25 ng Riken clone
DEPC-H_2O	37.2	
Total volume	*50*	

3. Conditions for the PCR reaction are as follows:

   PCR protocol.

   Step 1: 94 °C for 5 min. Run one cycle of **step 1**.

   Step 2: 94 °C for 5 min.

   Step 3: 58 °C for 1 min.

   Step 4 72 °C for 2 min.

   Run 35 cycles of **steps 2–4**.

   Step 5: 72 °C for 10 min. Run one cycle of **step 5**.
4. Gel purify the PCR amplification product. Run the plasmid digest on a 1% agarose gel containing ethidium bromide in Tris-acetate EDTA buffer.
5. A 1% agarose gel is prepared by dissolving 0.5 g of agarose in 50 ml of TAE buffer (*see* below). Add 2.5 μl of ethidium bromide to the gels to visualize the DNA bands by ultraviolet light (ethidium bromide is toxic, handle with care). Boil and cool before pouring.
6. Load samples in loading buffer (1.66 μl/10 μl sample). For the 1 kb DNA ladder, mix 600ul 1× TAE buffer, 300 μl 6× sample loading buffer, and 100 μl 1 kb DNA ladder stock.
7. To prepare 1 L of 10× stock solution of Tris-acetate EDTA buffer (TAE buffer), use 48.5 g Tris, 11.4 ml glacial acetic acid, and 20 ml 0.5 M EDTA (pH 8.0). Dissolve Tris in 800 ml of autoclaved deionized water. Add acetic acid and EDTA. Add deionized water to 1 L. Store at room temperature. Dilute stock solution 10:1 to make a 1× working solution.

8. View the gel using shortwave UV illumination and the excise gel fragment containing the PCR amplification product with a clean scalpel. Purify the excised *HindIII*-mouse progranulin cDNA-*Not1* PCR product with a Zymoclean gel purification kit according to manufacturer's recommended protocols.

Preparation of the *HindIII*-mouse progranulin cDNA-*Not1* PCR product for ligation.

It is necessary to remove the caps on the primers and expose the sticky ends of the restriction sites the *HindIII*-mouse progranulin cDNA-*Not1* PCR product prior to ligation with the *Tie2*-vector. This is achieved using *HindIII* and *NotI* restriction enzymes as follows:

1. Prepare the following reaction mix in an Eppendorf microfuge tube:

   30 μl *HindIII*-mouse progranulin cDNA-*Not1* PCR product (approximately 2.25 μg).

   15 U *HindIII*, 0.75 μl.

   15 U *NotI*, 1.5 μl.

   10 μg/μl, BSA 0.2 μl.

   10× NEBuffer2, 3.6 μl.
2. Incubate at 37 °C for 2 h. Inactivate enzymes by heating at 65 °C for 20 min.
3. Separate the *HindIII*-mouse progranulin cDNA-*NotI* fragment (1.8 kb) by agarose gel electrophoresis in a 1% gel followed by purification using the Zymoclean gel purification kit according to manufacturer's recommended protocols.
4. Analyze an aliquot of the product by agarose gel electrophoresis to confirm its purity.
5. Validate the sequence of the *mPgrn* insert by sequencing in both directions.

3.5.2 Insertion of Murine Pgrn Insert into Tie2 Vector

1. Open the *Tie2* vector with *HindIII/NotI* double digestion. The reaction mixture is as follows:

   Tie2 vector plasmid DNA 5ug in 15 μl water.

   HindIII, 0.75 μl (15 U).

   NotI, 1.5 μl (15 U).

   BSA (10 μg/ul) 0.2 μl.

   10× NEBuffer, 2, 2 μl (final conc. 1×).

   dH_2O, 0.55 μl.

   Total 20 μl.
2. Incubate at 37 °C for 2 h.
3. Inactivate enzymes by heating at 65 °C for 20 min.

4. Dephosphorylate the linearized *Tie2* vector to avoid self-annealing. Reaction components and conditions are as follows:

   Cleaved *Tie2* vector (5 μg), 20 μl.

   10× dephosphorylation buffer, 3 μl.

   2 U shrimp alkaline phosphatase, 2 μl (Promega).

   Water, 5 μl.

   Total 30 μl.

5. Incubate at 37 °C for 1 h. Inactivate shrimp alkaline phosphatase at 65 °C for 15 min.
6. Gel purify the dephosphorylated *Tie2* vector using the Zymoclean gel purification kit as above.

3.5.3 Ligation of HindIII-Mouse Progranulin cDNA-Not1 PCR Product into the Tie2-Vector

Control reactions are performed by omitting the *Pgrn* insert but including the ligase and including the *Pgrn* insert but omitting the ligase.

1. Add the opened *Tie2* vector and the *Pgrn* insert at an approximately 2:1 ratio by mass to obtain a final total DNA of 200 ng, for example, we used 130 and 70 ng of *Tie2* vector and *Pgrn* insert, respectively. Bring up to a volume of 9 μl in d. H_2O. Then add

   2 μl of 10× ligase buffer, 1 μl of T4 DNA ligase, and d.H_2O to a final reaction volume of 20 μl.

2. Incubate at room temperature (~24.5 °C) for 4 h, add 1 μl more ligase in reaction mix, continue to incubate at 16 °C overnight (~18 h), and inactivate enzyme at 65 °C for 20 min.
3. Transform 2 μl of ligated product in 50 μl of *E. coli* XTRACells-A using standard protocols. *Tie2* plasmid DNA (40 ng in 2 μl) can be used as transformation positive control.
4. Spread transformed cells on LB agar plates with 100 μg/ml ampicillin, and incubate at room temperature overnight. Pick colonies.
5. Grow picked colonies overnight at 30 °C in a gently shaking incubator in 3 ml LB broth plus ampicillin (100 μg/ml).
6. Purify plasmids from picked colonies using QIAprep Spin Miniprep Kit, or similar, according to manufacturer's protocol.
7. Identify *Tie2-Pgrn* plasmids by restriction enzyme mapping with *HindIII* and *NotI, SalI, SalI,* and *BamHI* (*see* reaction conditions for enzymatic digestion above).
8. To confirm that the *pgrn* insert is intact and undamaged, the *Tie2-progranulin* vector (*Tie2-Pgrn* vector) construct, 8.6 kb, was sequenced with the sequencing primers, mPgrnFwdSeq, mPgrnRvsSeq, and mPgrnInternalSeq.

For long-term storage of *Tie2-Pgrn* vector, prepare a glycerol stock using standard procedures.

3.5.4 Production of Tie2-Pgrn Founder Lines

1. Release the *Tie2-Pgrn* transgene, 5.8 kb, from the vector backbone, 2.4 kb by *SalI* digestion of the *Tie2-Pgrn* plasmid. The restriction digestion conditions are the same as above for other digestion reactions.
2. Gel purify using an extraction kit (Zymoclean), followed by an Elutip column (Schleicher & Schuell) following manufacturer's protocols.
3. The founder lines carrying the *Tie2-Pgrn* transgene are obtained by microinjection of the linearized *Tie2-Pgrn* transgene into pronuclei of fertilized eggs from B6C3F1 mice. This requires the use of a specialized transgenics facility and was performed as a paid service at the Transgenics Facility of the Clinical Research Institute of Montreal.
4. To identify the founder mice, use tail genomic DNA as the template for conventional PCR genotyping employing primers that selectively identify *Tie2-Pgrn* transgene. Details of the genotyping protocol are given below. Of 300 eggs that were microinjected, three *Tie2-Pgrn* transgenic founder mice were obtained from 47 live mice born. The presence and fidelity of *Tie2-Pgrn* transgene constructs in founder mice are confirmed by DNA sequence analysis.

3.5.5 Genotyping

Mice were normally genotyped at weaning using DNA from tail snips. Yolk sac genomic DNA was used to genotype embryos.

1. Cut ~0.5 cm of mouse tail or yolk sac; put into an autoclaved 1.5 ml Eppendorf microcentrifugation tube.
2. Add 500 μl of tail lysis buffer (10 mM Tris, pH 8.0, 100 mM NaCl, 10 mM EDTA, 0.5% SDS) with proteinase K (0.4 mg/ml).
3. Incubate overnight at 55 °C in a metal heating block.
4. Add 500 μl of phenol chloroform to lysate. Invert tubes ten times to mix.
5. Centrifuge at 20,000 × *g* (or top speed) in a desktop microcentrifuge for 15 min.
6. Transfer supernatant into fresh Eppendorf tubes, and add 1 ml of ice-cold EtOH. Vortex for ~5 s.
7. Precipitate DNA at −20 °C for 3 h or O/N. Centrifuge at 20,000 × *g* for 10 min.
8. Wash DNA pellet with 70% EtOH. Centrifuge at 20,000 × *g* for 5 min. Air dry.
9. Reconstitute DNA pellet with 50–100 μl of TE buffer, pH 8.0.

10. Incubate at 65 °C for 10 min.
11. The presence of the *Tie2-Pgrn* transgene is detected by PCR. Transgene-specific primers are used: genotyping forward primer, genotyping reverse primer 1, and genotyping reverse primer 2 (*see* above for details). Mouse *β-actin*, a housekeeper gene, is used as control.
12. The PCR reaction mix is as described above (generation of the *Tie2-Pgrn* vector) but replacing the Riken *Pgrn* plasmid with genomic DNA.
13. The PCR conditions are as follows:

    Step 1: 94 °C for 3 min. Run one cycle of **step 1**.

    Step 2: 94 °C for 30 s.

    Step 3: 69 °C for 45 s.

    Step 4: 72 °C for 1 min, Run 35 cycles of **steps 2–4**.

    Step 5: 72 °C for 7 min. Run one cycle of **step 5**.
14. Analyze the amplification products by electrophoresis in a 1% agarose gel as described above. The genotyping forward primer with reverse primer 1 yields a PCR product of 2.1 kb, and genotyping forward primer with reverse primer 2 yields a PCR product of 1.6 kb.

3.5.6 Backcrossing

Backcrossing is the interbreeding of a hybrid line, in this case, *Tie2-Pgrn* founder mice, which have a B6C3F1 genetic background, with an inbred parental line, (C57BL/6), to obtain offspring with a near pure genetic background. After sufficient backcrossing, the offspring possesses a genetic background of more than 99% that of the inbred parental lines (C57BL/6). Congenic strains were established by backcrossing *Tie2-Pgrn* positive donors with the inbred strain C57BL/6.

1. To develop congenic strains, *Tie2-Pgrn* positive donors were backcrossed with the inbred mouse strain C57BL/6. Male *Tie2-Pgrn* positive donors for breeding are preferred as one male can inseminate several females.
2. *Tie2-Pgrn* positive offspring in each litter were identified by genotyping (*see* above) soon after they were fully weaned.
3. After 5–10 generations of backcrossing, the mice are suitable for phenotypic analysis.

3.5.7 Estimation of Tie2-Pgrn Copy Number

The number of *Tie2-Pgrn* transgene copies in the transgenic lines is established by quantitative real-time PCR (SYBR Green I, Roche LightCycler). To selectively amplify *Tie2-Pgrn* transgene, we use primers *Tie2-Pgrn* Fwd and *Tie2-Pgrn* Rvs that span exons junctions. These primers do not amplify the endogenous progranulin gene in the genomic DNA of the transgenic mice. Mouse beta-actin primers mActin Fwd and mActin Rvs are used to normalize the quantification of the reaction. An RT-PCR strategy is used to determine copy number of *Tie2-Pgrn* transgene in our transgenic mice.

1. Calculate how much *Tie2-Pgrn* vector DNA plasmid should be spiked into a certain amount of gDNA in order to get one copy of transgene correspondingly and obtain a copy number standard curve. Calculations to obtain the standard curve are as follows:

   The mouse haploid genome has 3×10^9 bp; the transgenic founder mice are hemizygous for *Tie2-Pgrn* transgene. The real-time PCR product of *Tie2-Pgrn* transgene is 206 bp, the *Tie2-Pgrn* vector is 8157 bp, and the *Tie2-Pgrn* transgene is 5757 bp. To calculate the mass of *Tie2-Pgrn* vector DNA plasmid, 8157 bp, reflecting a single copy of *Tie2-Pgrn* transgene in 5 ng of genomic DNA, we used the following calculation: Mass = $8157/3 \times 10^9 \times 5/2 = 6797.5 \times 10^{-9}$ ng = 6.7975×10^{-18} g of *Tie2-Pgrn* vector plasmid per 5 ng of genomic DNA.
2. To quantify the copy number of *Tie2-Pgrn* transgenic mouse lines, generate a RT-PCR standard curve. PCR conditions were validated to obtain a high efficiency in both standard and unknown gDNA samples. The PCR amplification efficiency was from 1.96 to 1.98. The standard curves were prepared by "spiking" qRT-PCR reactions performed on 5 and 10 ng WT genomic DNA (gDNA) with the *Tie2-Pgrn* plasmid (plasmid DNA spiked) as in Table 2.
3. The qRT-PCR is performed using SYBR Green quantitative real-time PCR. The reaction mixes are as follows for the standard curve and the test samples. Include also negative controls with no spiked DNA to exclude the possibility of reagent contamination. The standard and sample reaction mixes are in Table 3.

Table 2

Preparation of DNA standards for the estimation of *Tie2-Pgrn* copy number

		5 ng gDNA per reaction	10 ng gDNA per reaction
	Copy number	**Spike with plasmid DNA**	**Spike with plasmid DNA**
Standard 1	1000	6.797 pg	13.594 pg
Standard 2	100	0.6797 pg	1.3594 pg
Standard 3	10	0.06797 pg	0.13594 pg
Standard 4	1	0.006797 pg	0.013594 pg

Table 3

qRT-PCR reaction mix for the estimation of *Tie2-Pgrn* copy number

PCR reaction	Vol. (μl) for Standard	Vol. (μl) for samples
PCR H_2O	8.3	10.3
$MgCl_2$ (3 mM)	2.0	2.0
Primermix (5 μM)	1.0	1.0
SYBR Green I	1.7	1.7
WT gDNA (5 or 10 ng)	2.0	–
Spiked DNA (serially diluted)	2.0	–
Transgenic gDNA (5 or 10 ng)	–	2
Total volume	*17*	*17*

Table 4

qRT-PCR conditions for the estimation of *Tie2-Pgrn* copy number

Steps	Temperature (°C)	Hold time (s)	Slope (C/s)	Acquisition mode
Denaturation (1 cycle)				
1	95	600	20	None
Amplification (35 cycles)				
1	95	10	20	None
2	66	5	20	None
3	72	10	20	None
4	83	1	20	Single
Melting (1 cycle)				
1	95	0	20	None
2	65	15	20	None
3	95	0	0.1	Continuous
Cooling (1 cycle)				
1	40	30	20	None

4. To perform qRT-PCR, use the conditions given in Table 4.
5. A melting curve of the PCR products is obtained at 60 °C for 30 s. The melting curve should show a single peak with no primer dimers.

3.5.8 Quantitative Measurement of Tie2-PGrn mRNA Transcripts

This part of the method is designed to detect and quantify the expression of the *Tie2-Pgrn* transgene mRNA by quantitative RT-PCR (*see* Fig. 5).

1. Extract total RNAs from mouse tissues (e.g., brain, lung, liver, kidney, and skin) using Qiagen's RNeasy Mini Kit (Cat. No. 74104) following the protocol supplied by the manufacturer. For more fibrous tissue such as the muscle and skin, it is better to use Qiagen's RNeasy Fibrous Tissue Mini Kit (Cat. No. 74704). (*See* **Note 4.**)
2. Generate single-stranded DNA from the total mRNA by reverse transcription. Use a single-strand synthesis kit such as Invitrogen's SuperScript III First-Strand Synthesis System for RT-PCR (Cat. No. 18080-051).

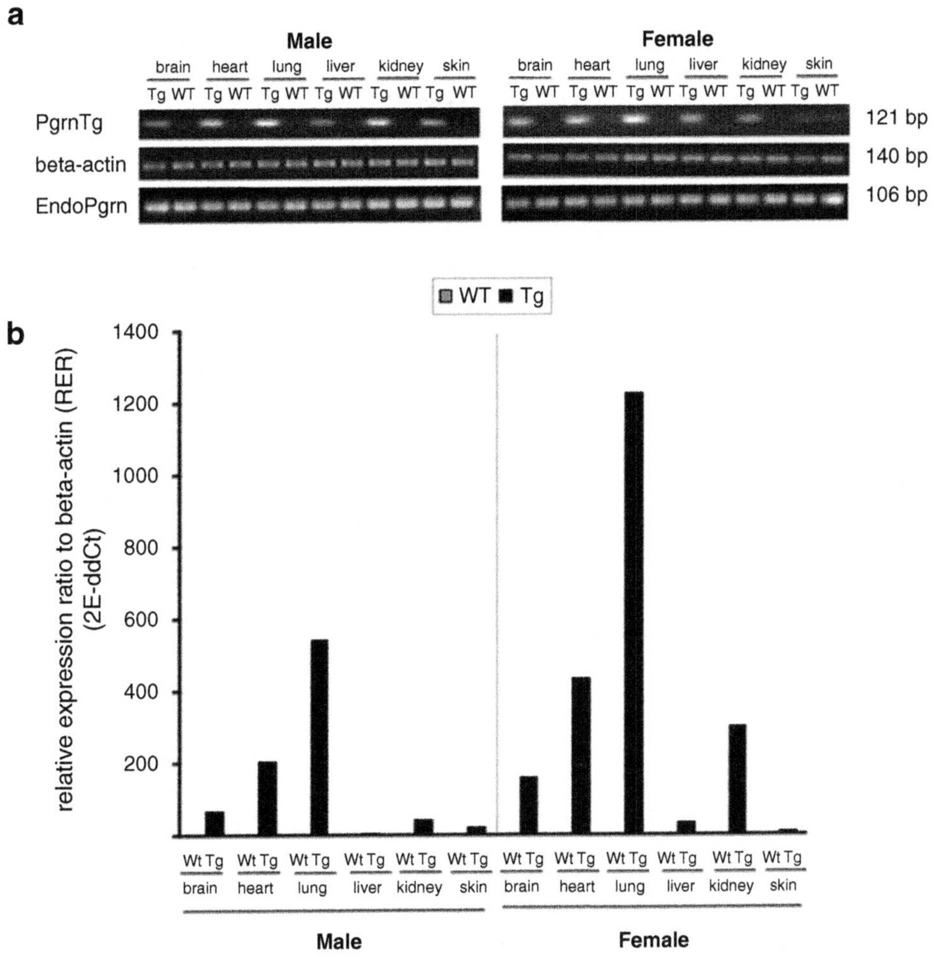

Fig. 5 (**a**) Reverse transcription PCR of tissue extracts from *Tie2-Pgrn* mice and WT littermates using *Tie2-Pgrn* transgene-specific primers analyzed in gel and (**b**) by quantitative real-time RT-PCR. Abbreviations: PgnTg, Tie2-PgrnTransgene; EndoPgrn, endogenous Pgrn. In (**b**), all values were normalized relative to beta-actin mRNA and are expressed as fold increase over the nonspecific background signal (partly taken from [23])

3. The PGRN primers used to detect *Tie2-Pgrn* mRNA expression and exclude endogenous *Pgrn* expression are Tie2-forward primer and Tie2-reverse primer (*see* above for details). To specifically detect endogenous *Pgrn* mRNA, use Pgrn-forward primer and Pgrn-reverse primer. Use mouse β-actin primers to provide control amplifications.
4. Perform qRT-PCR using the SYBR Green in a Roche LightCycler (*see* **Note 5**) method using the same reaction mixture as described for the determination of *Tie2-Pgrn* copy number but now using the reversed transcribed DNA obtained from tissues (**step 1**) as the template.
5. Use the PCR conditions in Table 4 to perform the qRT-PCR.
6. A melting curve of the PCR products is obtained at 60 °C for 30 s. The melting curve should show a single peak with no primer dimers. The negative control consists of PCR products without reverse-transcribed cDNA.
7. The relative expression ratio is calculated as $2^{\Delta \mathrm{Cp\ Tie2\text{-}Pgrn}}/2^{\Delta \mathrm{Cp\ beta\text{-}actin}}$. The reaction efficiency of the PCR is assumed to be 2; "ΔCp *Tie2-Pgrn*" is the difference in the crossing points (CP) between the transgenic mouse and its WT counterpart littermate obtained using the *Tie2-Pgrn* transgene-specific primers (the curve from the WT littermate is assumed to represent a background signal, not the presence of the *Tie2-PGrn* mRNA). "ΔCp beta-actin" is the difference in the crossing points for the amplification of beta-actin mRNA between the transgenic mouse and its WT counterpart littermate.

3.6 Immunohistochemistry

Tie2-Pgrn mice show vascular abnormalities that increase in severity as the *Tie2-Pgrn* copy number increases. These phenotypes are detailed in [23]. This section describes immunohistochemical methods to analyze the *Tie2-Pgrn* phenotype at the cellular level. Tissues from *Tie2-pgrn* positive and WT mice and embryos were fixed in 10% neutral buffered formalin and then embedded in paraffin. Sections were prepared at the Histology Core of the Bone Centre, McGill University including the staining of some sections by H&E stain. The primary antibodies used were:

Anti-mouse progranulin, dilution 1:70.

Anti-PECAM1/CD31 (platelet endothelial cell adhesion molecule), dilution 1:25, this stains endothelial cells.

Anti-alpha smooth muscle actin, dilution 1:100 which stains vascular smooth muscle cells.

Anti-desmin, dilution 1:100, which stains vascular pericytes.

Anti-laminin, dilution 1:50, which stains the basement membrane.

It is important to test the antibody dilution beforehand by staining surplus mouse tissue sections. Representative staining of *Tie2-Pgrn* section for progranulin and PECAM1 is shown in Fig. 6.

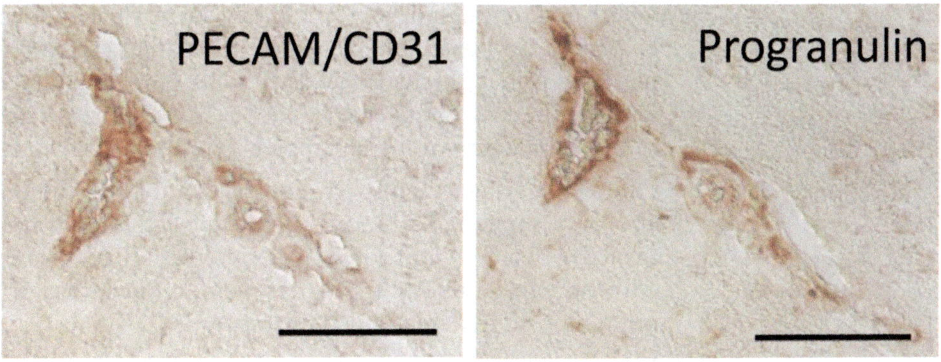

Fig. 6 Serial sections from the adult cerebellum from a *Tie2-pgrn* mouse stained immunostained for PECAM1, to identify endothelial cells, and for progranulin. The scale bar indicates 20 μm (taken from [23])

1. Prepare sodium citrate buffer, 2.94 g of sodium citrate in 1000 ml dH_2O + 0.05% Tween20 pH 6.0.
2. Prepare Tris-buffered saline, 0.025% Triton X-100 (TBST). Make10× Tris-buffered saline by mixing 24 g Tris base, 88 g NaCl, 900 ml distilled water, and pH to 7.6 with concentrated HCl; add distilled water for a final volume of 1 L. Make a 1× TBS solution by mixing 10× TBS with distilled water, and adjust pH to 7.6. Add 0.025% Triton X-100 and mix well.
3. Place slides in a glass slide holder.
4. Paraffin sections are cleared and rehydrated for staining by submerging in CitriSolv (*see* **Note 6**) and ethanol solutions as follows:

   CitriSolv, two times, 5 min.

   100% ethanol, two times, 3 min.

   95% ethanol for 3 min.

   70% ethanol for 3 min.

   50% ethanol for 3 min.
5. Transfer slides still in a glass slide holder to a beaker of hot sodium citrate buffer, and boil the slides for 20 min in a microwave oven. Cover with a lid to prevent evaporation. This step is for antigen retrieval.
6. Transfer slides to a container of TBST washing solution, and cool the slides for 5–10 min.
7. Remove the slides from the glass slide container. Wipe the slides gently with Kimwipe. Do not touch the tissue section. Place slides in a metal slide tray. Draw a circle around the tissue section of about 1 cm radius or according to the size of the section with a Dako pen. (*See* **Notes 7** and **8**.) Place slides in a metal slide tray.

8. Gently add enough peroxidase block (Dako dual endogenous enzyme block) blocking agent over the tissue section to completely cover it. Incubate slides for 30 min in a humidified chamber at room temperature by placing a sufficient amount of blocking agent over the sample to completely cover it.
9. Dilute the primary antibody in Dako antibody diluent (*see* above for suggested dilutions). Place enough diluted antibody over the tissue section to completely cover it. Incubate slides with primary antibody for 1 h at room temperature or overnight at 4 °C in a humidified chamber.
10. Wash the slides in TBST for three times, 30 min each wash at room temperature with gentle rocking.
11. Incubate the slides with Dako biotinylated link solution for 30 min at room temperature in a humidified chamber.
12. Wash the slides in TBST for three times, 30 min each wash at room temperature with gentle rocking.
13. Incubate the slide with Dako streptavidin horseradish peroxidase solution for 30 min at room temperature in a humidified chamber.
14. Wash the slides in TBST for three times, 30 min each wash at room temperature with gentle rocking.
15. Apply DAB substrate for about 1 min.
16. Wash in TBST for 10 min at RT with gentle rocking.
17. Mount the slides.

4 Notes

1. If long-term experiments are planned, it is important to cryopreserve a reserve stock of the cells. This serves two purposes, first as an insurance step if the cells become contaminated or in any other way damaged. Second, primary cell cultures senesce after multiple passages, to maintain the cells in early passage the cryopreserved stock can be thawed and regenerated. This should be done after approximately 7–8 passages.
2. Do not centrifuge living cells at more than 300 × *g*.
3. Tubes will form spontaneously even without growth factors stimulation, adding the growth factors accelerates tube formation. In addition, there is a tendency for the tubes to coalesce after long incubation and even to form "slug" like aggregates. It is important, therefore, to collect the data early, before unstimulated tube formation begins and before "slug" formation occurs so that the individual cellular chains are still easy to recognize. Shorter time periods also minimize the possibility that differential rates of proliferation between treatments might bias the results.

4. Other RNA purification protocols such as purification with Trizol solutions will work as well but require more manipulations.
5. Any reliable qRT-PCR system should work as well and should be used according to its availability.
6. CitriSolv is a safer alternative to xylene.
7. A Dako pen draws a hydrophobic boundary around the sample and prevents the staining solution from moving away from the sample during incubation with antibody.
8. It is critically important not to let the slides dry at any point. If you have a big section and the edges are starting to dry, add the appropriate reagent immediately, or dip them in TBST. During incubation with antibody, make sure the sections are well covered and will not dry out during the incubation period.

References

1. Chung AS, Ferrara N (2011) Developmental and pathological angiogenesis. Annu Rev Cell Dev Biol 27:563–584. https://doi.org/10.1146/annurev-cellbio-092910-154002
2. Bao P, Kodra A, Tomic-Canic M, Golinko MS, Ehrlich HP, Brem H (2009) The role of vascular endothelial growth factor in wound healing. J Surg Res 153(2):347–358. https://doi.org/10.1016/j.jss.2008.04.023
3. Johnson KE, Wilgus TA (2014) Vascular endothelial growth factor and angiogenesis in the regulation of cutaneous wound repair. Adv Wound Care (New Rochelle) 3(10):647–661. https://doi.org/10.1089/wound.2013.0517
4. Andres AC, Djonov V (2010) The mammary gland vasculature revisited. J Mammary Gland Biol Neoplasia 15(3):319–328. https://doi.org/10.1007/s10911-010-9186-9
5. Demir R, Seval Y, Huppertz B (2007) Vasculogenesis and angiogenesis in the early human placenta. Acta Histochem 109(4):257–265. https://doi.org/10.1016/j.acthis.2007.02.008
6. Demir R, Yaba A, Huppertz B (2010) Vasculogenesis and angiogenesis in the endometrium during menstrual cycle and implantation. Acta Histochem 112(3):203–214. https://doi.org/10.1016/j.acthis.2009.04.004
7. Cheng R, Ma JX (2015) Angiogenesis in diabetes and obesity. Rev Endocr Metab Disord 16(1):67–75. https://doi.org/10.1007/s11154-015-9310-7
8. Martínez CE, Smith PC, Palma Alvarado VA (2015) The influence of platelet-derived products on angiogenesis and tissue repair: a concise update. Front Physiol 6(290). https://doi.org/10.3389/fphys.2015.00290
9. Presta M, Dell'Era P, Mitola S, Moroni E, Ronca R, Rusnati M (2005) Fibroblast growth factor/fibroblast growth factor receptor system in angiogenesis. Cytokine Growth Factor Rev 16(2):159–178. https://doi.org/10.1016/j.cytogfr.2005.01.004
10. Mahabeleshwar GH, Feng W, Phillips DR, Byzova TV (2006) Integrin signaling is critical for pathological angiogenesis. J Exp Med 203(11):2495–2507. https://doi.org/10.1084/jem.20060807
11. Brooks PC, Clark RA, Cheresh DA (1994) Requirement of vascular integrin alpha v beta 3 for angiogenesis. Science 264(5158):569–571
12. Daniel R, He Z, Carmichael KP, Halper J, Bateman A (2000) Cellular localization of gene expression for progranulin. J Histochem Cytochem 48(7):999–1009. https://doi.org/10.1177/002215540004800713
13. He Z, Ong CH, Halper J, Bateman A (2003) Progranulin is a mediator of the wound response. Nat Med 9(2):225–229. https://doi.org/10.1038/nm816
14. Daniel R, Daniels E, He Z, Bateman A (2003) Progranulin (acrogranin/PC cell-derived

growth factor/granulin-epithelin precursor) is expressed in the placenta, epidermis, microvasculature, and brain during murine development. Dev Dyn 227(4):593–599. https://doi.org/10.1002/dvdy.10341

15. Desmarais JA, Cao M, Bateman A, Murphy BD (2008) Spatiotemporal expression pattern of progranulin in embryo implantation and placenta formation suggests a role in cell proliferation, remodeling, and angiogenesis. Reproduction 136(2):247–257. https://doi.org/10.1530/REP-08-0044
16. Davidson B, Alejandro E, Florenes VA, Goderstad JM, Risberg B, Kristensen GB, Trope CG, Kohn EC (2004) Granulin-epithelin precursor is a novel prognostic marker in epithelial ovarian carcinoma. Cancer 100(10):2139–2147. https://doi.org/10.1002/cncr.20219
17. Gonzalez EM, Mongiat M, Slater SJ, Baffa R, Iozzo RV (2003) A novel interaction between perlecan protein core and progranulin: potential effects on tumor growth. J Biol Chem 278(40):38113–38116. https://doi.org/10.1074/jbc.C300310200
18. Li LQ, Min LS, Jiang Q, Ping JL, Li J, Dai LC (2012) Progranulin expression in breast cancer with different intrinsic subtypes. Pathol Res Pract 208(4):210–216. https://doi.org/10.1016/j.prp.2012.02.001
19. Chen XY, Li JS, Liang QP, He DZ, Zhao J (2008) Expression of PC cell-derived growth factor and vascular endothelial growth factor in esophageal squamous cell carcinoma and their clinicopathologic significance. Chin Med J 121(10):881–886
20. Wang M, Li G, Yin J, Lin T, Zhang J (2012) Progranulin overexpression predicts overall survival in patients with glioblastoma. Med Oncol 29(4):2423–2431. https://doi.org/10.1007/s12032-011-0131-6
21. Yang D, Wang LL, Dong TT, Shen YH, Guo XS, Liu CY, Liu J, Zhang P, Li J, Sun YP (2015) Progranulin promotes colorectal cancer proliferation and angiogenesis through TNFR2/Akt and ERK signaling pathways. Am J Cancer Res 5(10):3085–3097
22. Tangkeangsirisin W, Serrero G (2004) PC cell-derived growth factor (PCDGF/GP88, progranulin) stimulates migration, invasiveness and VEGF expression in breast cancer cells. Carcinogenesis 25(9):1587–1592. https://doi.org/10.1093/carcin/bgh171
23. Toh H, Cao M, Daniels E, Bateman A (2013) Expression of the growth factor progranulin in endothelial cells influences growth and development of blood vessels: a novel mouse model. PLoS One 8(5):e64989. https://doi.org/10.1371/journal.pone.0064989
24. Kawase R, Ohama T, Matsuyama A, Matsuwaki T, Okada T, Yamashita T, Yuasa-Kawase M, Nakaoka H, Nakatani K, Inagaki M, Tsubakio-Yamamoto K, Masuda D, Nakagawa-Toyama Y, Nishida M, Ohmoto Y, Nishihara M, Komuro I, Yamashita S (2013) Deletion of progranulin exacerbates atherosclerosis in ApoE knockout mice. Cardiovasc Res 100(1):125–133. https://doi.org/10.1093/cvr/cvt178
25. Jackman K, Kahles T, Lane D, Garcia-Bonilla L, Abe T, Capone C, Hochrainer K, Voss H, Zhou P, Ding A, Anrather J, Iadecola C (2013) Progranulin deficiency promotes post-ischemic blood-brain barrier disruption. J Neurosci 33(50):19579–19589. https://doi.org/10.1523/JNEUROSCI.4318-13.2013
26. Shakado S, Sakisaka S, Noguchi K, Yoshitake M, Harada M, Mimura Y, Sata M, Tanikawa K (1995) Effects of extracellular matrices on tube formation of cultured rat hepatic sinusoidal endothelial cells. Hepatology 22(3):969–973
27. Francescone RA 3rd, Faibish M, Shao R (2011) A Matrigel-based tube formation assay to assess the vasculogenic activity of tumor cells. J Vis Exp (55). https://doi.org/10.3791/3040
28. Hendrix MJ, Seftor EA, Hess AR, Seftor RE (2003) Vasculogenic mimicry and tumour-cell plasticity: lessons from melanoma. Nat Rev Cancer 3(6):411–421. https://doi.org/10.1038/nrc1092
29. Nicosia RF (2009) The aortic ring model of angiogenesis: a quarter century of search and discovery. J Cell Mol Med 13(10):4113–4136. https://doi.org/10.1111/j.1582-4934.2009.00891.x
30. Schlaeger TM, Bartunkova S, Lawitts JA, Teichmann G, Risau W, Deutsch U, Sato TN (1997) Uniform vascular-endothelial-cell-specific gene expression in both embryonic and adult transgenic mice. Proc Natl Acad Sci U S A 94(7):3058–3063
31. Nishida M, Carley WW, Gerritsen ME, Ellingsen O, Kelly RA, Smith TW (1993) Isolation and characterization of human and rat cardiac microvascular endothelial cells. Am J Phys 264(2 Pt 2):H639–H652

Index

Andrew Bateman et al. (eds.), *Progranulin: Methods and Protocols*, Methods in Molecular Biology, vol. 1806, https://doi.org/10.1007/978-1-4939-8559-3, © Springer Science+Business Media, LLC, part of Springer Nature 2018

S

T

U

W

X

Y

Z

Printed by Printforce, the Netherlands